PROBABILITY, RANDOM VARIABLES, AND RANDOM SIGNAL PRINCIPLES

McGraw-Hill Series in Electrical Engineering

Consulting Editor
Stephen W. Director, Carnegie-Mellon University

Circuits and Systems
Communications and Signal Processing
Control Theory
Electronics and Electronic Circuits
Power and Energy
Electromagnetics
Computer Engineering
Introductory
Radar and Antennas
VLSI

Previous Consulting Editors

Ronald N. Bracewell, Colin Cherry, James F. Gibbons, Willis W. Harman,
Hubert Heffner, Edward W. Herold, John G. Linvill, Simon Ramo,
Ronald A. Rohrer, Anthony E. Siegman, Charles Susskind, Frederick E. Terman,
John G. Truxal, Ernst Weber, and John R. Whinnery

Communications and Signal Processing

Consulting Editor
Stephen W. Director, Carnegie-Melon University

Antoniou: *Digital Filters: Analysis and Design*
Candy: *Signal Processing: The Model-Based Approach*
Carlson: *Communications Systems: An Introduction to Signals and Noise in Electrical Communication*
Cherin: *An Introduction to Optical Fibers*
Cooper and McGillem: *Modern Communications and Spread Spectrum*
Davenport: *Probability and Random Processes: An Introduction for Applied Scientists and Engineers*
Drake: *Fundamentals of Applied Probability Theory*
Guiasu: *Information Theory with New Applications*
Keiser: *Optical Fiber Communications*
Melsa and Cohn: *Decision and Estimation Theory*
Papoulis: *Probability, Random Variables, and Stochastic Processes*
Papoulis: *Signal Analysis*
Papoulis: *The Fourier Integral and Its Applications*
Peebles: *Probability, Random Variables, and Random Signal Principles*
Proakis: *Digital Communications*
Schwartz: *Information Transmission, Modulation, and Noise*
Schwartz and Shaw: *Signal Processing*
Shooman: *Probabilistic Reliability: An Engineering Approach*
Smith: *Modern Communication Circuits*
Taub and Schilling: *Principles of Communication Systems*
Viterbi and Omura: *Principles of Digital Communication and Coding*

PROBABILITY, RANDOM VARIABLES, AND RANDOM SIGNAL PRINCIPLES

Second Edition

Peyton Z. Peebles, Jr., Ph.D.

Professor of Electrical Engineering
University of Florida

McGraw-Hill Book Company

New York   St. Louis   San Francisco   Auckland   Bogotá   Hamburg
Johannesburg   London   Madrid   Mexico   Milan   Montreal   New Delhi
Panama   Paris   São Paulo   Singapore   Sydney   Tokyo   Toronto

To MY MOTHER
Maida Erlene Denton Dials

AND STEPFATHER
Ralph Phillip Dials

This book was set in Times Roman by Santype International Limited;
the editor was Sanjeev Rao; the cover was designed by Rafael Hernandez;
the production supervisors were Leroy A. Young and Fred Schulte.
Project supervision was done by Albert Harrison, Harley Editorial Services.
R. R. Donnelley & Sons Company was printer and binder.

PROBABILITY, RANDOM VARIABLES, AND RANDOM SIGNAL PRINCIPLES

1234567890 DOCDOC 8943210987

ISBN 0-07-049219-0

Library of Congress Cataloging-in-Publication Data

Peebles, Peyton Z.
    Probability, random variables, and random signal
principles.

    (McGraw-Hill series in electrical engineering.
Communications and information theory)
        Bibliography: p.
        1. Probabilities.   2. Random variables.   3. Signal
theory (Telecommunication)   I. Title.   II. Series.
TA340.P43   1987   519.2   86-10550
ISBN 0-07-049219-0
ISBN 0-07-049220-4 (solutions manual)

CONTENTS

Preface to the Second Edition   xiii

Preface to the First Edition   xv

Chapter 1   Probability   1

    1.0   Introduction to Book and Chapter   1
    1.1   Set Definitions   3
    1.2   Set Operations   5
        Venn Diagram / Equality and Difference / Union and
        Intersection / Complement / Algebra of Sets / De
        Morgan's Laws / Duality Principle
    1.3   Probability Introduced through Sets   8
        Experiments and Sample Spaces / Discrete and
        Continuous Sample Spaces / Events / Probability
        Definition and Axioms / Mathematical Model of
        Experiments
    1.4   Joint and Conditional Probability   13
        Joint Probability / Conditional Probability / Total
        Probability / Bayes' Theorem
    1.5   Independent Events   18
        Two Events / Multiple Events / Properties of Independent
        Events
   *1.6   Combined Experiments   22
        *Combined Sample Space / *Events on the Combined Space /
        *Probabilities
    1.7   Bernoulli Trials   25
        Problems   28
        Additional Problems   32

    * Star indicates more advanced material.

Chapter 2 The Random Variable   34

    2.0   Introduction   34
    2.1   The Random Variable Concept   34
            Definition of a Random Variable / Conditions for a
            Function to Be a Random Variable / Discrete and
            Continuous Random Variables / Mixed Random Variable
    2.2   Distribution Function   37
    2.3   Density Function   40
            Existence / Properties of Density Functions
    2.4   The Gaussian Random Variable   43
    2.5   Other Distribution and Density Examples   46
            Binomial / Poisson / Uniform / Exponential / Rayleigh
    2.6   Conditional Distribution and Density Functions   51
            Conditional Distribution / Properties of Conditional
            Distribution / Conditional Density / Properties of
            Conditional Density / ★Methods of Defining
            Conditioning Event
          Problems   57
          Additional Problems   63

Chapter 3 Operations on One Random Variable—
            Expectation   66

    3.0   Introduction   66
    3.1   Expectation   66
            Expected Value of a Random Variable / Expected Value
            of a Function of a Random Variable / ★Conditional
            Expected Value
    3.2   Moments   69
            Moments about the Origin / Central Moments /
            Variance and Skew
  ★3.3   Functions That Give Moments   71
            ★Characteristic Function / ★Moment Generating Function
    3.4   Transformations of a Random Variable   74
            Monotonic Transformations of a Continuous Random
            Variable / Nonmonotonic Transformations of a
            Continuous Random Variable / Transformation of a
            Discrete Random Variable
          Problems   80
          Additional Problems   84

Chapter 4 Multiple Random Variables   87

    4.0   Introduction   87
    4.1   Vector Random Variables   88
    4.2   Joint Distribution and Its Properties   89
            Joint Distribution Function / Properties of the Joint
            Distribution / Marginal Distribution Functions

4.3	Joint Density and Its Properties	93
	Joint Density Function / Properties of the Joint Density / Marginal Density Functions	
4.4	Conditional Distribution and Density	96
	Conditional Distribution and Density—Point Conditioning / ★Conditional Distribution and Density—Interval Conditioning	
4.5	Statistical Independence	100
4.6	Distribution and Density of a Sum of Random Variables	102
	Sum of Two Random Variables / ★Sum of Several Random Variables	
★4.7	Central Limit Theorem	105
	★Unequal Distributions / ★Equal Distributions	
	Problems	109
	Additional Problems	113

Chapter 5   Operations on Multiple Random Variables   117

5.0	Introduction	117
5.1	Expected Value of a Function of Random Variables	117
	Joint Moments about the Origin / Joint Central Moments	
★5.2	Joint Characteristic Functions	122
5.3	Jointly Gaussian Random Variables	124
	Two Random Variables / ★N Random Variables / ★Some Properties of Gaussian Random Variables	
★5.4	Transformations of Multiple Random Variables	128
★5.5	Linear Transformation of Gaussian Random Variables	130
★5.6	Complex Random Variables	133
	Problems	134
	Additional Problems	139

Chapter 6   Random Processes   142

6.0	Introduction	142
6.1	The Random Process Concept	142
	Classification of Processes / Deterministic and Nondeterministic Processes	
6.2	Stationarity and Independence	147
	Distribution and Density Functions / Statistical Independence / First-Order Stationary Processes / Second-Order and Wide-Sense Stationarity / N-Order and Strict-Sense Stationarity / Time Averages and Ergodicity	
6.3	Correlation Functions	152
	Autocorrelation Function and Its Properties / Cross-Correlation Function and Its Properties / Covariance Functions	

6.4	Measurement of Correlation Functions	156
6.5	Gaussian Random Processes	158
*6.6	Complex Random Processes	160
	Problems	162
	Additional Problems	168

Chapter 7   Spectral Characteristics of Random Processes   172

7.0	Introduction	172
7.1	Power Density Spectrum and Its Properties	173
	The Power Density Spectrum / Properties of the Power Density Spectrum / Bandwidth of the Power Density Spectrum	
7.2	Relationship between Power Spectrum and Autocorrelation Function	179
7.3	Cross-Power Density Spectrum and Its Properties	182
	The Cross-Power Density Spectrum / Properties of the Cross-Power Density Spectrum	
*7.4	Relationship between Cross-Power Spectrum and Cross-Correlation Function	185
7.5	Some Noise Definitions and other Topics	187
	White and Colored Noise / Product Device Response to a Random Signal	
*7.6	Power Spectrums of Complex Processes	193
	Problems	194
	Additional Problems	200

Chapter 8   Linear Systems with Random Inputs   205

8.0	Introduction	205
8.1	Linear System Fundamentals	205
	The General Linear System / Linear Time-Invariant Systems / Time-Invariant System Transfer Function / Idealized Systems / Causal and Stable Systems	
8.2	Random Signal Response of Linear Systems	210
	System Response—Convolution / Mean and Mean-Squared Value of System Response / Autocorrelation Function of Response / Cross-Correlation Functions of Input and Output	
8.3	System Evaluation Using Random Noise	214
8.4	Spectral Characteristics of System Response	215
	Power Density Spectrum of Response / Cross-Power Density Spectrums of Input and Output	
8.5	Noise Bandwidth	217
*8.6	Bandpass, Band-Limited, and Narrowband Processes	219
	*Band-Limited Processes / *Narrowband Processes / *Properties of Band-Limited Processes / *Proof of Properties of Band-Limited Processes	

8.7 Modeling of Noise Sources   226
Resistive (Thermal) Noise Source / Arbitrary Noise
Sources, Effective Noise Temperature / An Antenna as a
Noise Source

8.8 Incremental Modeling of Noisy Networks   231
Available Power Gain / Equivalent Networks, Effective
Input Noise Temperature / Spot Noise Figures

8.9 Modeling of Practical Noisy Networks   236
Average Noise Figures / Average Noise
Temperatures / Modeling of Attenuators /
Model of Example System

Problems   240
Additional Problems   251

Chapter 9   Optimum Linear Systems   257

9.0 Introduction   257
9.1 Systems That Maximize Signal-to-Noise Ratio   258
Matched Filter for Colored Noise / Matched Filter for
White Noise

9.2 Systems That Minimize Mean-Squared Error   262
Wiener Filters / Minimum Mean-Squared Error

9.3 Optimization by Parameter Selection   266
Problems   267
Additional Problems   271

Chapter 10   Some Practical Applications of the Theory   275

10.0 Introduction   275
10.1 Noise in an Amplitude Modulation Communication System   275
AM System and Waveforms / Noise Performance

10.2 Noise in a Frequency Modulation Communication System   280
FM System and Waveforms / FM System Performance

10.3 Noise in a Simple Control System   285
Transfer Function / Error Function / Wiener Filter Application

10.4 Noise in a Phase-Locked Loop   288
Phase Detector / Loop Transfer Function / Loop Noise
Performance

10.5 Characteristics of Random Computer-Type Waveform   296
Process Description / Power Spectrum / Autocorrelation
Function

10.6 Envelope and Phase of a Sinusoidal Signal Plus Noise   298
Waveforms / Probability Density of the
Envelope / Probability Density of Phase

10.7 Radar Detection Using a Single Observation   301
False Alarm Probability and Threshold / Detection
Probability

Problems   305

Appendix A   Review of the Impulse Function   310

Appendix B   Gaussian Distribution Function   313

Appendix C   Useful Mathematical Quantities   315
    Trigonometric Identities   315
    Indefinite Integrals   316
        Rational Algebraic Functions / Trigonometric
        Functions / Exponential Functions
    Definite Integrals   318
    Finite Series   319

Appendix D   Review of Fourier Transforms   320
    Existence   321
    Properties   321
        Linearity / Time and Frequency Shifting / Scaling /
        Duality / Differentiation / Integration / Conjugation /
        Convolution / Correlation / Parseval's Theorem
    Multidimensional Fourier Transforms   323
    Problems   324

Appendix E   Table of Useful Fourier Transforms   328

Appendix F   Some Probability Densities and Distributions   330
    Discrete Functions   331
        Bernoulli / Binomial / Pascal / Poisson
    Continuous Functions   333
        Arcsine / Beta / Cauchy / Chi-Square with N Degrees
        of Freedom / Erlang / Exponential / Gamma /
        Gaussian-Univariate / Gaussian-Bivariate / Laplace /
        Log-Normal / Rayleigh / Rice / Uniform / Weibull

Bibliography   339

Index   341

PREFACE TO THE
SECOND EDITION

Because the first edition of this book was well received by the academic and engineering community, a special attempt was made in the second edition to include only those changes that seemed to clearly improve the book's use in the classroom. Most of the modifications were included only after obtaining input from several users of the book.

Except for a few minor corrections and additions, just six significant changes were made. Only two, a new section on the central limit theorem and one on gaussian random processes, represent modification of the original text. A third change, a new chapter (10) added at the end of the book, serves to illustrate a number of the book's theoretical principles by applying them to problems encountered in practice. A fourth change is the addition of Appendix F, which is a convenient list of some useful probability densities that are often encountered.

The remaining two changes are probably the most significant, especially for instructors using the book. First, the number of examples that illustrate the topics discussed has been increased by about 30 percent (over 85 examples are now included). These examples were carefully scattered throughout the text in an effort to include at least one in each section where practical to do so. Second, over 220 new student exercises (problems) have been added at the ends of the chapters (a 54 percent increase).

The book now contains 630 problems and a complete solutions manual is available to instructors from the publisher. This addition was in response to instructors that had used most of the exercises in the first edition. For these instructors' convenience in identifying the new problems, they are listed in each chapter as "Additional Problems."

All other aspects of the book, such as its purpose (a textbook), intended audience (juniors, seniors, first-year graduate students), level, and style of presentation, remain as before.

I would like to thank D. I. Starry for her excellent work in typing the manuscript and the University of Florida for making her services available. Finally, I am again indebted to my wife, Barbara, for her selfless efforts in helping me proofread the book. If the number of in-print errors is small, it is greatly due to her work.

Peyton Z. Peebles, Jr.

PREFACE TO THE FIRST EDITION

This book has been written specifically as a textbook with the purpose of introducing the principles of probability, random variables, and random signals to either junior or senior engineering students.

The *level* of material included in the book has been selected to apply to a typical undergraduate program. However, a small amount of more advanced material is scattered throughout to serve as stimulation for the more advanced student, or to fill out course content in schools where students are at a more advanced level. (Such topics are keyed by a star ★.) The *amount* of material included has been determined by my desire to fit the text to courses of up to one semester in length. (More is said below about course structure.)

The *need* for the book is easily established. The engineering applications of probability concepts have historically been taught at the graduate level, and many excellent texts exist at that level. In recent times, however, many colleges and universities are introducing these concepts into the undergraduate curricula, especially in electrical engineering. This fact is made possible, in part, by refinements and simplifications in the theory such that it can now be grasped by junior or senior engineering students. Thus, there is a definite need for a text that is clearly written in a manner appealing to such students. I have tried to respond to this need by paying careful attention to the organization of the contents, the development of discussions in simple language, and the inclusion of text examples and many problems at the end of each chapter. The book contains over 400 problems and a solutions manual for all problems is available to instructors from the publisher.

Many of the examples and problems have purposely been made very simple in an effort to instill a sense of accomplishment in the student, which, hopefully, will provide the encouragement to go on to the more challenging problems. Although emphasis is placed on examples and problems of electrical engineering, the concepts and theory are applicable to all areas of engineering.

The International System of Units (SI) has been used primarily throughout the text. However, because technology is presently in a transitional stage with regard to measurements, some of the more established customary units (gallons, °F, etc.) are also utilized; in such instances, values in SI units follow in parentheses.

The *student background* required to study the book is only that typical of junior or senior engineering students. Specifically, it is assumed the student has been introduced to multivariable calculus, Fourier series, Fourier transforms, impulse functions, and some linear system theory (transfer function concepts, especially). I recognize, however, that students tend to forget a fair amount of what is initially taught in many of these areas, primarily through lack of opportunity to apply the material in later courses. Therefore, I have inserted short reviews of some of these required topics. These reviews are occasionally included in the text, but, for the most part, exist in appendixes at the end of the book.

The *order of the material* is dictated by the main topic. Chapter 1 introduces probability from the axiomatic definition using set theory. In my opinion this approach is more modern and mathematically correct than other definitions. It also has the advantage of creating a better base for students desiring to go on to graduate work. Chapter 2 introduces the theory of a single random variable. Chapter 3 introduces operations on one random variable that are based on statistical expectation. Chapter 4 extends the theory to several random variables, while Chapter 5 defines operations with several variables. Chapters 6 and 7 introduce random processes. Definitions based on temporal characterizations are developed in Chapter 6. Spectral characterizations are included in Chapter 7.

The remainder of the text is concerned with the response of linear systems with random inputs. Chapter 8 contains the general theory, mainly for linear time-invariant systems; while Chapter 9 considers specific optimum systems that either maximize system output signal-to-noise ratio or minimize a suitably defined average error.

Finally, the book closes with a number of appendixes that contain material helpful to the student in working problems, in reviewing background topics, and in the interpretation of the text.

The book can profitably be used in curricula based on either the quarter or the semester system. At the University of Tennessee, a *one-quarter undergraduate course* at the junior level has been successfully taught that covers Chapters 1 through 8, except for omitting Sections 2.6, 3.4, 4.4, 8.7 through 8.9, and all starred material. The class met three hours per week.

A *one-semester undergraduate course* (three hours per week) can readily be structured to cover Chapters 1 through 9, omitting all starred material except that in Sections 3.3, 5.3, 7.4, and 8.6.

Although the text is mainly developed for the undergraduate, I have also successfully used it in a *one-quarter graduate course* (first-year, three hours per week) that covers Chapters 1 through 7, including all starred material.

It should be possible to cover the entire book, including all starred material, in a *one-semester graduate course* (first-year, three hours per week).

I am indebted to many people who have helped make the book possible. Drs. R. C. Gonzalez and M. O. Pace read portions of the manuscript and suggested a number of improvements. Dr. T. V. Blalock taught from an early version of the manuscript, independently worked a number of the problems, and provided various improvements. I also extend my appreciation to the Advanced Book Program of Addison-Wesley Publishing Company for allowing me to adapt and use several of the figures from my earlier book *Communication System Principles* (1976), and to Dr. J. M. Googe, head of the electrical engineering department of the University of Tennessee, for his support and encouragement of this project. Typing of the bulk of the manuscript was ably done by Ms. Belinda Hudgens; other portions and various corrections were typed by Kymberly Scott, Sandra Wilson, and Denise Smiddy. Finally, I thank my wife, Barbara, for her aid in proofreading the entire book.

Peyton Z. Peebles, Jr.

PROBABILITY

1.0 INTRODUCTION TO BOOK AND CHAPTER

The primary goals of this book are to introduce the reader to the principles of random signals and to provide tools whereby one can deal with systems involving such signals. Toward these goals, perhaps the first thing that should be done is define what is meant by random signal. A *random signal* is a time waveform† that can be characterized only in some probabilistic manner. In general, it can be either a desired or undesired waveform.

The reader has no doubt heard background hiss while listening to an ordinary broadcast radio receiver. The waveform causing the hiss, when observed on an oscilloscope, would appear as a randomly fluctuating voltage with time. It is undesirable, since it interferes with our ability to hear the radio program, and is called *noise*.

Undesired random waveforms (noise) also appear in the outputs of other types of systems. In a radio astronomer's receiver, noise interferes with the desired signal from outer space (which itself is a random, but desirable, signal). In a television system, noise shows up in the form of picture interference often called "snow." In a sonar system, randomly generated sea sounds give rise to a noise that interferes with the desired echoes.

The number of desirable random signals is almost limitless. For example, the bits in a computer bit stream appear to fluctuate randomly with time between the

---

† We shall usually assume random signals to be voltage-time waveforms. However, the theory to be developed throughout the book will apply, in most cases, to random functions other than voltage, of arguments other than time.

zero and one states, thereby creating a random signal. In another example, the output voltage of a wind-powered generator would be random because wind speed fluctuates randomly. Similarly, the voltage from a solar detector varies randomly due to the randomness of cloud and weather conditions. Still other examples are: the signal from an instrument designed to measure instantaneous ocean wave height; the space-originated signal at the output of the radio astronomer's antenna (the relative intensity of this signal from space allows the astronomer to form radio maps of the heavens); and the voltage from a vibration analyzer attached to an automobile driving over rough terrain.

In Chapters 8 and 9 we shall study methods of characterizing systems having random input signals. However, from the above examples, it is obvious that random signals only represent the behavior of more fundamental underlying random phenomena. Phenomena associated with the desired signals of the last paragraph are: information source for computer bit stream; wind speed; various weather conditions such as cloud density and size, cloud speed, etc.; ocean wave height; sources of outer space signals; and terrain roughness. All these phenomena must be described in some probabilistic way.

Thus, there are actually two things to be considered in characterizing random signals. One is how to describe any one of a variety of random phenomena; another is how to bring time into the problem so as to create the random signal of interest. To accomplish the first item, we shall introduce mathematical concepts in Chapters 2, 3, 4, and 5 (random variables) that are sufficiently general they can apply to any suitably defined random phenomena. To accomplish the second item, we shall introduce another mathematical concept, called a random process, in Chapters 6 and 7. All these concepts are based on probability theory.

The purpose of this chapter is to introduce the elementary aspects of probability theory on which all of our later work is based. Several approaches exist for the definition and discussion of probability. Only two of these are worthy of modern-day consideration, while all others are mainly of historical interest and are not commented on further here. Of the more modern approaches, one uses the relative frequency definition of probability. It gives a degree of physical insight which is popular with engineers, and is often used in texts having principal topics other than probability theory itself (for example, see Peebles, 1976).†

The second approach to probability uses the axiomatic definition. It is the most mathematically sound of all approaches and is most appropriate for a text having its topics based principally on probability theory. The axiomatic approach also serves as the best basis for readers wishing to proceed beyond the scope of this book to more advanced theory. Because of these facts, we adopt the axiomatic approach in this book.

Prior to the introduction of the axioms of probability, it is necessary that we first develop certain elements of set theory.‡

---

† References are quoted by name and date of publication. They are listed at the end of the book.

‡ Our treatment is limited to the level required to introduce the desired probability concepts. For additional details the reader is referred to McFadden (1963), or Milton and Tsokos (1976).

1.1 SET DEFINITIONS

A *set* is a collection of objects. The objects are called *elements* of the set and may be anything whatsoever. We may have a set of voltages, a set of airplanes, a set of chairs, or even a set of sets, called a *class* of sets. A set is usually denoted by a capital letter while an element is represented by a lower-case letter. Thus, if a is an element of set A, then we write

$$a \in A \tag{1.1-1}$$

If a is not an element of A, we write

$$a \notin A \tag{1.1-2}$$

A set is specified by the content of two braces: $\{\cdot\}$. Two methods exist for specifying content, the tabular method and the rule method. In the tabular method the elements are enumerated explicitly. For example, the set of all integers between 5 and 10 would be $\{6, 7, 8, 9\}$. In the rule method, a set's content is determined by some rule, such as: {integers between 5 and 10}.† The rule method is usually more convenient to use when the set is large. For example, {integers from 1 to 1000 inclusive} would be cumbersome to write explictly using the tabular method.

A set is said to be *countable* if its elements can be put in one-to-one correspondence with the natural numbers, which are the integers 1, 2, 3, etc. If a set is not countable it is called *uncountable*. A set is said to be *empty* if it has no elements. The empty set is given the symbol $\emptyset$ and is often called the *null set*.

A *finite set* is one that is either empty or has elements that can be counted, with the counting process terminating. In other words, it has a finite number of elements. If a set is not finite it is called *infinite*. An infinite set having countable elements is called *countably infinite*.

If every element of a set A is also an element in another set B, A is said to be contained in B. A is known as a *subset* of B and we write

$$A \subseteq B \tag{1.1-3}$$

If at least one element exists in B which is not in A, then A is a *proper subset* of B, denoted by (Thomas, 1969)

$$A \subset B \tag{1.1-4}$$

The null set is clearly a subset of all other sets.

Two sets, A and B, are called *disjoint* or *mutually exclusive* if they have no common elements.

---

† Sometimes notations such as $\{I \mid 5 < I < 10, I \text{ an integer}\}$ or $\{I: 5 < I < 10, I \text{ an integer}\}$ are seen in the literature.

Example 1.1-1 To illustrate the topics discussed above, we identify the sets listed below.

$A = \{1, 3, 5, 7\}$ $D = \{0.0\}$

$B = \{1, 2, 3, \ldots\}$ $E = \{2, 4, 6, 8, 10, 12, 14\}$

$C = \{0.5 < c \le 8.5\}$ $F = \{-5.0 < f \le 12.0\}$

The set A is tabularly specified, countable, and finite. B is also tabularly specified and countable, but is infinite. Set C is rule-specified, uncountable, and infinite, since it contains *all* numbers greater than 0.5 but not exceeding 8.5. Similarly, sets D and E are countably finite, while set F is uncountably infinite. It should be noted that D is *not* the null set; it has one element, the number zero.

Set A is contained in sets B, C, and F. Similarly, $C \subset F$, $D \subset F$, and $E \subset B$. Sets B and F are not subsets of any of the other sets or of each other. Sets A, D, and E are mutually exclusive of each other. The reader may wish to identify which of the remaining sets are also mutually exclusive.

The largest or all-encompassing set of objects under discussion in a given situation is called the *universal set,* denoted S. All sets (of the situation considered) are subsets of the universal set. An example will help clarify the concept of a universal set.

Example 1.1-2 Suppose we consider the problem of rolling a die. We are interested in the numbers that show on the upper face. Here the universal set is $S = \{1, 2, 3, 4, 5, 6\}$. In a gambling game, suppose a person wins if the number comes up odd. This person wins for any number in the set $A = \{1, 3, 5\}$. Another person might win if the number shows four or less; that is, for any number in the set $B = \{1, 2, 3, 4\}$.

Observe that both A and B are subsets of S. For any universal set with N elements, there are 2^N possible subsets of S. (The reader should check this for a few values of N.) For the present example, $N = 6$ and $2^N = 64$, so that there are 64 ways one can define "winning" with one die.

It should be noted that winning or losing in the above gambling game is related to a set. The game itself is partially specified by its universal set (other games typically have a different universal set). These facts are not just coincidence, and we shall shortly find that sets form the basis on which our study of probability is constructed.

(a)

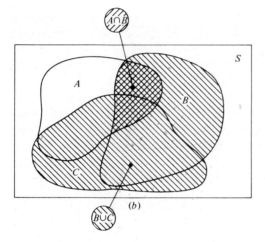

(b)

Figure 1.2-1 Venn diagrams. (a) Illustration of subsets and mutually exclusive sets, and (b) illustration of intersection and union of sets. [*Adapted from Peebles,(1976) with permission of publishers Addison-Wesley, Advanced Book Program.*]

1.2 SET OPERATIONS

In working with sets, it is helpful to introduce a geometrical representation that enables us to associate a physical picture with sets.

Venn Diagram

Such a representation is the Venn diagram.† Here sets are represented by closed-plane figures. Elements of the sets are represented by the enclosed points (area). The universal set S is represented by a rectangle as illustrated in Figure 1.2-1a. Three sets A, B, and C are shown. Set C is disjoint from both A and B, while set B is a subset of A.

Equality and Difference

Two sets A and B are *equal* if all elements in A are present in B and all elements in B are present in A; that is, if $A \subseteq B$ and $B \subseteq A$. For equal sets we write $A = B$.

The *difference* of two sets A and B, denoted $A - B$, is the set containing all

† After John Venn (1834–1923), an Englishman.

elements of A that are not present in B. For example, with $A = \{0.6 < a \leq 1.6\}$ and $B = \{1.0 \leq b \leq 2.5\}$, then $A - B = \{0.6 < c < 1.0\}$ or $B - A = \{1.6 < d \leq 2.5\}$. Note that $A - B \neq B - A$.

Union and Intersection

The *union* (call it C) of two sets A and B is written

$$C = A \cup B \tag{1.2-1}$$

It is the set of all elements of A or B or both. The union is sometimes called the *sum* of two sets.

The *intersection* (call it D) of two sets A and B is written

$$D = A \cap B \tag{1.2-2}$$

It is the set of all elements common to both A and B. Intersection is sometimes called the *product* of two sets. For mutually exclusive sets A and B, $A \cap B = \emptyset$. Figure 1.2-1*b* illustrates the Venn diagram area to be associated with the intersection and union of sets.

By repeated application of (1.2-1) or (1.2-2), the union and intersection of N sets A_n, $n = 1, 2, \ldots, N$, become

$$C = A_1 \cup A_2 \cup \cdots \cup A_N = \bigcup_{n=1}^{N} A_n \tag{1.2-3}$$

$$D = A_1 \cap A_2 \cap \cdots \cap A_N = \bigcap_{n=1}^{N} A_n \tag{1.2-4}$$

Complement

The *complement* of a set A, denoted by $\bar{A}$, is the set of all elements not in A. Thus,

$$\bar{A} = S - A \tag{1.2-5}$$

It is also easy to see that $\bar{\emptyset} = S$, $\bar{S} = \emptyset$, $A \cup \bar{A} = S$, and $A \cap \bar{A} = \emptyset$.

Example 1.2-1 We illustrate intersection, union, and complement by taking an example with the four sets

$S = \{1 \leq \text{integers} \leq 12\}$ $\qquad$ $B = \{2, 6, 7, 8, 9, 10, 11\}$

$A = \{1, 3, 5, 12\}$ $\qquad$ $C = \{1, 3, 4, 6, 7, 8\}$

Applicable unions and intersections here are:

$A \cup B = \{1, 2, 3, 5, 6, 7, 8, 9, 10, 11, 12\}$ $\qquad$ $A \cap B = \emptyset$

$A \cup C = \{1, 3, 4, 5, 6, 7, 8, 12\}$ $\qquad$ $A \cap C = \{1, 3\}$

$B \cup C = \{1, 2, 3, 4, 6, 7, 8, 9, 10, 11\}$ $\qquad$ $B \cap C = \{6, 7, 8\}$

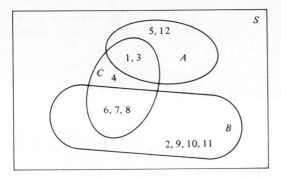

Figure 1.2-2 Venn diagram applicable to Example 1.2-1.

Complements are:

$$\bar{A} = \{2, 4, 6, 7, 8, 9, 10, 11\}$$
$$\bar{B} = \{1, 3, 4, 5, 12\}$$
$$\bar{C} = \{2, 5, 9, 10, 11, 12\}$$

The various sets are illustrated in Figure 1.2-2.

Algebra of Sets

All subsets of the universal set form an algebraic system for which a number of theorems may be stated (Thomas, 1969). Three of the most important of these relate to laws involving unions and intersections. The *commutative law* states that

$$A \cap B = B \cap A \qquad (1.2\text{-}6)$$
$$A \cup B = B \cup A \qquad (1.2\text{-}7)$$

The *distributive law* is written as

$$A \cap (B \cup C) = (A \cap B) \cup (A \cap C) \qquad (1.2\text{-}8)$$
$$A \cup (B \cap C) = (A \cup B) \cap (A \cup C) \qquad (1.2\text{-}9)$$

The *associative law* is written as

$$(A \cup B) \cup C = A \cup (B \cup C) = A \cup B \cup C \qquad (1.2\text{-}10)$$
$$(A \cap B) \cap C = A \cap (B \cap C) = A \cap B \cap C \qquad (1.2\text{-}11)$$

These are just restatements of (1.2-3) and (1.2-4).

De Morgan's Laws

By use of a Venn diagram we may readily prove *De Morgan's laws*†, which state that the complement of a union (intersection) of two sets A and B equals the intersection (union) of the complements $\bar{A}$ and $\bar{B}$. Thus,

$$\overline{(A \cup B)} = \bar{A} \cap \bar{B} \qquad (1.2\text{-}12)$$
$$\overline{(A \cap B)} = \bar{A} \cup \bar{B} \qquad (1.2\text{-}13)$$

† After Augustus De Morgan (1806–1871), an English mathematician.

From the last two expressions one can show that if in an identity we replace unions by intersections, intersections by unions, and sets by their complements, then the identity is preserved (Papoulis, 1965, p. 23).

Example 1.2-2 We verify De Morgan's law (1.2-13) by using the example sets $A = \{2 < a \le 16\}$ and $B = \{5 < b \le 22\}$ when $S = \{2 < s \le 24\}$. First, if we define $C = A \cap B$, the reader can readily see from Venn diagrams that $C = A \cap B = \{5 < c \le 16\}$ so $\bar{C} = \overline{A \cap B} = \{2 < c \le 5, \ 16 < c \le 24\}$. This result is the left side of (1.2-13).

Second, we compute $\bar{A} = S - A = \{16 < a \le 24\}$ and $\bar{B} = S - B = \{2 < b \le 5, \ 22 < b \le 24\}$. Thus, $C = \bar{A} \cup \bar{B} = \{2 < c \le 5, \ 16 < c \le 24\}$. This result is the right side of (1.2-13) and De Morgan's law is verified.

Duality Principle

This principle (Papoulis, 1965) states: if in an identity we replace unions by intersections, intersections by unions, S by $\emptyset$, and $\emptyset$ by S, then the identity is preserved. For example, since

$$A \cap (B \cup C) = (A \cap B) \cup (A \cap C) \tag{1.2-14}$$

is a valid identity from (1.2-8), it follows that

$$A \cup (B \cap C) = (A \cup B) \cap (A \cup C) \tag{1.2-15}$$

is also valid, which is just (1.2-9).

1.3 PROBABILITY INTRODUCED THROUGH SETS

Basic to our study of probability is the idea of a physical *experiment*. In this section we develop a mathematical model of an experiment. Of course, we are interested only in experiments that are regulated in some probabilistic way. A single performance of the experiment is called a *trial* for which there is an *outcome*.

Experiments and Sample Spaces

Although there exists a precise mathematical procedure for defining an experiment, we shall rely on reason and examples. This simplified approach will ultimately lead us to a valid mathematical model for any real experiment.† To

---

† Most of our early definitions involving probability are rigorously established only through concepts beyond our scope. Although we adopt a simplified development of the theory, our final results are no less valid or useful than if we had used the advanced concepts.

illustrate, one experiment might consist of rolling a single die and observing the number that shows up. There are six such numbers and they form all the possible outcomes in the experiment. If the die is "unbiased" our intuition tells us that each outcome is equally likely to occur and the *likelihood* of any one occurring is $\frac{1}{6}$ (later we call this number the *probability* of the outcome). This experiment is seen to be governed, in part, by two *sets*. One is the set of all possible outcomes, and the other is the set of the likelihoods of the outcomes. Each set has six elements. For the present, we consider only the set of outcomes.

The set of all possible outcomes in any given experiment is called the *sample space* and it is given the symbol *S*. In effect, the sample space is a universal set for the given experiment. *S* may be different for different experiments, but all experiments are governed by some sample space. The definition of sample space forms the first of three elements in our mathematical model of experiments. The remaining elements are *events* and *probability*, as discussed below.

Discrete and Continuous Sample Spaces

In the earlier die-tossing experiment, *S* was a finite set with six elements. Such sample spaces are said to be *discrete* and finite. The sample space can also be discrete and *infinite* for some experiments. For example, *S* in the experiment "choose randomly a positive integer" is the countably infinite set $\{1, 2, 3, \ldots\}$.

Some experiments have an uncountably infinite sample space. An illustration would be the experiment "obtain a number by spinning the pointer on a wheel of chance numbered from 0 to 12." Here any number *s* from 0 to 12 can result and $S = \{0 < s \leq 12\}$. Such a sample space is called *continuous*.

Events

In most situations, we are interested in some *characteristic* of the outcomes of our experiment as opposed to the outcomes themselves. In the experiment "draw a card from a deck of 52 cards," we might be more interested in whether we draw a spade as opposed to having any interest in individual cards. To handle such situations we define the concept of an event.

An *event* is defined as a subset of the sample space. Because an event is a set, all the earlier definitions and operations applicable to sets will apply to events. For example, if two events have no common outcomes they are *mutually exclusive*.

In the above card experiment, 13 of the 52 possible outcomes are spades. Since any one of the spade outcomes satisfies the event "draw a spade," this event is a set with 13 elements. We have earlier stated that a set with *N* elements can have as many as 2^N subsets (events defined on a sample space having *N* possible outcomes). In the present example, $2^N = 2^{52} \approx 4.5(10^{15})$ events.

As with the sample space, events may be either discrete or continuous. The card event "draw a spade" is a discrete, finite event. An example of a discrete, countably infinite event would be "select an odd integer" in the experiment "randomly select a positive integer." The event has a countably infinite number of elements: $\{1, 3, 5, 7, \ldots\}$. However, events defined on a coutably infinite sample space do not *have* to be countably infinite. The event $\{1, 3, 5, 7\}$ is clearly not infinite but applies to the integer selection experiment.

Events defined on continuous sample spaces are usually continuous. In the experiment "choose randomly a number a from 6 to 13," the sample space is $S = \{6 \leq s \leq 13\}$. An event of interest might correspond to the chosen number falling between 7.4 and 7.6; that is, the event (call it A) is $A = \{7.4 < a < 7.6\}$.

Discrete events may also be defined on continuous sample spaces. An example of such an event is $A = \{6.13692\}$ for the sample space $S = \{6 \leq s \leq 13\}$ of the previous paragraph. We comment later on this type of event.

The above definition of an event as a subset of the sample space forms the second of three elements in our mathematical model of experiments. The third element involves defining probability.

Probability Definition and Axioms

To each event defined on a sample space S, we shall assign a nonnegative number called *probability*. Probability is therefore a function; it is a function of the events defined. We adopt the notation $P(A)$† for "the probability of event A." When an event is stated explicitly as a set by using braces, we employ the notation $P\{\cdot\}$ instead of $P(\{\cdot\})$.

The assigned probabilities are chosen so as to satisfy three *axioms*. Let A be any event defined on a sample space S. Then the first two axioms are

axiom 1: $$P(A) \geq 0 \qquad (1.3\text{-}1a)$$

axiom 2: $$P(S) = 1 \qquad (1.3\text{-}1b)$$

The first only represents our desire to work with nonnegative numbers. The second axiom recognizes that the sample space itself is an event, and, since it is the all encompassing event, it should have the highest possible probability, which is selected as unity. For this reason, S is known as the *certain event*. Alternatively, the null set $\varnothing$ is an event with no elements; it is known as the *impossible event* and its probability is 0.

The third axiom applies to N events A_n, $n = 1, 2, \ldots, N$, where N may possibly be infinite, defined on a sample space S, and having the property $A_m \cap A_n = \varnothing$ for all $m \neq n$. It is

axiom 3: $$P\left(\bigcup_{n=1}^{N} A_n\right) = \sum_{n=1}^{N} P(A_n) \qquad \text{if} \qquad A_m \cap A_n = \varnothing \qquad (1.3\text{-}1c)$$

for all $m \neq n = 1, 2, \ldots, N$, with N possibly infinite. The axiom states that the

---

† Occasionally it will be convenient to use brackets, such as $P[A]$ when A is itself an event such as $C - (B \cap D)$.

probability of the event equal to the union of any number of mutually exclusive events is equal to the sum of the individual event probabilities.

An example should help give a physical picture of the meaning of the above axioms.

Example 1.3-1 Let an experiment consist of obtaining a number x by spinning the pointer on a "fair" wheel of chance that is labeled from 0 to 100 points. The sample space is $S = \{0 < x \leq 100\}$. We reason that probability of the pointer falling between any two numbers $x_2 \geq x_1$ should be $(x_2 - x_1)/100$ since the wheel is fair. As a check on this assignment, we see that the event $A = \{x_1 < x \leq x_2\}$ satisfies axiom 1 for all x_1 and x_2, and axiom 2 when $x_2 = 100$ and $x_1 = 0$.

Now suppose we break the wheel's periphery into N contiguous segments $A_n = \{x_{n-1} < x \leq x_n\}$, $x_n = (n)100/N$, $n = 1, 2, \ldots, N$, with $x_0 = 0$. Then $P(A_n) = 1/N$, and, for any N,

$$P\left(\bigcup_{n=1}^{N} A_n\right) = \sum_{n=1}^{N} P(A_n) = \sum_{n=1}^{N} \frac{1}{N} = 1 = P(S)$$

from axiom 3.

Example 1.3-1 allows us to return to our earlier discussion of discrete events defined on continuous sample spaces. If the interval $x_n - x_{n-1}$ is allowed to approach zero ($\rightarrow 0$), the probability $P(A_n) \rightarrow P(x_n)$; that is, $P(A_n)$ becomes the probability of the pointer falling exactly on the point x_n. Since $N \rightarrow \infty$ in this situation, $P(A_n) \rightarrow 0$. Thus, the probability of a discrete event defined on a continuous sample space is 0. This fact is true in general.

A consequence of the above statement is that events can occur even if their probability is 0. Intuitively, any number can be obtained from the wheel of chance, but that precise number may never occur again. The infinite sample space has only one outcome satisfying such a discrete event, so its probability is 0. Such events are *not* the same as the impossible event which has *no* elements and *cannot* occur. The converse situation can also happen where events with probability 1 may *not* occur. An example for the wheel of chance experiment would be the event $A = \{$all numbers except the number $x_n\}$. Events with probability 1 (that may not occur) are not the same as the certain event which *must* occur.

Mathematical Model of Experiments

The axioms of probability, introduced above, complete our mathematical model of an experiment. We pause to summarize. Given some real physical experiment having a set of particular outcomes possible, we first defined a *sample space* to mathematically represent the physical outcomes. Second, it was recognized that certain characteristics of the outcomes in the real experiment were of interest, as opposed to the outcomes themselves; *events* were defined to mathematically represent these characteristics. Finally, *probabilities* were assigned to the defined events to mathematically account for the random nature of the experiment.

Thus, a real experiment is defined mathematically by three things: (1) assignment of a sample space; (2) definition of events of interest; and (3) making probability assignments to the events such that the axioms are satisfied. Establishing the correct model for an experiment is probably the single most difficult step in solving probability problems.

Example 1.3-2 An experiment consists of observing the sum of the numbers showing up when two dice are thrown. We develop a model for this experiment.

The sample space consists of $6^2 = 36$ points as shown in Figure 1.3-1. Each possible outcome corresponds to a sum having values from 2 to 12.

Suppose we are mainly interested in three events defined by $A = \{\text{sum} = 7\}$, $B = \{8 < \text{sum} \leq 11\}$, and $C = \{10 < \text{sum}\}$. In assigning probabilities to these events, it is first convenient to define 36 *elementary events* $A_{ij} = \{\text{sum for outcome } (i, j) = i + j\}$, where i represents the row and j represents the column locating a particular possible outcome in Figure 1.3-1. An elementary event has only one element.

For probability assignments, intuition indicates that each possible outcome has the same likelihood of occurrence if the dice are fair, so $P(A_{ij}) = \frac{1}{36}$. Now because the events A_{ij}, i and $j = 1, 2, \ldots, N = 6$, are mutually exclusive, they must satisfy axiom 3. But since the events A, B, and C are simply the unions of appropriate elementary events, their probabilities are derived from axiom 3. From Figure 1.3-1 we easily find

$$P(A) = P\left(\bigcup_{i=1}^{6} A_{i, 7-i}\right) = \sum_{i=1}^{6} P(A_{i, 7-i}) = 6\left(\frac{1}{36}\right) = \frac{1}{6}$$

$$P(B) = 9\left(\frac{1}{36}\right) = \frac{1}{4}$$

$$P(C) = 3\left(\frac{1}{36}\right) = \frac{1}{12}$$

Figure 1.3-1 Sample space applicable to Example 1.3-2.

As a matter of interest, we also observe the probabilities of the events $B \cap C$ and $B \cup C$ to be $P(B \cap C) = 2(\frac{1}{36}) = \frac{1}{18}$ and $P(B \cup C) = 10(\frac{1}{36}) = \frac{5}{18}$.

1.4 JOINT AND CONDITIONAL PROBABILITY

In some experiments, such as in Example 1.3-2 above, it may be that some events are not mutually exclusive because of common elements in the sample space. These elements correspond to the simultaneous or *joint* occurrence of the non-exclusive events. For two events A and B, the common elements from the event $A \cap B$.

Joint Probability

The probability $P(A \cap B)$ is called the *joint probability* for two events A and B which intersect in the sample space. A study of a Venn diagram will readily show that

$$P(A \cap B) = P(A) + P(B) - P(A \cup B) \tag{1.4-1}$$

Equivalently,

$$P(A \cup B) = P(A) + P(B) - P(A \cap B) \le P(A) + P(B) \tag{1.4-2}$$

In other words, the probability of the union of two events never exceeds the sum of the event probabilities. The equality holds only for mutually exclusive events because $A \cap B = \emptyset$, and therefore, $P(A \cap B) = P(\emptyset) = 0$.

Conditional Probability

Given some event B with nonzero probability

$$P(B) > 0 \tag{1.4-3}$$

we define the *conditional probability* of an event A, given B, by

$$P(A \mid B) = \frac{P(A \cap B)}{P(B)} \tag{1.4-4}$$

The probability $P(A \mid B)$ simply reflects the fact that the probability of an event A may depend on a second event B. If A and B are mutually exclusive, $A \cap B = \emptyset$, and $P(A \mid B) = 0$.

Conditional probability is a defined quantity and cannot be proven. However, as a probability it must satisfy the three axioms given in (1.3-1). $P(A \mid B)$ obviously satisfies axiom 1 by its definition because $P(A \cap B)$ and $P(B)$ are non-negative numbers. The second axiom is shown to be satisfied by letting $S = A$:

$$P(S \mid B) = \frac{P(S \cap B)}{P(B)} = \frac{P(B)}{P(B)} = 1 \tag{1.4-5}$$

The third axiom may be shown to hold by considering the union of A with an event C, where A and C are mutually exclusive. If $P(A \cup C | B) = P(A | B) + P(C | B)$ is true, then axiom 3 holds. Since $A \cap C = \varnothing$ then events $A \cap B$ and $B \cap C$ are mutually exclusive (use a Venn diagram to verify this fact) and

$$P[(A \cup C) \cap B] = P[(A \cap B) \cup (C \cap B)] = P(A \cap B) + P(C \cap B)$$

(1.4-6)

Thus, on substitution into (1.4-4)

$$P[(A \cup C) | B] = \frac{P[(A \cup C) \cap B]}{P(B)} = \frac{P(A \cap B)}{P(B)} + \frac{P(C \cap B)}{P(B)}$$

$$= P(A | B) + P(C | B)$$

(1.4-7)

and axiom 3 holds.

Example 1.4-1 In a box there are 100 resistors having resistance and tolerance as shown in Table 1.4-1. Let a resistor be selected from the box and assume each resistor has the same likelihood of being chosen. Define three events: A as "draw a 47-Ω resistor," B as "draw a resistor with 5% tolerance," and C as "draw a 100-Ω resistor." From the table, the applicable probabilities are†

$$P(A) = P(47 \ \Omega) = \frac{44}{100}$$

$$P(B) = P(5\%) = \frac{62}{100}$$

$$P(C) = P(100 \ \Omega) = \frac{32}{100}$$

The joint probabilities are

$$P(A \cap B) = P(47 \ \Omega \cap 5\%) = \frac{28}{100}$$

$$P(A \cap C) = P(47 \ \Omega \cap 100 \ \Omega) = 0$$

$$P(B \cap C) = P(5\% \cap 100 \ \Omega) = \frac{24}{100}$$

† It is reasonable that probabilities are related to the *number* of resistors in the box that satisfy an event, since each resistor is equally likely to be selected. An alternative approach would be based on elementary events similar to that used in Example 1.3-2. The reader may view the latter approach as more rigorous but less readily applied.

Table 1.4-1 Numbers of resistors in a box having given resistance and tolerance.

Resistance (Ω)	Tolerance		
	5%	10%	Total
22	10	14	24
47	28	16	44
100	24	8	32
Total	62	38	100

By using (1.4-4) the conditional probabilities become

$$P(A\,|\,B) = \frac{P(A \cap B)}{P(B)} = \frac{28}{62}$$

$$P(A\,|\,C) = \frac{P(A \cap C)}{P(C)} = 0$$

$$P(B\,|\,C) = \frac{P(B \cap C)}{P(C)} = \frac{24}{32}$$

$P(A\,|\,B) = P(47\,\Omega\,|\,5\%)$ is the probability of drawing a 47-Ω resistor given that the resistor drawn is 5%. $P(A\,|\,C) = P(47\,\Omega\,|\,100\,\Omega)$ is the probability of drawing a 47-Ω resistor given that the resistor drawn is 100 Ω; this is clearly an impossible event so the probability of it is 0. Finally, $P(B\,|\,C) = P(5\%\,|\,100\,\Omega)$ is the probability of drawing a resistor of 5% tolerance given that the resistor is 100 Ω.

Total Probability

The probability $P(A)$ of any event A defined on a sample space S can be expressed in terms of conditional probabilities. Suppose we are given N mutually exclusive events B_n, $n = 1, 2, \ldots, N$, whose union equals S as illustrated in Figure 1.4-1. These events satisfy

$$B_m \cap B_n = \varnothing \qquad m \neq n = 1, 2, \ldots, N \tag{1.4-8}$$

$$\bigcup_{n=1}^{N} B_n = S \tag{1.4-9}$$

We shall prove that

$$P(A) = \sum_{n=1}^{N} P(A\,|\,B_n)P(B_n) \tag{1.4-10}$$

which is known as the *total probability* of event A.

$$\bigcup_{n=1}^{N} B_n = S, \quad B_m \cap B_n = \emptyset \qquad \text{for all } m \neq n$$

Figure 1.4-1 Venn diagram of N mutually exclusive events B_n and another event A.

Since $A \cap S = A$, we may start the proof using (1.4-9) and (1.2-8):

$$A \cap S = A \cap \left(\bigcup_{n=1}^{N} B_n \right) = \bigcup_{n=1}^{N} (A \cap B_n) \tag{1.4-11}$$

Now the events $A \cap B_n$ are mutually exclusive as seen from the Venn diagram (Fig. 1.4-1). By applying axiom 3 to these events, we have

$$P(A) = P(A \cap S) = P\left[\bigcup_{n=1}^{N} (A \cap B_n) \right] = \sum_{n=1}^{N} P(A \cap B_n) \tag{1.4-12}$$

where (1.4-11) has been used. Finally, (1.4-4) is substituted into (1.4-12) to obtain (1.4-10).

Bayes' Theorem†

The definition of conditional probability, as given by (1.4-4), applies to any two events. In particular, let B_n be one of the events defined above in the subsection on total probability. Equation (1.4-4) can be written

$$P(B_n | A) = \frac{P(B_n \cap A)}{P(A)} \tag{14-13}$$

if $P(A) \neq 0$, or, alternatively,

$$P(A | B_n) = \frac{P(A \cap B_n)}{P(B_n)} \tag{1.4-14}$$

if $P(B_n) \neq 0$. One form of Bayes' theorem is obtained by equating these two expressions:

$$P(B_n | A) = \frac{P(A | B_n)P(B_n)}{P(A)} \tag{1.4-15}$$

† The theorem is named for Thomas Bayes (1702–1761), an English philosopher.

Another form derives from a substitution of $P(A)$ as given by (1.4-10),

$$P(B_n \mid A) = \frac{P(A \mid B_n)P(B_n)}{P(A \mid B_1)P(B_1) + \cdots + P(A \mid B_N)P(B_N)} \qquad (1.4\text{-}16)$$

for $n = 1, 2, \ldots, N$.

An example will serve to illustrate Bayes' theorem and conditional probability.

Example 1.4-2 An elementary binary communication system consists of a transmitter that sends one of two possible symbols (a 1 or a 0) over a channel to a receiver. The channel occasionally causes errors to occur so that a 1 shows up at the receiver as a 0, and vice versa.

The sample space has two elements (0 or 1). We denote by B_i, $i = 1, 2$, the events "the symbol before the channel is 1," and "the symbol before the channel is 0," respectively. Furthermore, define A_i, $i = 1, 2$, as the events "the symbol after the channel is 1," and "the symbol after the channel is 0," respectively. The probabilities that the symbols 1 and 0 are selected for transmission are assumed to be

$$P(B_1) = 0.6 \qquad \text{and} \qquad P(B_2) = 0.4$$

Conditional probabilities describe the effect the channel has on the transmitted symbols. The reception probabilities given a 1 was transmitted are assumed to be

$$P(A_1 \mid B_1) = 0.9$$
$$P(A_2 \mid B_1) = 0.1$$

The channel is presumed to affect 0s in the same manner so

$$P(A_1 \mid B_2) = 0.1$$
$$P(A_2 \mid B_2) = 0.9$$

In either case, $P(A_1 \mid B_i) + P(A_2 \mid B_i) = 1$ because A_1 and A_2 are mutually exclusive and are the only "receiver" events (other than the uninteresting events $\varnothing$ and S) possible. The channel is often shown diagrammatically as illustrated in Figure 1.4-2. Because of its form it is usually called a *binary symmetric channel*.

From (1.4-10) we obtain the "received" symbol probabilities

$$P(A_1) = P(A_1 \mid B_1)P(B_1) + P(A_1 \mid B_2)P(B_2)$$
$$= 0.9(0.6) + 0.1(0.4) = 0.58$$
$$P(A_2) = P(A_2 \mid B_1)P(B_1) + P(A_2 \mid B_2)P(B_2)$$
$$= 0.1(0.6) + 0.9(0.4) = 0.42$$

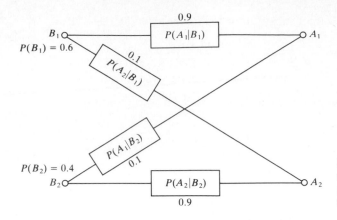

Figure 1.4-2 Binary symmetric communication system diagrammatical model applicable to Example 1.4-2.

From either (1.4-15) or (1.4-16) we have

$$P(B_1 \mid A_1) = \frac{P(A_1 \mid B_1)P(B_1)}{P(A_1)} = \frac{0.9(0.6)}{0.58} = \frac{0.54}{0.58} \approx 0.931$$

$$P(B_2 \mid A_2) = \frac{P(A_2 \mid B_2)P(B_2)}{P(A_2)} = \frac{0.9(0.4)}{0.42} = \frac{0.36}{0.42} \approx 0.857$$

$$P(B_1 \mid A_2) = \frac{P(A_2 \mid B_1)P(B_1)}{P(A_2)} = \frac{0.1(0.6)}{0.42} = \frac{0.06}{0.42} \approx 0.143$$

$$P(B_2 \mid A_1) = \frac{P(A_1 \mid B_2)P(B_2)}{P(A_1)} = \frac{0.1(0.4)}{0.58} = \frac{0.04}{0.58} \approx 0.069$$

These last two numbers are probabilities of system error while $P(B_1 \mid A_1)$ and $P(B_2 \mid A_2)$ are probabilities of correct system transmission of symbols.

In Bayes' theorem (1.4-16), the probabilities $P(B_n)$ are usually referred to as *a priori probabilities*, since they apply to the events B_n before the performance of the experiment. Similarly, the probabilities $P(A \mid B_n)$ are numbers typically known prior to conducting the experiment. Example 1.4-2 described such a case. The conditional probabilities are sometimes called *transition probabilities* in a communications context. On the other hand, the probabilities $P(B_n \mid A)$ are called *a posteriori probabilities*, since they apply after the experiment's performance when some event A is obtained.

1.5 INDEPENDENT EVENTS

In this section we introduce the concept of statistically independent events. Although a given problem may involve any number of events in general, it is most instructive to consider first the simplest possible case of two events.

Two Events

Let two events A and B have nonzero probabilities of occurrence; that is, assume $P(A) \neq 0$ and $P(B) \neq 0$. We call the events *statistically independent* if the probability of occurrence of one event is not affected by the occurrence of the other event. Mathematically, this statement is equivalent to requiring

$$P(A \mid B) = P(A) \qquad (1.5\text{-}1)$$

for statistically independent events. We also have

$$P(B \mid A) = P(B) \qquad (1.5\text{-}2)$$

for statistically independent events. By substitution of (1.5-1) into (1.4-4), independence† also means that the probability of the joint occurrence (intersection) of two events must equal the product of the two event probabilities:

$$P(A \cap B) = P(A)P(B) \qquad (1.5\text{-}3)$$

Not only is (1.5-3) [or (1.5-1)] necessary for two events to be independent but it is sufficient. As a consequence, (1.5-3) can, and often does, serve as a test of independence.

Statistical independence is fundamental to much of our later work. When events are independent it will often be found that probability problems are greatly simplified.

It has already been stated that the joint probability of two mutually exclusive events is 0:

$$P(A \cap B) = 0 \qquad (1.5\text{-}4)$$

If the two events have nonzero probabilities of occurrence, then, by comparison of (1.5-4) with (1.5-3), we easily establish that two events cannot be both mutually exclusive and statistically independent. Hence, in order for two events to be independent they *must* have an intersection $A \cap B \neq \emptyset$.

If a problem involves more than two events, those events satisfying either (1.5-3) or (1.5-1) are said to be *independent by pairs*.

Example 1.5-1 In an experiment, one card is selected from an ordinary 52-card deck. Define events A as "select a king," B as "select a jack or queen," and C as "select a heart." From intuition, these events have probabilities $P(A) = \frac{4}{52}$, $P(B) = \frac{8}{52}$, and $P(C) = \frac{13}{52}$.

It is also easy to state joint probabilities. $P(A \cap B) = 0$ (it is not possible to simultaneously select a king and a jack or queen), $P(A \cap C) = \frac{1}{52}$, and $P(B \cap C) = \frac{2}{52}$.

---

† We shall often use only the word independence to mean statistical independence.

We determine whether A, B, and C are independent by pairs by applying (1.5-3):

$$P(A \cap B) = 0 \neq P(A)P(B) = \frac{32}{52^2}$$

$$P(A \cap C) = \frac{1}{52} = P(A)P(C) = \frac{1}{52}$$

$$P(B \cap C) = \frac{2}{52} = P(B)P(C) = \frac{2}{52}$$

Thus, A and C are independent as a pair, as are B and C. However, A and B are not independent, as we might have guessed from the fact that A and B are mutually exclusive.

In many practical problems, statistical independence of events is often *assumed*. The justification hinges on there being no apparent physical connection between the mechanisms leading to the events. In other cases, probabilities assumed for elementary events may lead to independence of other events defined from them (Cooper and McGillem, 1971, p. 24).

Multiple Events

When more than two events are involved, independence by pairs is not sufficient to establish the events as statistically independent, even if *every* pair satisfies (1.5-3).

In the case of three events A_1, A_2, and A_3, they are said to be independent if, and only if, they are independent by all pairs and are also independent as a triple; that is, they must satisfy the *four* equations:

$$P(A_1 \cap A_2) = P(A_1)P(A_2) \tag{1.5-5a}$$

$$P(A_1 \cap A_3) = P(A_1)P(A_3) \tag{1.5-5b}$$

$$P(A_2 \cap A_3) = P(A_2)P(A_3) \tag{1.5-5c}$$

$$P(A_1 \cap A_2 \cap A_3) = P(A_1)P(A_2)P(A_3) \tag{1.5-5d}$$

The reader may wonder if satisfaction of (1.5-5d) might be sufficient to guarantee independence by pairs, and therefore, satisfaction of all four conditions? The answer is no, and supporting examples are relatively easy to construct. The reader might try this exercise.

More generally, for N events $A_1, A_2, \ldots, A_N$ to be called statistically independent, we require that all the conditions

$$P(A_i \cap A_j) = P(A_i)P(A_j)$$

$$P(A_i \cap A_j \cap A_k) = P(A_i)P(A_j)P(A_k)$$

$$\vdots \qquad (1.5\text{-}6)$$

$$P(A_1 \cap A_2 \cap \cdots \cap A_N) = P(A_1)P(A_2) \cdots P(A_N)$$

be satisfied for all $1 \le i < j < k < \cdots \le N$. There are $2^N - N - 1$ of these conditions (Davenport, 1970, p. 83).

Example 1.5-2 Consider drawing four cards from an ordinary 52-card deck. Let events A_1, A_2, A_3, A_4 define drawing an ace on the first, second, third, and fourth cards, respectively. Consider two cases. First, draw the cards assuming each is replaced after the draw. Intuition tells us that these events are independent so $P(A_1 \cap A_2 \cap A_3 \cap A_4) = P(A_1)P(A_2)P(A_3)P(A_4) = (4/52)^4 \approx 3.50(10^{-5})$.

On the other hand, suppose we keep each card after it is drawn. We now expect these are not independent events. In the general case we may write

$$P(A_1 \cap A_2 \cap A_3 \cap A_4)$$

$$= P(A_1)P(A_2 \cap A_3 \cap A_4 | A_1)$$

$$= P(A_1)P(A_2|A_1)P(A_3 \cap A_4 | A_1 \cap A_2)$$

$$= P(A_1)P(A_2|A_1)P(A_3|A_1 \cap A_2)P(A_4|A_1 \cap A_2 \cap A_3)$$

$$= \frac{4}{52} \cdot \frac{3}{51} \cdot \frac{2}{50} \cdot \frac{1}{49} \approx 3.69(10^{-6})$$

Thus, we have approximately 9.5-times better chance of drawing four aces when cards are replaced than when kept. This is an intuitively satisfying result since replacing the ace drawn raises chances for an ace on the succeeding draw.

Properties of Independent Events

Many properties of independent events may be summarized by the statement: If N events $A_1, A_2, \ldots, A_N$ are independent, then any one of them is independent of any event formed by unions, intersections, and complements of the others (Papoulis, 1965, p. 42). Several examples of the application of this statement are worth listing for illustration.

For two independent events A_1 and A_2 it results that A_1 is independent of $\bar{A}_2$, $\bar{A}_1$ is independent of A_2, and $\bar{A}_1$ is independent of $\bar{A}_2$. These statements are proved as a problem at the end of this chapter.

For three independent events A_1, A_2, and A_3 any one is independent of the joint occurrence of the other two. For example

$$P[A_1 \cap (A_2 \cap A_3)] = P(A_1)P(A_2)P(A_3) = P(A_1)P(A_2 \cap A_3) \qquad (1.5\text{-}7)$$

with similar statements possible for the other cases $A_2 \cap (A_1 \cap A_3)$ and $A_3 \cap (A_1 \cap A_2)$. Any one event is also independent of the union of the other two. For example

$$P[A_1 \cap (A_2 \cup A_3)] = P(A_1)P(A_2 \cup A_3) \qquad (1.5\text{-}8)$$

This result and (1.5-7) do not necessarily hold if the events are only independent by pairs.

*1.6 COMBINED EXPERIMENTS

All of our work up to this point is related to outcomes from a single experiment. Many practical problems arise where such a constrained approach does not apply. One example would be the simultaneous measurement of wind speed and barometric pressure at some location and instant in time. *Two* experiments are actually being conducted; one has the outcome " speed"; the other outcome is "pressure." Still another type of problem involves conducting the *same* experiment several times, such as flipping a coin N times. In this case there are N performances of the same experiment. To handle these situations we introduce the concept of a combined experiment.

A *combined experiment* consists of forming a *single* experiment by suitably combining individual experiments, which we now call *subexperiments*. Recall that an experiment is defined by specifying three quantities. They are: (1) the applicable sample space, (2) the events defined on the sample space, and (3) the probabilities of the events. We specify these three quantities below, beginning with the sample space, for a combined experiment.

*Combined Sample Space

Consider only two subexperiments first. Let S_1 and S_2 be the sample spaces of the two subexperiments and let s_1 and s_2 represent the elements of S_1 and S_2 respectively. We form a new space S, called the *combined sample space*,† whose elements are all the ordered pairs (s_1, s_2). Thus, if S_1 has M elements and S_2 has N elements, then S will have MN elements. The combined sample space is denoted

$$S = S_1 \times S_2 \qquad (1.6\text{-}1)$$

---

† Also called the *cartesian product space* in some texts.

Example 1.6-1 If S_1 corresponds to flipping a coin, then $S_1 = \{H, T\}$, where H is the element "heads" and T represents "tails." Let $S_2 = \{1, 2, 3, 4, 5, 6\}$ corresponding to rolling a single die. The combined sample space $S = S_1 \times S_2$ becomes

$$S = \{(H, 1), (H, 2), (H, 3), (H, 4), (H, 5), (H, 6),$$
$$(T, 1), (T, 2), (T, 3), (T, 4), (T, 5), (T, 6)\}$$

In the new space, elements are considered to be single objects, each object being a pair of items.

Example 1.6-2 We flip a coin twice, each flip being taken as one sub-experiment. The applicable sample spaces are now

$$S_1 = \{H, T\}$$
$$S_2 = \{H, T\}$$
$$S = \{(H, H), (H, T), (T, H), (T, T)\}$$

In this last example, observe that the element (H, T) is considered different from the element (T, H); this fact emphasizes the elements of S are *ordered* pairs of objects.

The more general situation of N subexperiments is a direct extension of the above concepts. For N sample spaces S_n, $n = 1, 2, \ldots, N$, having elements s_n, the combined sample space S is denoted

$$S = S_1 \times S_2 \times \cdots \times S_N \tag{1.6-2}$$

and it is the set of all ordered N-tuples

$$(s_1, s_2, \ldots, s_N) \tag{1.6-3}$$

★Events on the Combined Space

Events may be defined on the combined sample space through their relationship with events defined on the subexperiment sample spaces. Consider two subexperiments with sample spaces S_1 and S_2. Let A be any event defined on S_1 and B be any event defined on S_2, then

$$C = A \times B \tag{1.6-4}$$

is an event defined on S consisting of all pairs (s_1, s_2) such that

$$s_1 \in A \quad \text{and} \quad s_2 \in B \tag{1.6-5}$$

Since elements of A correspond to elements of the event $A \times S_2$ defined on S, and elements of B correspond to the event $S_1 \times B$ defined on S, we easily find that

$$A \times B = (A \times S_2) \cap (S_1 \times B) \tag{1.6-6}$$

Thus, the event defined by the subset of S given by $A \times B$ is the intersection of the subsets $A \times S_2$ and $S_1 \times B$. We consider all subsets of S of the form $A \times B$ as events. All intersections and unions of such events are also events (Papoulis, 1965, p. 50).

Example 1.6-3 Let $S_1 = \{0 \leq x \leq 100\}$ and $S_2 = \{0 \leq y \leq 50\}$. The combined sample space is the set of all pairs of numbers (x, y) with $0 \leq x \leq 100$ and $0 \leq y \leq 50$ as illustrated in Figure 1.6-1. For events

$$A = \{x_1 < x < x_2\}$$
$$B = \{y_1 < y < y_2\}$$

where $0 \leq x_1 < x_2 \leq 100$ and $0 \leq y_1 < y_2 \leq 50$, the events $S_1 \times B$ and $A \times S_2$ are horizontal and vertical strips as shown. The event

$$A \times B = \{x_1 < x < x_2\} \times \{y_1 < y < y_2\}$$

is the rectangle shown. An event $S_1 \times \{y = y_1\}$ would be a horizontal line.

In the more general case of N subexperiments with sample spaces S_n on which events A_n are defined, the events on the combined sample space S will all be sets of the form

$$A_1 \times A_2 \times \cdots \times A_N \tag{1.6-7}$$

and unions and intersections of such sets (Papoulis, 1965, pp. 53–54).

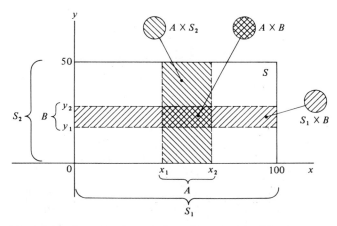

Figure 1.6-1 A combined sample space for two subexperiments.

★Probabilities

To complete the definition of a combined experiment we must assign probabilities to the events defined on the combined sample space S. Consider only two subexperiments first. Since all events defined on S will be unions and intersections of events of the form $A \times B$, where $A \subset S_1$ and $B \subset S_2$, we only need to determine $P(A \times B)$ for any A and B. We shall only consider the case where

$$P(A \times B) = P(A)P(B) \tag{1.6-8}$$

Subexperiments for which (1.6-8) is valid are called *independent experiments*.

To see what elements of S correspond to elements of A and B, we only need substitute S_2 for B or S_1 for A in (1.6-8):

$$P(A \times S_2) = P(A)P(S_2) = P(A) \tag{1.6-9}$$

$$P(S_1 \times B) = P(S_1)P(B) = P(B) \tag{1.6-10}$$

Thus, elements in the set $A \times S_2$ correspond to elements of A, and those of $S_1 \times B$ correspond to those of B.

For N independent experiments, the generalization of (1.6-8) becomes

$$P(A_1 \times A_2 \times \cdots \times A_N) = P(A_1)P(A_2) \cdots P(A_N) \tag{1.6-11}$$

where $A_n \subset S_n, n = 1, 2, \ldots, N$.

With independent experiments, the above results show that probabilities for events defined on S are completely determined from probabilities of events defined in the subexperiments.

1.7 BERNOULLI TRIALS

We shall close this chapter on probability by considering a very practical problem. It involves any experiment for which there are only two possible outcomes on any trial. Examples of such an experiment arc numerous: flipping a coin, hitting or missing the target in artillery, passing or failing an exam, receiving a 0 or a 1 in a computer bit stream, or winning or losing in a game of chance, are just a few.

For this type of experiment, we let A be the elementary event having one of the two possible outcomes as its element. $\bar{A}$ is the only other possible elementary event. Specifically, we shall repeat the basic experiment N times and determine the probability that A is observed exactly k times out of the N trials. Such repeated experiments are called *Bernoulli trials*.† Those readers familiar with combined experiments will recognize this experiment as the combination of N identical subexperiments. For readers who omitted the section on combined experiments, we shall develop the problem so that the omission will not impair their understanding of the material.

---

† After the Swiss mathematician Jacob Bernoulli (1654–1705).

Assume that elementary events are statistically independent for every trial. Let event A occur on any given trial with probability

$$P(A) = p \tag{1.7-1}$$

The event $\bar{A}$ then has probability

$$P(\bar{A}) = 1 - p \tag{1.7-2}$$

After N trials of the basic experiment, one *particular* sequence of outcomes has A occurring k times, followed by $\bar{A}$ occurring $N - k$ times.† Because of assumed statistical independence of trials, the probability of this one sequence is

$$\underbrace{P(A)P(A) \cdots P(A)}_{k \text{ terms}} \underbrace{P(\bar{A})P(\bar{A}) \cdots P(\bar{A})}_{N - k \text{ terms}} = p^k(1 - p)^{N-k} \tag{1.7-3}$$

Now there are clearly other particular sequences that will yield k events A and $N - k$ events $\bar{A}$.‡ The probability of each of these sequences is given by (1.7-3). Since the sum of all such probabilities will be the desired probability of A occurring exactly k times in N trials, we only need find the number of such sequences. Some thought will reveal that this is the number of ways of taking k objects at a time from N objects. From combinatorial analysis, the number is known to be

$$\binom{N}{k} = \frac{N!}{k!(N - k)!} \tag{1.7-4}$$

The quantity $\binom{N}{k}$ is called the *binomial coefficient*. It is sometimes given the symbol C_k^N.

From the product of (1.7-4) and (1.7-3) we finally obtain

$$P\{A \text{ occurs exactly } k \text{ times}\} = \binom{N}{k} p^k(1 - p)^{N-k} \tag{1.7-5}$$

Example 1.7-1 A submarine attempts to sink an aircraft carrier. It will be successful only if two or more torpedoes hit the carrier. If the sub fires three torpedoes and the probability of a hit is 0.4 for each torpedo, what is the probability that the carrier will be sunk?

---

† This particular sequence corresponds to one N-dimensional element in the combined sample space S.

‡ All such sequences define all the elements of S that satisfy the event $\{A$ occurs exactly k times in N trials$\}$ defined on the combined sample space.

Define the event $A = \{\text{torpedo hits}\}$. Then $P(A) = 0.4$, and $N = 3$. Probabilities are found from (1.7-5):

$$P\{\text{exactly no hits}\} = \binom{3}{0}(0.4)^0(1 - 0.4)^3 = 0.216$$

$$P\{\text{exactly one hit}\} = \binom{3}{1}(0.4)^1(1 - 0.4)^2 = 0.432$$

$$P\{\text{exactly two hits}\} = \binom{3}{2}(0.4)^2(1 - 0.4)^1 = 0.288$$

$$P\{\text{exactly three hits}\} = \binom{3}{3}(0.4)^3(1 - 0.4)^0 = 0.064$$

The answer we desire is

$$P\{\text{carrier sunk}\} = P\{\text{two or more hits}\}$$
$$= P\{\text{exactly two hits}\} + P\{\text{exactly three hits}\}$$
$$= 0.352$$

Example 1.7-2 In a culture used for biological research the growth of unavoidable bacteria occasionally spoils results of an experiment that requires at least three out of four cultures to be unspoiled to obtain a single datum point. Experience has shown that about 6 of every 100 cultures are randomly spoiled by the bateria. If the experiment requires three simultaneously derived, unspoiled data points for success, we find the probability of success for any given set of 12 cultures (three data points of four cultures each).

We treat individual datum points first as a Bernoulli trial problem with $N = 4$ and $p = P\{\text{good culture}\} = {}^{94}\!/_{100} = 0.94$. Here

$$P\{\text{valid datum point}\} = P\{3 \text{ good cultures}\} + P\{4 \text{ good cultures}\}$$

$$= \binom{4}{3}(0.94)^3(1 - 0.94)^1 + \binom{4}{4}(0.94)^4(1 - 0.94)^0 \approx 0.98$$

Finally, we treat the required three data points as a Bernoulli trial problem with $N = 3$ and $p = P\{\text{valid datum point}\} = 0.98$. Now

$$P\{\text{successful experiment}\} = P\{3 \text{ valid data points}\}$$

$$= \binom{3}{3}(0.98)^3(1 - 0.98)^0 \approx 0.941.$$

Thus, the given experiment will be successful about 94.1 percent of the time.

PROBLEMS

1-1 Specify the following sets by the rule method.

$$A = \{1, 2, 3\}, \; B = \{8, 10, 12, 14\}, \; C = \{1, 3, 5, 7, \ldots\}$$

1-2 Use the tabular method to specify a class of sets for the sets of Problem 1-1.

1-3 State whether the following sets are countable or uncountable, or, finite or infinite. $A = \{1\}$, $B = \{x = 1\}$, $C = \{0 < \text{integers}\}$, $D = \{\text{children in public school No. 5}\}$, $E = \{\text{girls in public school No. 5}\}$, $F = \{\text{girls in class in public school No. 5 at 3:00 A.M.}\}$, $G = \{\text{all lengths not exceeding one meter}\}$, $H = \{-25 \le x \le -3\}$, $I = \{-2, -1, 1 \le x \le 2\}$.

1-4 For each set of Problem 1-3, determine if it is equal to, or a subset of, any of the other sets.

1-5 State every possible subset of the set of letters $\{a, b, c, d\}$.

1-6 A thermometer measures temperatures from -40 to $130°F$ (-40 to $54.4°C$).

(*a*) State a universal set to describe temperature measurements. Specify subsets for:

(*b*) Temperature measurements not exceeding water's freezing point, and

(*c*) Measurements exceeding the freezing point but not exceeding $100°F$ ($37.8°C$).

★**1-7** Prove that a set with N elements has 2^N subsets.

1-8 A random noise voltage at a given time may have any value from -10 to 10 V.

(*a*) What is the universal set describing noise voltage?

(*b*) Find a set to describe the voltages available from a half-wave rectifier for positive voltages that has a linear output–input voltage characteristic.

(*c*) Repeat parts (*a*) and (*b*) if a dc voltage of -3 V is added to the random noise.

1-9 Show that $C \subset A$ if $C \subset B$ and $B \subset A$.

1-10 Two sets are given by $A = \{-6, -4, -0.5, 0, 1.6, 8\}$ and $B = \{-0.5, 0, 1, 2, 4\}$. Find:

    (*a*) $A - B$ (*b*) $B - A$ (*c*) $A \cup B$ (*d*) $A \cap B$

1-11 A univeral set is given as $S = \{2, 4, 6, 8, 10, 12\}$. Define two subsets as $A = \{2, 4, 10\}$ and $B = \{4, 6, 8, 10\}$. Determine the following:

    (*a*) $\bar{A} = S - A$ (*b*) $A - B$ and $B - A$ (*c*) $A \cup B$ (*d*) $A \cap B$
    (*e*) $\bar{A} \cap B$

1-12 Using Venn diagrams for three sets A, B, and C, shade the areas corresponding to the sets:

    (*a*) $(A \cup B) - C$ (*b*) $\bar{B} \cap A$ (*c*) $A \cap B \cap C$ (*d*) $\overline{(A \cup B)} \cap C$

1-13 Sketch a Venn diagram for three events where $A \cap B \ne \varnothing$, $B \cap C \ne \varnothing$, $C \cap A \ne \varnothing$, but $A \cap B \cap C = \varnothing$.

1-14 Use Venn diagrams to show that the following identities are true:

(a) $(\overline{A \cup B}) \cap C = C - [(A \cap C) \cup (B \cap C)]$

(b) $(A \cup B \cup C) - (A \cap B \cap C) = (\bar{A} \cap B) \cup (\bar{B} \cap C) \cup (\bar{C} \cap A)$

(c) $(\overline{A \cap B \cap C}) = \bar{A} \cup \bar{B} \cup \bar{C}$

1-15 Use Venn diagrams to prove De Morgan's laws $(\overline{A \cup B}) = \bar{A} \cap \bar{B}$ and $(\overline{A \cap B}) = \bar{A} \cup \bar{B}$.

1-16 A universal set is $S = \{-20 < s \le -4\}$. If $A = \{-10 \le s \le -5\}$ and $B = \{-7 < s < -4\}$, find:

(a) $A \cup B$

(b) $A \cap B$

(c) A third set C such that the sets $A \cap C$ and $B \cap C$ are as large as possible while the smallest element in C is -9.

(d) What is the set $A \cap B \cap C$?

1-17 Use De Morgan's laws to show that:

(a) $\overline{A \cap (B \cup C)} = (\bar{A} \cup \bar{B}) \cap (\bar{A} \cup \bar{C})$

(b) $(\overline{A \cap B \cap C}) = \bar{A} \cup \bar{B} \cup \bar{C}$

In each case check your results using a Venn diagram.

1-18 A die is tossed. Find the probabilities of the events $A = \{$odd number shows up$\}$, $B = \{$number larger than 3 shows up$\}$, $A \cup B$, and $A \cap B$.

1-19 In a game of dice, a "shooter" can win outright if the sum of the two numbers showing up is either 7 or 11 when two dice are thrown. What is his probability of winning outright?

1-20 A pointer is spun on a fair wheel of chance having its periphery labeled from 0 to 100.

(a) What is the sample space for this experiment?

(b) What is the probability that the pointer will stop between 20 and 35?

(c) What is the probability that the wheel will stop on 58?

1-21 An experiment has a sample space with 10 equally likely elements $S = \{a_1, a_2, \ldots, a_{10}\}$. Three events are defined as $A = \{a_1, a_5, a_9\}$, $B = \{a_1, a_2, a_6, a_9\}$, and $C = \{a_6, a_9\}$. Find the probabilities of:

(a) $A \cup C$

(b) $B \cup \bar{C}$

(c) $A \cap (B \cup C)$

(d) $\overline{A \cup B}$

(e) $(A \cup B) \cap C$

1-22 Let A be an arbitrary event. Show that $P(\bar{A}) = 1 - P(A)$.

1-23 An experiment consists of rolling a single die. Two events are defined as: $A = \{$a 6 shows up$\}$ and $B = \{$a 2 or a 5 shows up$\}$.

(a) Find $P(A)$ and $P(B)$.

(b) Define a third event C so that $P(C) = 1 - P(A) - P(B)$.

1-24 In a box there are 500 colored balls: 75 black, 150 green, 175 red, 70 white, and 30 blue. What are the probabilities of selecting a ball of each color?

1-25 A single card is drawn from a 52-card deck.

(a) What is the probability that the card is a jack?

(b) What is the probability the card will be a 5 or smaller?

(c) What is the probability that the card is a red 10?

1-26 Two cards are drawn from a 52-card deck (the first is not replaced).

(a) Given the first card is a queen, what is the probability that the second is also a queen?

(b) Repeat part (a) for the first card a queen and the second card a 7.

(c) What is the probability that both cards will be a queen?

1-27 An ordinary 52-card deck is thoroughly shuffled. You are dealt four cards up. What is the probability that all four cards are sevens?

1-28 For the resistor selection experiment of Example 1.4-1, define event D as "draw a 22-Ω resistor," and E as "draw a resistor with 10% tolerance." Find $P(D)$, $P(E)$, $P(D \cap E)$, $P(D \mid E)$, and $P(E \mid D)$.

1-29 For the resistor selection experiment of Example 1.4-1, define two mutually exclusive events B_1 and B_2 such that $B_1 \cup B_2 = S$.

(a) Use the total probability theorem to find the probability of the event "select a 22-Ω resistor," denoted D.

(b) Use Bayes' theorem to find the probability that the resistor selected had 5% tolerance, given it was 22 Ω.

1-30 In three boxes there are capacitors as shown in Table P1-30. An experiment consists of first randomly selecting a box, assuming each has the same likelihood of selection, and then selecting a capacitor from the chosen box.

(a) What is the probability of selecting a 0.01-μF capacitor, given that box 2 is selected?

(b) If a 0.01-μF capacitor is selected, what is the probability it came from box 3? (Hint: Use Bayes' and total probability theorems.)

Table P1-30 Capacitors

Value (μF)	Number in box			Totals
	1	2	3	
0.01	20	95	25	140
0.1	55	35	75	165
1.0	70	80	145	295
Totals	145	210	245	600

1-31 For Problem 1-30, list the nine conditional probabilities of capacitor selection, given certain box selections.

1-32 Rework Example 1.4-2 if $P(B_1) = 0.6$, $P(B_2) = 0.4$, $P(A_1 \mid B_1) = P(A_2 \mid B_2) = 0.95$, and $P(A_2 \mid B_1) = P(A_1 \mid B_2) = 0.05$.

1-33 Rework Example 1.4-2 if $P(B_1) = 0.7$, $P(B_2) = 0.3$, $P(A_1|B_1) = P(A_2|B_2) = 1.0$, and $P(A_2|B_1) = P(A_1|B_2) = 0$. What type of channel does this system have?

1-34 A company sells high fidelity amplifiers capable of generating 10, 25, and 50 W of audio power. It has on hand 100 of the 10-W units, of which 15% are defective, 70 of the 25-W units with 10% defective, and 30 of the 50-W units with 10% defective.

(a) What is the probability that an amplifier sold from the 10-W units is defective?

(b) If each wattage amplifier sells with equal likelihood, what is the probability of a randomly selected unit being 50 W and defective?

(c) What is the probability that a unit randomly selected for sale is defective?

1-35 A missile can be accidentally launched if two relays A and B both have failed. The probabilities of A and B failing are known to be 0.01 and 0.03 respectively. It is also known that B is more likely to fail (probability 0.06) if A has failed.

(a) What is the probability of an accidental missile launch?

(b) What is the probability that A will fail if B has failed?

(c) Are the events "A fails" and "B fails" statistically independent?

1-36 Determine whether the three events A, B, and C of Example 1.4-1 are statistically independent.

1-37 List the various equations that four events A_1, A_2, A_3, and A_4 must satisfy if they are to be statistically independent.

★**1-38** Given that two events A_1 and A_2 are statistically independent, show that:

(a) A_1 is independent of $\bar{A}_2$

(b) $\bar{A}_1$ is independent of A_2

(c) $\bar{A}_1$ is independent of $\bar{A}_2$

★**1-39** An experiment consists of randomly selecting one of five cities on Florida's west coast for a vacation. Another experiment consists of selecting at random one of four acceptable motels in which to stay. Define sample spaces S_1 and S_2 for the two experiments and a combined space $S = S_1 \times S_2$ for the combined experiment having the two subexperiments.

★**1-40** Sketch the area in the combined sample space of Example 1.6-3 corresponding to the event $A \times B$ where:

(a) $A = \{10 < x \le 15\}$ and $B = \{20 < y \le 50\}$

(b) $A = \{x = 40\}$ and $B = \{5 < y \le 40\}$

1-41 A production line manufactures 5-gal (18.93-liter) gasoline cans to a volume tolerance of 5%. The probability of any one can being out of tolerance is 0.03. If four cans are selected at random:

(a) What is the probability they are all out of tolerance?

(b) What is the probability of exactly two being out?

(c) What is the probability that all are in tolerance?

1-42 Spacecraft are expected to land in a prescribed recovery zone 80% of the time. Over a period of time, six spacecraft land.

(*a*) Find the probability that none lands in the prescribed zone.

(*b*) Find the probability that at least one will land in the prescribed zone.

(*c*) The landing program is called successful if the probability is 0.9 that three or more out of six spacecraft will land in the prescribed zone. Is the program successful?

1-43 In the submarine problem of Example 1.7-1, find the probabilities of sinking the carrier when fewer ($N = 2$) or more ($N = 4$) torpedoes are fired.

ADDITIONAL PROBLEMS

1-44 Use the tabular method to define a set A that contains all integers with magnitudes not exceeding 7. Define a second set B having odd integers larger than -2 and not larger than 5. Determine if $A \subset B$ and if $B \subset A$.

1-45 A set A has three elements a_1, a_2, and a_3. Determine all possible subsets of A.

1-46 Shade Venn diagrams to illustrate each of the following sets: (*a*) $(A \cup \bar{B}) \cap \bar{C}$, (*b*) $(\overline{A \cap B}) \cup \bar{C}$, (*c*) $(A \cup \bar{B}) \cup (C \cap D)$, (*d*) $(A \cap B \cap \bar{C}) \cup (\bar{B} \cap C \cap D)$.

1-47 A universal set S is comprised of all points in a rectangular area defined by $0 \le x \le 3$ and $0 \le y \le 4$. Define three sets by $A = \{y \le 3(x-1)/2\}$, $B = \{y \ge 1\}$, and $C = \{y \ge 3 - x\}$. Shade in Venn diagrams corresponding to the sets (*a*) $A \cap B \cap C$, and (*b*) $C \cap B \cap \bar{A}$.

1-48 The take-off-roll distance for aircraft at a certain airport can be any number from 80 m to 1750 m. Propeller aircraft require from 80 m to 1050 m while jets use from 950 m to 1750 m. The overall runway is 2000 m.

(*a*) Determine sets A, B, and C defined as "propeller aircraft take-off distances," "jet aircraft take-off distances," and "runway length safety margin," respectively.

(*b*) Determine the set $A \cap B$ and give its physical significance.

(*c*) What is the meaning of the set $\overline{A \cup B}$?

(*d*) What are the meanings of the sets $\overline{A \cup B \cup C}$ and $A \cup B$?

1-49 Prove that DeMorgan's law (1.2-13) can be extended to N events A_i, $i = 1$, 2, ..., N as follows

$$\overline{(A_1 \cap A_2 \cap \cdots \cap A_N)} = (\bar{A}_1 \cup \bar{A}_2 \cup \cdots \cup \bar{A}_N).$$

1-50 Work Problem 1-49 for (1.2-12) to prove

$$\overline{(A_1 \cup A_2 \cup \cdots \cup A_N)} = (\bar{A}_1 \cap \bar{A}_2 \cap \cdots \cap \bar{A}_N).$$

1-51 A pair of fair dice are thrown in a gambling problem. Person A wins if the sum of numbers showing up is six or less *and* one of the dice shows four. Person B wins if the sum is five or more *and* one of the dice shows a four. Find: (*a*) The probability that A wins, (*b*) the probability of B winning, and (*c*) the probability that both A and B win.

1-52 You (person A) and two others (B and C) each toss a fair coin in a two-step gambling game. In step 1 the person whose toss is not a match to either of the other two is "odd man out." Only the remaining two whose coins match go on to step 2 to resolve the ultimate winner.

(a) What is the probability you will advance to step 2 after the first toss?

(b) What is the probability you will be out after the first toss?

(c) What is the probability that no one will be out after the first toss?

*1-53 The communication system of Example 1.4-2 is to be extended to the case of three transmitted symbols 0, 1, and 2. Define appropriate events A_i and B_i, $i = 1$, 2, 3, to represent symbols after and before the channel, respectively. Assume channel transition probabilities are all equal at $P(A_i | B_j) = 0.1$, $i \neq j$, and are $P(A_i | B_j) = 0.8$ for $i = j = 1$, 2, 3, while symbol transmission probabilities are $P(B_1) = 0.5$, $P(B_2) = 0.3$, and $P(B_3) = 0.2$.

(a) Sketch the diagram analogous to Fig. 1.4-2.

(b) Compute received symbol probabilities $P(A_1)$, $P(A_2)$, and $P(A_3)$.

(c) Compute the a posteriori probabilities for this system.

(d) Repeat parts (b) and (c) for all transmission symbol probabilities equal. Note the effect.

1-54 Show that there are $2^N - N - 1$ equations required in (1.5-6). (*Hint:* Recall that the binomial coefficient is the number of combinations of N things taken n at a time.)

1-55 A student is known to arrive late for a class 40% of the time. If the class meets five times each week find: (a) the probability the student is late for at least three classes in a given week, and (b) the probability the student will not be late at all during a given week.

1-56 An airline in a small city has five departures each day. It is known that any given flight has a probability of 0.3 of departing late. For any given day find the probabilities that: (a) no flights depart late, (b) all flights depart late, and (c) three or more depart on time.

1-57 The local manager of the airline of Problem 1-56 desires to make sure that 90% of flights leave on time. What is the largest probability of being late that the individual flights can have if the goal is to be achieved? Will the operation have to be improved significantly?

1-58 A man wins in a gambling game if he gets two heads in five flips of a biased coin. The probability of getting a head with the coin is 0.7.

(a) Find the probability the man will win. Should he play this game?

(b) What is his probability of winning if he wins by getting at least four heads in five flips? Should he play this new game?

*1-59 A rifleman can achieve a "marksman" award if he passes a test. He is allowed to fire six shots at a target's bull's eye. If he hits the bull's eye with at least five of his six shots he wins a set. He becomes a marksman only if he can repeat the feat three times straight, that is, if he can win three straight sets. If his probability is 0.8 of hitting a bull's eye on any one shot, find the probabilities of his: (a) winning a set, and (b) becoming a marksman.

TWO

THE RANDOM VARIABLE

2.0 INTRODUCTION

In the previous chapter we introduced the concept of an event to describe characteristics of outcomes of an experiment. Events allowed us more flexibility in determining properties of an experiment than could be obtained by considering only the outcomes themselves. An event could be almost anything from "descriptive," such as "draw a spade," to numerical, such as "the outcome is 3."

In this chapter, we introduce a new concept that will allow events to be defined in a more consistent manner; they will always be numerical. The new concept is that of a *random variable*, and it will constitute a powerful tool in the solution of practical probabilistic problems.

2.1 THE RANDOM VARIABLE CONCEPT

Definition of a Random Variable

We define a real *random variable*† as a real *function* of the elements of a sample space S. We shall represent a random variable by a capital letter (such as W, X, or Y) and any particular value of the random variable by a lowercase letter (such

† Complex random variables are considered in Chapter 5.

as w, x, or y). Thus, given an experiment defined by a sample space S with elements s, we assign to every s a real number

$$X(s) \tag{2.1-1}$$

according to some rule and call $X(s)$ a random variable.

A random variable X can be considered to be a function that maps all elements of the sample space into points on the real line or some parts thereof. We illustrate, by two examples, the mapping of a random variable.

Example 2.1-1 An experiment consists of rolling a die and flipping a coin. The applicable sample space is illustrated in Figure 2.1-1. Let the random variable be a function X chosen such that (1) a coin head (H) outcome corresponds to positive values of X that are equal to the numbers that show up on the die, and (2) a coin tail (T) outcome corresponds to negative values of X that are equal in magnitude to *twice* the number that shows on the die. Here X maps the sample space of 12 elements into 12 values of X from -12 to 6 as shown in Figure 2.1-1.

Example 2.1-2 Figure 2.1-2 illustrates an experiment where the pointer on a wheel of chance is spun. The possible outcomes are the numbers from 0 to 12 marked on the wheel. The sample space consists of the numbers in the set $\{0 < s \le 12\}$. We define a random variable by the function

$$X = X(s) = s^2$$

Points in S now map onto the real line as the set $\{0 < x \le 144\}$.

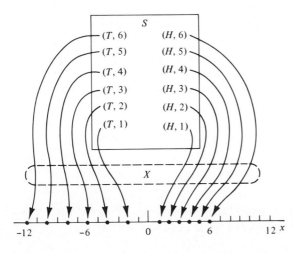

Figure 2.1-1 A random variable mapping of a sample space.

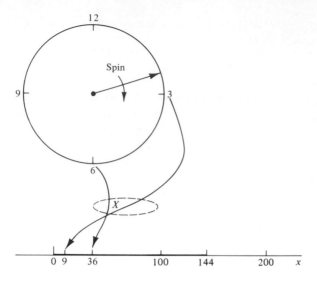

Figure 2.1-2 Mapping applicable to Example 2.1-2.

As seen in these two examples, a random variable is a function that maps each point in S into some point on the real line. It is not necessary that the sample-space points map uniquely, however. More than one point in S may map into a single value of X. For example, in the extreme case, we might map all six points in the sample space for the experiment "throw a die and observe the number that shows up" into the one point $X = 2$.

Conditions for a Function to be a Random Variable

Thus, a random variable may be almost any function we wish. We shall, however, require that it not be multivalued. That is, every point in S must correspond to only one value of the random variable.

Moreover, we shall require that two additional conditions be satisfied in order that a function X be a random variable (Papoulis, 1965, p. 88). First, the set $\{X \leq x\}$ shall be an event for any real number x. The satisfaction of this condition will be no trouble in practical problems. This set corresponds to those points s in the sample space for which the random variable $X(s)$ does not exceed the number x. The probability of this event, denoted by $P\{X \leq x\}$, is equal to the sum of the probabilities of all the elementary events corresponding to $\{X \leq x\}$.

The second condition we require is that the probabilities of the events $\{X = \infty\}$ and $\{X = -\infty\}$ be 0:

$$P\{X = -\infty\} = 0 \qquad P\{X = \infty\} = 0 \qquad (2.1\text{-}2)$$

This condition does not prevent X from being either $-\infty$ or ∞ for some values of s; it only requires that the probability of the set of those s be zero.

Discrete and Continuous Random Variables

A *discrete random variable* is one having only discrete values. Example 2.1-1 illustrated a discrete random variable. The sample space for a discrete random variable can be discrete, continuous, or even a mixture of discrete and continuous points. For example, the "wheel of chance" of Example 2.1-2 has a continuous sample space, but we could define a discrete random variable as having the value 1 for the set of outcomes $\{0 < s \leq 6\}$ and -1 for $\{6 < s \leq 12\}$. The result is a discrete random variable defined on a continuous sample space.

A *continuous random variable* is one having a continuous range of values. It cannot be produced from a discrete sample space because of our requirement that all random variables be single-valued functions of all sample-space points. Similarly, a purely continuous random variable cannot result from a mixed sample space because of the presence of the discrete portion of the sample space. The random variable of Example 2.1-2 is continuous.

Mixed Random Variable

A *mixed random variable* is one for which some of its values are discrete and some are continuous. The mixed case is usually the least important type of random variable, but it occurs in some problems of practical significance.

2.2 DISTRIBUTION FUNCTION

The probability $P\{X \leq x\}$ is the probability of the event $\{X \leq x\}$. It is a number that depends on x; that is, it is a function of x. We call this function, denoted $F_X(x)$, the *cumulative probability distribution function* of the random variable X. Thus,

$$F_X(x) = P\{X \leq x\} \tag{2.2-1}$$

We shall often call $F_X(x)$ just the *distribution function* of X. The argument x is any real number ranging from $-\infty$ to ∞.

The distribution function has some specific properties derived from the fact that $F_X(x)$ is a probability. These are:†

(1)	$F_X(-\infty) = 0$	(2.2-2a)
(2)	$F_X(\infty) = 1$	(2.2-2b)
(3)	$0 \leq F_X(x) \leq 1$	(2.2-2c)
(4)	$F_X(x_1) \leq F_X(x_2)$ if $x_1 < x_2$	(2.2-2d)
(5)	$P\{x_1 < X \leq x_2\} = F_X(x_2) - F_X(x_1)$	(2.2-2e)
(6)	$F_X(x^+) = F_X(x)$	(2.2-2f)

---

† We use the notation x^+ to imply $x + \varepsilon$ where $\varepsilon > 0$ is infinitesimally small; that is, $\varepsilon \to 0$.

The first three of these properties are easy to justify, and the reader should justify them as an exercise. The fourth states that $F_X(x)$ is a nondecreasing function of x. The fifth property states that the probability that X will have values larger than some number x_1 but not exceeding another number x_2 is equal to the difference in $F_X(x)$ evaluated at the two points. It is justified from the fact that the events $\{X \leq x_1\}$ and $\{x_1 < X \leq x_2\}$ are mutually exclusive, so the probability of the event $\{X \leq x_2\} = \{X \leq x_1\} \cup \{x_1 < X \leq x_2\}$ is the sum of the probabilities $P\{X \leq x_1\}$ and $P\{x_1 < X \leq x_2\}$. The sixth property states that $F_X(x)$ is a function continuous from the right.

Properties 1, 2, 4, and 6 may be used as tests to determine if some function, say $G_X(x)$, could be a valid distribution function. If so, all four tests must be passed.

If X is a discrete random variable, consideration of its distribution function defined by (2.2-1) shows that $F_X(x)$ must have a stairstep form, such as shown in Figure 2.2-1a. The amplitude of a step will equal the probability of occurrence of the value of X where the step occurs. If the values of X are denoted x_i, we may write $F_X(x)$ as

$$F_X(x) = \sum_{i=1}^{N} P\{X = x_i\}u(x - x_i) \tag{2.2-3}$$

where $u(\cdot)$ is the unit-step function defined by†

$$u(x) = \begin{cases} 1 & x \geq 0 \\ 0 & x < 0 \end{cases} \tag{2.2-4}$$

and N may be infinite for some random variables. By introducing the shortened notation

$$P(x_i) = P\{X = x_i\} \tag{2.2-5}$$

(2.2-3) can be written as

$$F_X(x) = \sum_{i=1}^{N} P(x_i)u(x - x_i) \tag{2.2-6}$$

We next consider an example that illustrates the distribution function of a discrete random variable.

Example 2.2-1 Let X have the discrete values in the set $\{-1, -0.5, 0.7, 1.5, 3\}$. The corresponding probabilities are assumed to be $\{0.1, 0.2, 0.1, 0.4, 0.2\}$. Now $P\{X < -1\} = 0$ because there are no sample space points in the set $\{X < -1\}$. Only when $X = -1$ do we obtain one outcome. Thus, there is an immediate jump in probability of 0.1 in the function $F_X(x)$ at the point $x = -1$. For $-1 < x < -0.5$, there are no additional sample space points so $F_X(x)$ remains constant at the value 0.1. At $x = -0.5$ there is another jump of

---

† This definition differs slightly from (A-5) by including the equality so that $u(x)$ satisfies (2.2-2f).

Figure 2.2-1 Distribution function (*a*) and density function (*b*) applicable to the dscrete random variable of Example 2.2-1. [*Adapted from Peebles (1976) with permission of publishers Addison-Wesley, Advanced Book Program.*]

0.2 in $F_X(x)$. This process continues until all points are included. $F_X(x)$ then equals 1.0 for all x above the last point. Figure 2.2-1*a* illustrates $F_X(x)$ for this discrete random variable.

A continuous random variable will have a continuous distribution function. We consider an example for which $F_X(x)$ is the continuous function shown in Figure 2.2-2*a*.

Example 2.2-2 We return to the fair wheel-of-chance experiment. Let the wheel be numbered from 0 to 12 as shown in Figure 2.1-2. Clearly the probability of the event $\{X \leq 0\}$ is 0 because there are no sample space points in this set. For $0 < x \leq 12$ the probability of $\{0 < X \leq x\}$ will increase linearly with x for a fair wheel. Thus, $F_X(x)$ will behave as shown in Figure 2.2-2*a*.

The distribution function of a mixed random variable will be a sum of two parts, one of stairstep form, the other continuous.

(a)

(b)

Figure 2.2-2 Distribution function (a) and density function (b) applicable to the continuous random variable of Example 2.2-2. [*Adapted from Peebles (1976) with permission of publishers Addison-Wesley, Advanced Book Program.*]

2.3 DENSITY FUNCTION

The *probability density function*, denoted by $f_X(x)$, is defined as the derivative of the distribution function:

$$f_X(x) = \frac{dF_X(x)}{dx} \tag{2.3-1}$$

We often call $f_X(x)$ just the *density function* of the random variable X.

Existence

If the derivative of $F_X(x)$ exists then $f_X(x)$ exists and is given by (2.3-1). There may, however, be places where $dF_X(x)/dx$ is not defined. For example, a continuous random variable will have a continuous distribution $F_X(x)$, but $F_X(x)$ may have corners (points of abrupt change in slope). The distribution shown in Figure 2.2-2a is such a function. For such cases, we plot $f_X(x)$ as a function with step-type discontinuities (such as in Figure 2.2-2b). We shall assume that the number of points where $F_X(x)$ is not differentiable is countable.

For discrete random variables having a stairstep form of distribution function, we introduce the concept of the *unit-impulse function* $\delta(x)$ to describe the derivative of $F_X(x)$ at its stairstep points. The unit-impulse function and its properties are reviewed in Appendix A. It is shown there that $\delta(x)$ may be defined by its integral property

$$\phi(x_0) = \int_{-\infty}^{\infty} \phi(x)\delta(x - x_0) \, dx \tag{2.3-2}$$

where $\phi(x)$ is any function continuous at the point $x = x_0$; $\delta(x)$ can be interpreted as a "function" with infinite amplitude, area of unity, and zero duration. The unit-impulse and the unit-step functions are related by

$$\delta(x) = \frac{du(x)}{d(x)} \tag{2.3-3}$$

or

$$\int_{-\infty}^{x} \delta(\xi) \, d\xi = u(x) \tag{2.3-4}$$

The more general impulse function is shown symbolically as a vertical arrow occurring at the point $x = x_0$ and having an amplitude equal to the amplitude of the step function for which it is the derivative.

We return to the case of a discrete random variable and differentiate $F_X(x)$, as given by (2.2-6), to obtain

$$f_X(x) = \sum_{i=1}^{N} P(x_i)\delta(x - x_i) \tag{2.3-5}$$

Thus, the density function for a discrete random variable exists in the sense that we use impulse functions to describe the derivative of $F_X(x)$ at its stairstep points. Figure 2.2-1*b* is an example of the density function for the random variable having the function of Figure 2.2-1*a* as its distribution.

A physical interpretation of (2.3-5) is readily achieved. Clearly, the probability of X having one of its particular values, say x_i, is a number $P(x_i)$. If this probability is assigned to the *point* x_i, then the *density* of probability is infinite because a point has no "width" on the x axis. The infinite "amplitude" of the impulse function describes this infinite density. The "size" of the density of probability at $x = x_i$ is accounted for by the scale factor $P(x_i)$ giving $P(x_i)\delta(x - x_i)$ for the density at the point $x = x_i$.

Properties of Density Functions

Several properties that $f_X(x)$ satisfies may be stated:

$$(1) \quad 0 \le f_X(x) \qquad \text{all } x \tag{2.3-6a}$$

$$(2) \quad \int_{-\infty}^{\infty} f_X(x) \, dx = 1 \tag{2.3-6b}$$

$$(3) \quad F_X(x) = \int_{-\infty}^{x} f_X(\xi) \, d\xi \tag{2.3-6c}$$

$$(4) \quad P\{x_1 < X \le x_2\} = \int_{x_1}^{x_2} f_X(x) \, dx \tag{2.3-6d}$$

Proofs of these properties are left to the reader as exercises. Properties 1 and 2 require that the density function be nonnegative and have an area of unity. These two properties may also be used as tests to see if some function, say $g_X(x)$, can be a valid probability density function. Both tests must be satisfied for validity. Property 3 is just another way of writing (2.3-1) and serves as the link between $F_X(x)$ and $f_X(x)$. Property 4 relates the probability that X will have values from x_1 to, and including, x_2 to the density function.

Example 2.3-1 Let us test the function $g_X(x)$ shown in Figure 2.3-1a to see if it can be a valid density function. It obviously satisfies property 1 since it is nonnegative. Its area is $a\alpha$ which must equal unity to satisfy property 2. Therefore $a = 1/\alpha$ is necessary if $g_X(x)$ is to be a density.

Suppose $a = 1/\alpha$. To find the applicable distribution function we first write

$$
g_X(x) = \begin{cases}
0 & x_0 - \alpha > x \ge x_0 + \alpha \\[2mm]
\dfrac{1}{\alpha^2}(x - x_0 + \alpha) & x_0 - \alpha \le x < x_0 \\[2mm]
\dfrac{1}{\alpha} - \dfrac{1}{\alpha^2}(x - x_0) & x_0 \le x < x_0 + \alpha
\end{cases}
$$

Next, by using (2.3-6c), we obtain

$$
G_X(x) = \begin{cases}
0 & x_0 - \alpha > x \\[2mm]
\displaystyle\int_{x_0 - \alpha}^{x} g_X(\xi) \, d\xi = \dfrac{1}{2\alpha^2}(x - x_0 + \alpha)^2 & x_0 - \alpha \le x < x_0 \\[4mm]
\dfrac{1}{2} + \displaystyle\int_{x_0}^{x} g_X(\xi) \, d\xi = \dfrac{1}{2} + \dfrac{1}{\alpha}(x - x_0) - \dfrac{1}{2\alpha^2}(x - x_0)^2 & \\[2mm]
& x_0 \le x < x_0 + \alpha \\[2mm]
1 & x_0 + \alpha \le x
\end{cases}
$$

This function is plotted in Figure 2.3-1b.

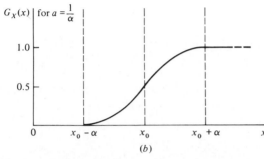

Figure 2.3-1 A possible probability density function (*a*) and a distribution function (*b*) applicable to Example 2.3-1.

Example 2.3-2 Suppose a random variable is known to have the triangular probability density of the preceding example with $x_0 = 8$, $\alpha = 5$ and $a = 1/\alpha = \frac{1}{5}$. From the earlier work

$$f_X(x) = \begin{cases} 0 & 3 > x \geq 13 \\ (x - 3)/25 & 3 \leq x < 8 \\ 0.2 - (x - 8)/25 & 8 \leq x < 13 \end{cases}$$

We shall use this probability density in (2.3-6d) to find the probability that X has values greater than 4.5 but not greater than 6.7. The probability is

$$P\{4.5 < X \leq 6.7\} = \int_{4.5}^{6.7} [(x - 3)/25] \, dx$$

$$= \frac{1}{25}\left[\frac{x^2}{2} - 3x\right]\Bigg|_{4.5}^{6.7} = 0.2288$$

Thus, the event $\{4.5 < X \leq 6.7\}$ has a probability of 0.2288 or 22.88%.

2.4 THE GAUSSIAN RANDOM VARIABLE

A random variable X is called *gaussian*† if its density function has the form

$$f_X(x) = \frac{1}{\sqrt{2\pi\sigma_X^2}} e^{-(x-a_X)^2/2\sigma_X^2} \tag{2.4-1}$$

† After the German mathematician Johann Friedrich Carl Gauss (1777–1855). The gaussian density is often called the *normal density*.

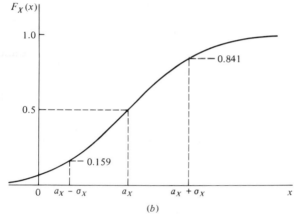

Figure **2.4-1** Density (*a*) and distribution (*b*) functions of a gaussian random variable.

where $\sigma_X > 0$ and $-\infty < a_X < \infty$ are real constants. This function is sketched in Figure 2.4-1*a*. Its maximum value $(2\pi\sigma_X^2)^{-1/2}$ occurs at $x = a_X$. Its "spread" about the point $x = a_X$ is related to σ_X. The function decreases to 0.607 times its maximum at $x = a_X + \sigma_X$ and $x = a_X - \sigma_X$.

The gaussian density is the most important of all densities. It enters into nearly all areas of engineering and science. We shall encounter the gaussian random variable frequently in later work when we discuss some important types of noise.

The distribution function is found from (2.3-6*c*) using (2.4-1). The integral is

$$F_X(x) = \frac{1}{\sqrt{2\pi\sigma_X^2}} \int_{-\infty}^{x} e^{-(\xi - a_X)^2/2\sigma_X^2} \, d\xi \qquad (2.4-2)$$

This integral has no known closed-form solution and must be evaluated by numerical methods. To make the results generally available, we could develop a set of tables of $F_X(x)$ for various x with a_X and σ_X as parameters. However, this approach has limited value because there is an infinite number of possible combinations of a_X and σ_X, which requires an infinite number of tables. A better approach is possible where only one table of $F_X(x)$ is developed that corresponds to normalized (specific) values of a_X and σ_X. We then show that the one table can be used in the general case where a_X and σ_X can be arbitrary.

We start by first selecting the normalized case where $a_X = 0$ and $\sigma_X = 1$. Denote the corresponding distribution function by $F(x)$. From (2.4-2), $F(x)$ is

$$F(x) = \frac{1}{\sqrt{2\pi}} \int_{-\infty}^{x} e^{-\xi^2/2} \, d\xi \qquad (2.4\text{-}3)$$

which is a function of x only. This function is tabularized in Appendix B for $x \geq 0$. For negative values of x we use the relationship

$$F(-x) = 1 - F(x) \qquad (2.4\text{-}4)$$

To show that the general distribution function $F_X(x)$ of (2.4-2) can be found in terms of $F(x)$ of (2.4-3), we make the variable change

$$u = (\xi - a_X)/\sigma_X \qquad (2.4\text{-}5)$$

in (2.4-2) to obtain

$$F_X(x) = \frac{1}{\sqrt{2\pi}} \int_{-\infty}^{(x-a_X)/\sigma_X} e^{-u^2/2} \, du \qquad (2.4\text{-}6)$$

From (2.4-3), this expression is clearly equivalent to

$$F_X(x) = F\left(\frac{x - a_X}{\sigma_X}\right) \qquad (2.4\text{-}7)$$

Figure 2.4-1b depicts the behavior of $F_X(x)$.

We consider two examples to illustrate the application of (2.4-7).

Example 2.4-1 We find the probability of the event $\{X \leq 5.5\}$ for a gaussian random variable having $a_X = 3$ and $\sigma_X = 2$.

Here $(x - a_X)/\sigma_X = (5.5\text{-}3)/2 = 1.25$. From (2.4-7) and the definition of $F_X(x)$

$$P\{X \leq 5.5\} = F_X(5.5) = F(1.25)$$

By using the table in Appendix B

$$P\{X \leq 5.5\} = F(1.25) = 0.8944$$

Example 2.4-2 Assume that the height of clouds above the ground at some location is a gaussian random variable X with $a_X = 1830$ m and $\sigma_X = 460$ m. We find the probability that clouds will be higher than 2750 m (about 9000 ft). From (2.4-7) and Appendix B:

$$P\{X > 2750\} = 1 - P\{X \le 2750\} = 1 - F_X(2750)$$

$$= 1 - F\left(\frac{2750 - 1830}{460}\right) = 1 - F(2.0)$$

$$= 1 - 0.9772 = 0.0228$$

The probability that clouds are higher than 2750 m is therefore about 2.20 percent if their behavior is as assumed.

2.5 OTHER DISTRIBUTION AND DENSITY EXAMPLES

Many distribution functions are important enough to have been given names. We give five examples. The first two are for discrete random variables; the remaining three are for continuous random variables. Other distributions are listed in Appendix F.

Binomial

Let $0 < p < 1$, and $N = 1, 2, \ldots$, then the function

$$f_X(x) = \sum_{k=0}^{N} \binom{N}{k} p^k (1 - p)^{N-k} \delta(x - k) \tag{2.5-1}$$

is called the *binomial density function*. The quantity $\binom{N}{k}$ is the binomial coefficient defined in (1.7-4) as

$$\binom{N}{k} = \frac{N!}{k!(N-k)!} \tag{2.5-2}$$

The binomial density can be applied to the Bernoulli trial experiment of Chapter 1. It applies to many games of chance, detection problems in radar and sonar, and many experiments having only two possible outcomes on any given trial.

By integration of (2.5-1), the *binomial distribution function* is found:

$$F_X(x) = \sum_{k=0}^{N} \binom{N}{k} p^k (1 - p)^{N-k} u(x - k) \tag{2.5-3}$$

Figure 2.5-1 illustrates the binomial density and distribution functions for $N = 6$ and $p = 0.25$.

Figure 2.5-1 Binomial density (a) and distribution (b) functions for the case $N = 6$ and $p = 0.25$.

Poisson

The *Poisson*† random variable X has a density and distribution given by

$$f_X(x) = e^{-b} \sum_{k=0}^{\infty} \frac{b^k}{k!} \delta(x - k) \qquad (2.5\text{-}4)$$

$$F_X(x) = e^{-b} \sum_{k=0}^{\infty} \frac{b^k}{k!} u(x - k) \qquad (2.5\text{-}5)$$

where $b > 0$ is a real constant. When plotted, these functions appear quite similar to those for the binomial random variable (Figure 2.5-1). In fact, if $N \to \infty$ and $p \to 0$ for the binomial case in such a way that $Np = b$, a constant, the Poisson case results.

The Poisson random variable applies to a wide variety of counting-type applications. It describes the number of defective units in a sample taken from a production line, the number of telephone calls made during a period of time, the

---

† After the French mathematician Siméon Denis Poisson (1781–1840).

number of electrons emitted from a small section of a cathode in a given time interval, etc. If the time interval of interest has duration T, and the events being counted are known to occur at an average rate λ and have a Poisson distribution, then b in (2.5-4) is given by

$$b = \lambda T \tag{2.5-6}$$

We illustrate these points by means of an example.

Example 2.5-1 Assume automobile arrivals at a gasoline station are Poisson and occur at an average rate of 50/h. The station has only one gasoline pump. If all cars are assumed to require one minute to obtain fuel, what is the probability that a waiting line will occur at the pump?

A waiting line will occur if two or more cars arrive in any one-minute interval. The probability of this event is one minus the probability that either none or one car arrives. From (2.5-6), with $\lambda = {}^{50}\!/_{60}$ cars/minute and $T = 1$ minute, we have $b = {}^{5}\!/_{6}$. On using (2.5-5)

$$\text{Probability of a waiting line} = 1 - F_X(1) - F_X(0)$$

$$= 1 - e^{-5/6}\left[1 + \frac{5}{6}\right] = 0.2032$$

We therefore expect a line at the pump about 20.32% of the time.

Uniform

The *uniform* probability density and distribution functions are defined by:

$$f_X(x) = \begin{cases} 1/(b-a) & a \leq x \leq b \\ 0 & \text{elsewhere} \end{cases} \tag{2.5-7}$$

$$F_X(x) = \begin{cases} 0 & x < a \\ (x-a)/(b-a) & a \leq x < b \\ 1 & b \leq x \end{cases} \tag{2.5-8}$$

for real constants $-\infty < a < \infty$ and $b > a$. Figure 2.5-2 illustrates the behavior of the above two functions.

The uniform density finds a number of practical uses. A particularly important application is in the quantization of signal samples prior to encoding in digital communication systems. Quantization amounts to "rounding off" the actual sample to the nearest of a large number of discrete "quantum levels." The errors introduced in the round-off process are uniformly distributed.

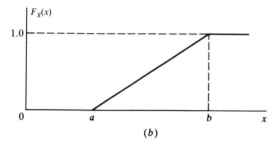

Figure 2.5-2 Uniform probability density function (*a*) and its distribution function (*b*).

Exponential

The *exponential* density and distribution functions are:

$$f_X(x) = \begin{cases} \dfrac{1}{b} e^{-(x-a)/b} & x > a \\ 0 & x < a \end{cases} \tag{2.5-9}$$

$$F_X(x) = \begin{cases} 1 - e^{-(x-a)/b} & x > a \\ 0 & x < a \end{cases} \tag{2.5-10}$$

for real numbers $-\infty < a < \infty$ and $b > 0$. These functions are plotted in Figure 2.5-3.

The exponential density is useful in describing raindrop sizes when a large number of rainstorm measurements are made. It is also known to approximately describe the fluctuations in signal strength received by radar from certain types of aircraft as illustrated by the following example.

Example 2.5-2 The power reflected from an aircraft of complicated shape that is received by a radar can be described by an exponential random variable P. The density of P is therefore

$$f_P(p) = \begin{cases} \dfrac{1}{P_0} e^{-p/P_0} & p > 0 \\ 0 & p < 0 \end{cases}$$

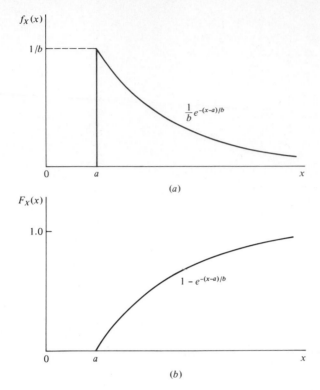

Figure 2.5-3 Exponential density (a) and distribution (b) functions.

where P_0 is the average amount of received power. At some given time P may have a value different from its average value and we ask: what is the probability that the received power is larger than the power received on the average?

We must find $P\{P > P_0\} = 1 - P\{P \leq P_0\} = 1 - F_P(P_0)$. From (2.5-10)

$$P\{P > P_0\} = 1 - (1 - e^{-P_0/P_0}) = e^{-1} \approx 0.368$$

In other words, the received power is larger than its average value about 36.8 per cent of the time.

Rayleigh

The *Rayleigh*† density and distribution functions are:

$$f_X(x) = \begin{cases} \dfrac{2}{b}(x-a)e^{-(x-a)^2/b} & x \geq a \\ 0 & x < a \end{cases} \qquad (2.5\text{-}11)$$

$$F_X(x) = \begin{cases} 1 - e^{-(x-a)^2/b} & x \geq a \\ 0 & x < a \end{cases} \qquad (2.5\text{-}12)$$

† Named for the English physicist John William Strutt, Lord Rayleigh (1842–1919).

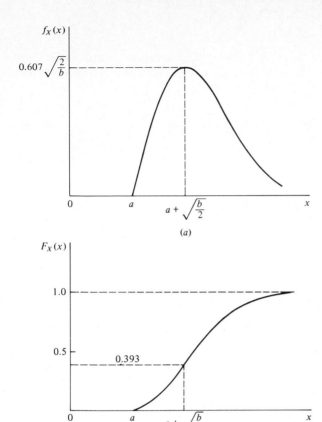

(a)

(b)

Figure 2.5-4 Rayleigh density (a) and distribution (b) functions.

for real constants $-\infty < a < \infty$ and $b > 0$. These functions are plotted in Figure 2.5-4.

The Rayleigh density describes the envelope of one type of noise when passed through a bandpass filter. It also is important in analysis of errors in various measurement systems.

2.6 CONDITIONAL DISTRIBUTION AND DENSITY FUNCTIONS

The concept of conditional probability was introduced in Chapter 1. Recall that, for two events A and B where $P(B) \neq 0$, the conditional probability of A given B had occurred was

$$P(A \mid B) = \frac{P(A \cap B)}{P(B)} \tag{2.6-1}$$

In this section we extend the conditional probability concept to include random variables.

Conditional Distribution

Let A in (2.6-1) be identified as the event $\{X \leq x\}$ for the random variable X. The resulting probability $P\{X \leq x \,|\, B\}$ is defined as the *conditional distribution function* of X, which we denote $F_X(x \,|\, B)$. Thus

$$F_X(x \,|\, B) = P\{X \leq x \,|\, B\} = \frac{P\{X \leq x \cap B\}}{P(B)} \qquad (2.6\text{-}2)$$

where we use the notation $\{X \leq x \cap B\}$ to imply the joint event $\{X \leq x\} \cap B$. This joint event consists of all outcomes s such that

$$X(s) \leq x \qquad \text{and} \qquad s \in B \qquad (2.6\text{-}3)$$

    The conditional distribution (2.6-2) applies to discrete, continuous, or mixed random variables.

Properties of Conditional Distribution

All the properties of ordinary distributions apply to $F_X(x \,|\, B)$. In other words, it has the following characteristics:

(1) $F_X(-\infty \,|\, B) = 0$ (2.6-4a)

(2) $F_X(\infty \,|\, B) = 1$ (2.6-4b)

(3) $0 \leq F_X(x \,|\, B) \leq 1$ (2.6-4c)

(4) $F_X(x_1 \,|\, B) \leq F_X(x_2 \,|\, B)$ if $x_1 < x_2$ (2.6-4d)

(5) $P\{x_1 < X \leq x_2 \,|\, B\} = F_X(x_2 \,|\, B) - F_X(x_1 \,|\, B)$ (2.6-4e)

(6) $F_X(x^+ \,|\, B) = F_X(x \,|\, B)$ (2.6-4f)

These characteristics have the same general meanings as described earlier following (2.2-2).

Conditional Density

In a manner similar to the ordinary density function, we define *conditional density function* of the random variable X as the derivative of the conditional distribution function. If we denote this density by $f_X(x \,|\, B)$, then

$$f_X(x \,|\, B) = \frac{dF_X(x \,|\, B)}{dx} \qquad (2.6\text{-}5)$$

If $F_X(x \,|\, B)$ contains step discontinuities, as when X is a discrete or mixed random variable, we assume that impulse functions are present in $f_X(x \,|\, B)$ to account for the derivatives at the discontinuities.

Properties of Conditional Density

Because conditional density is related to conditional distribution through the derivative, it satisfies the same properties as the ordinary density function. They are:

$$(1) \quad f_X(x \,|\, B) \geq 0 \qquad\qquad (2.6\text{-}6a)$$

$$(2) \quad \int_{-\infty}^{\infty} f_X(x \,|\, B) \, dx = 1 \qquad\qquad (2.6\text{-}6b)$$

$$(3) \quad F_X(x \,|\, B) = \int_{-\infty}^{x} f_X(\xi \,|\, B) \, d\xi \qquad\qquad (2.6\text{-}6c)$$

$$(4) \quad P\{x_1 < X \leq x_2 \,|\, B\} = \int_{x_1}^{x_2} f_X(x \,|\, B) \, dx \qquad\qquad (2.6\text{-}6d)$$

We take an example to illustrate conditional density and distribution.

Example 2.6-1 Two boxes have red, green, and blue balls in them; the number of balls of each color is given in Table 2.6-1. Our experiment will be to select a box and then a ball from the selected box. One box (number 2) is slightly larger than the other, causing it to be selected more frequently. Let B_2 be the event "select the larger box" while B_1 is the event "select the smaller box." Assume $P(B_1) = \frac{2}{10}$ and $P(B_2) = \frac{8}{10}$. (B_1 and B_2 are mutually exclusive and $B_1 \cup B_2$ is the certain event, since some box must be selected; therefore, $P(B_1) + P(B_2)$ must equal unity.)

Now define a discrete random variable X to have values $x_1 = 1$, $x_2 = 2$, and $x_3 = 3$ when a red, green, or blue ball is selected, and let B be an event equal to either B_1 or B_2. From Table 2.6-1:

$$P(X = 1 \,|\, B = B_1) = \frac{5}{100} \qquad P(X = 1 \,|\, B = B_2) = \frac{80}{150}$$

$$P(X = 2 \,|\, B = B_1) = \frac{35}{100} \qquad P(X = 2 \,|\, B = B_2) = \frac{60}{150}$$

$$P(X = 3 \,|\, B = B_1) = \frac{60}{100} \qquad P(X = 3 \,|\, B = B_2) = \frac{10}{150}$$

Table 2.6-1 Numbers of colored balls in two boxes

x_i	Ball color	Box 1	Box 2	Totals
1	Red	5	80	85
2	Green	35	60	95
3	Blue	60	10	70
Totals		100	150	250

The conditional probability density $f_X(x \mid B_1)$ becomes

$$f_X(x \mid B_1) = \frac{5}{100}\,\delta(x-1) + \frac{35}{100}\,\delta(x-2) + \frac{60}{100}\,\delta(x-3)$$

By direct integration of $f_X(x \mid B_1)$:

$$F_X(x \mid B_1) = \frac{5}{100}\,u(x-1) + \frac{35}{100}\,u(x-2) + \frac{60}{100}\,u(x-3)$$

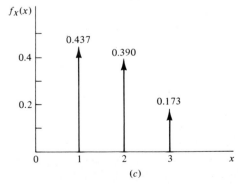

Figure 2.6-1 Distributions (a) and densities (b) and (c) applicable to Example 2.6-1.

For comparison, we may find the density and distribution of X by determining the probabilities $P(X = 1)$, $P(X = 2)$, and $P(X = 3)$. These are found from the total probability theorem embodied in (1.4-10):

$$P(X = 1) = P(X = 1 \mid B_1)P(B_1) + P(X = 1 \mid B_2)P(B_2)$$

$$= \frac{5}{100}\left(\frac{2}{10}\right) + \frac{80}{150}\left(\frac{8}{10}\right) = 0.437$$

$$P(X = 2) = \frac{35}{100}\left(\frac{2}{10}\right) + \frac{60}{150}\left(\frac{8}{10}\right) = 0.390$$

$$P(X = 3) = \frac{60}{100}\left(\frac{2}{10}\right) + \frac{10}{150}\left(\frac{8}{10}\right) = 0.173$$

Thus

$$f_X(x) = 0.437\,\delta(x - 1) + 0.390\,\delta(x - 2) + 0.173\,\delta(x - 3)$$

and

$$F_X(x) = 0.437u(x - 1) + 0.390u(x - 2) + 0.173u(x - 3)$$

These distributions and densities are plotted in Figure 2.6-1.

---

*Methods of Defining Conditioning Event

The preceding example illustrates how the conditioning event B can be defined from some characteristic of the physical experiment. There are several other ways of defining B (Cooper and McGillem, 1971, p. 61). We shall consider two of these in detail.

In one method, event B is defined in terms of the random variable X. We discuss this case further in the next paragraph. In another method, event B may depend on some random variable other than X. We discuss this case further in Chapter 4.

One way to define event B in terms of X is to let

$$B = \{X \le b\} \tag{2.6-7}$$

where b is some real number $-\infty < b < \infty$. After substituting (2.6-7) in (2.6-2), we get†

$$F_X(x \mid X \le b) = P\{X \le x \mid X \le b\} = \frac{P\{X \le x \cap X \le b\}}{P\{X \le b\}} \tag{2.6-8}$$

---

† Notation used has allowed for deletion of some braces for convenience. Thus, $F_X(x \mid \{X \le b\})$ is written $F_X(x \mid X \le b)$ and $P(\{X \le x\} \cap \{X \le b\})$ becomes $P\{X \le x \cap X \le b\}$.

for all events $\{X \leq b\}$ for which $P\{X \leq b\} \neq 0$. Two cases must be considered; one is where $b \leq x$; the second is where $x < b$. If $b \leq x$, the event $\{X \leq b\}$ is a subset of the event $\{X \leq x\}$, so $\{X \leq x\} \cap \{X \leq b\} = \{X \leq b\}$. Equation (2.6-8) becomes

$$F_X(x \mid X \leq b) = \frac{P\{X \leq x \cap X \leq b\}}{P\{X \leq b\}} = \frac{P\{X \leq b\}}{P\{X \leq b\}} = 1 \qquad b \leq x \qquad (2.6\text{-}9)$$

When $x < b$ the event $\{X \leq x\}$ is a subset of the event $\{X \leq b\}$, so $\{X \leq x\} \cap \{X \leq b\} = \{X \leq x\}$ and (2.6-8) becomes

$$F_X(x \mid X \leq b) = \frac{P\{X \leq x \cap X \leq b\}}{P\{X \leq b\}} = \frac{P\{X \leq x\}}{P\{X \leq b\}} = \frac{F_X(x)}{F_X(b)} \qquad x < b \qquad (2.6\text{-}10)$$

By combining the last two expressions, we obtain

$$F_X(x \mid X \leq b) = \begin{cases} \dfrac{F_X(x)}{F_X(b)} & x < b \\[2mm] 1 & b \leq x \end{cases} \qquad (2.6\text{-}11)$$

(a)

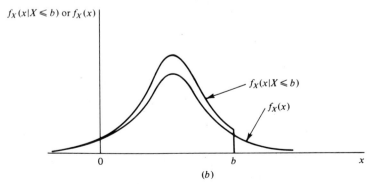

(b)

Figure 2.6-2 Possible distribution functions (a) and density functions (b) applicable to a conditioning event $B = \{X \leq b\}$.

The conditional density function derives from the derivative of (2.6-11):

$$f_X(x \mid X \le b) = \begin{cases} \dfrac{f_X(x)}{F_X(b)} = \dfrac{f_X(x)}{\int_{-\infty}^{b} f_X(x)\,dx} & x < b \\ 0 & x \ge b \end{cases} \tag{2.6-12}$$

Figure 2.6-2 sketches possible functions representing (2.6-11) and (2.6-12).

From our assumption that the conditioning event has nonzero probability, we have $0 < F_X(b) \le 1$, so the expression of (2.6-11) shows that the conditional distribution function is never smaller than the ordinary distribution function:

$$F_X(x \mid X \le b) \ge F_X(x) \tag{2.6-13}$$

A similar statement holds for the conditional density function of (2.6-12) wherever it is nonzero:

$$f_X(x \mid X \le b) \ge f_X(x) \qquad x < b \tag{2.6-14}$$

The principal results (2.6-11) and (2.6-12) can readily be extended to the more general event $B = \{a < X \le b\}$ (see Problem 2-39).

Example 2.6-2 The radial "miss-distance" of landings from parachuting sky divers, as measured from a target's center, is a Rayleigh random variable with $b = 800$ m^2 and $a = 0$. From (2.5-12) we have

$$F_X(x) = [1 - e^{-x^2/800}]u(x)$$

The target is a circle of 50-m radius with a bull's eye of 10-m radius. We find the probability of a parachuter hitting the bull's eye given that the landing is on the target.

The required probability is given by (2.6-11) with $x = 10$ and $b = 50$:

$$P(\text{bull's eye} \mid \text{landing on target}) = F_X(10)/F_X(50)$$

$$= (1 - e^{-100/800})/(1 - e^{-2500/800}) = 0.1229$$

Parachuter accuracy is such that about 12.29% of landings falling on the target will actually hit the bull's eye.

PROBLEMS

2-1 The sample space for an experiment is $S = \{0, 1, 2.5, 6\}$. List all possible values of the following random variables:

(a) $X = 2s$
(b) $X = 5s^2 - 1$
(c) $X = \cos(\pi s)$
(d) $X = (1 - 3s)^{-1}$

2-2 Work Problem 2-1 for $S = \{-2 < s \le 5\}$.

2-3 Given that a random variable X has the following possible values, state if X is discrete, continuous, or mixed.

(a) $\{-20 < x < -5\}$

(b) $\{10, 12 < x \le 14, 15, 17\}$

(c) $\{-10 \text{ for } s > 2 \text{ and } 5 \text{ for } s \le 2, \text{ where } 1 < s \le 6\}$

(d) $\{4, 3.1, 1, -2\}$

2-4 A random variable X is a function. So is probability P. Recall that the *domain* of a function is the set of values its argument may take on while its *range* is the set of corresponding values of the function. In terms of sets, events, and sample spaces, state the domain and range for X and P.

2-5 A man matches coin flips with a friend. He wins $2 if coins match and loses $2 if they do not match. Sketch a sample space showing possible outcomes for this experiment and illustrate how the points map onto the real line x that defines the values of the random variable $X =$ "dollars won on a trial." Show a second mapping for a random variable $Y =$ "dollars won by the friend on a trial."

2-6 Temperature in a given city varies randomly during any year from -21 to $49°C$. A house in the city has a thermostat that assumes only three positions: 1 represents "call for heat below $18.3°C$," 2 represents "dead or idle zone," and 3 represents "call for air conditioning above $21.7°C$." Draw a sample space for this problem showing the mapping necessary to define a random variable $X =$ "thermostat setting."

2-7 A random voltage can have any value defined by the set $S = \{a \le s \le b\}$. A quantizer divides S into 6 equal-sized contiguous subsets and generates a voltage random variable X having values $\{-4, -2, 0, 2, 4, 6\}$. Each value of X is equal to the midpoint of the subset of S from which it is mapped.

(a) Sketch the sample space and the mapping to the line x that defines the values of X.

(b) Find a and b.

*__2-8__ A random signal can have any voltage value (at a given time) defined by the set $S = \{a_0 < s \le a_N\}$, where a_0 and a_N are real numbers and N is any integer $N \ge 1$. A voltage quantizer divides S into N equal-sized contiguous subsets and converts the signal level into one of a set of discrete levels a_n, $n = 1, 2, \ldots, N$, that correspond to the "input" subsets $\{a_{n-1} < s \le a_n\}$. The set $\{a_1, a_2, \ldots, a_N\}$ can be taken as the discrete values of an "output" random variable X of the quantizer. If the smallest "input" subset is defined by $\Delta = a_1 - a_0$ and other subsets by $a_n - a_{n-1} = 2^{n-1}\Delta$, determine Δ and the quantizer levels a_n in terms of a_0, a_N, and N.

2-9 An honest coin is tossed three times.

(a) Sketch the applicable sample space S showing all possible elements. Let X be a random variable that has values representing the number of heads obtained on any triple toss. Sketch the mapping of S onto the real line defining X.

(b) Find the probabilities of the values of X.

2-10 Work Problem 2-9 for a biased coin for which $P\{\text{head}\} = 0.6$.

2-11 Resistor R_2 in Figure P2-11 is randomly selected from a box of resistors containing 180-Ω, 470-Ω, 1000-Ω, and 2200-Ω resistors. All resistor values have the same likelihood of being selected. The voltage E_2 is a discrete random variable. Find the set of values E_2 can have and give their probabilities.

$R_1 = 820 \ \Omega$

$E_1 = 12 \ V$

R_2

E_2

Figure P2-11

2-12 Bolts made on a production line are nominally designed to have a 760-mm length. A go-no-go testing device eliminates all bolts less than 650 mm and over 920 mm in length. The surviving bolts are then made available for sale and their lengths are known to be described by a uniform probability density function. A certain buyer orders all bolts that can be produced with a $\pm 5\%$ tolerance about the nominal length. What fraction of the production line's output is he purchasing?

2-13 Find and sketch the density and distribution functions for the random variables of parts (a), (b), and (c) in Problem 2-1 if the sample space elements have equal likelihoods of occurrence.

2-14 If temperature in Problem 2-6 is uniformly distributed, sketch the density and distribution functions of the random variable X.

2-15 For the uniform random variable defined by (2.5-7) find:
 (a) $P\{0.9a + 0.1b < X \le 0.7a + 0.3b\}$
 (b) $P\{(a + b)/2 < X \le b\}$

2-16 Determine which of the following are valid distribution functions:

(a) $G_X(x) = \begin{cases} 1 - e^{-x/2} & x \ge 0 \\ 0 & x < 0 \end{cases}$

(b) $G_X(x) = \begin{cases} 0 & x < 0 \\ 0.5 + 0.5 \sin [\pi(x - 1)/2] & 0 \le x < 2 \\ 1 & x \ge 2 \end{cases}$

(c) $G_X(x) = \dfrac{x}{a} [u(x - a) - u(x - 2a)]$

2-17 Determine the real constant a, for arbitrary real constants m and $0 < b$, such that

$$f_X(x) = ae^{-|x - m|/b}$$

is a valid density function (called the Laplace† density).

† After the French mathematician Marquis Pierre Simon de Laplace (1749–1827).

2-18 An intercom system master station provides music to six hospital rooms. The probability that any one room will be switched on and draw power at any time is 0.4. When on, a room draws 0.5 W.

(a) Find and plot the density and distribution functions for the random variable " power delivered by the master station."

(b) If the master-station amplifier is overloaded when more than 2 W is demanded, what is its probability of overload?

*2-19** The amplifier in the master station of Problem 2-18 is replaced by a 4-W unit that must now supply 12 rooms. Is the probability of overload better than if two independent 2-W units supplied six rooms each?

2-20 Justify that a distribution function $F_X(x)$ satisfies (2.2-2a, b, c).

2-21 Use the definition of the impulse function to evaluate the following integrals.

(*Hint:* Refer to Appendix A.)

(a) $\displaystyle\int_3^4 (3x^2 + 2x - 4)\delta(x - 3.2)\, dx$

(b) $\displaystyle\int_{-\infty}^{\infty} \cos{(6\pi x)}\delta(x - 1)\, dx$

(c) $\displaystyle\int_{-\infty}^{\infty} \frac{24\delta(x - 2)\, dx}{x^4 + 3x^2 + 2}$

(d)† $\displaystyle\int_{-\infty}^{\infty} \delta(x - x_0)e^{-j\omega x}\, dx$

(e) $\displaystyle\int_{-3}^{3} u(x - 2)\delta(x - 3)\, dx$

2-22 Show that the properties of a density function $f_X(x)$, as given by (2.3-6), are valid.

2-23 For the random variable defined in Example 2.3-1, find:
(a) $P\{x_0 - 0.6\alpha < X \le x_0 + 0.3\alpha\}$
(b) $P\{X = x_0\}$

2-24 A random variable X is gaussian with $a_X = 0$ and $\sigma_X = 1$.
(a) What is the probability that $|X| > 2$?
(b) What is the probability that $X > 2$?

2-25 Work Problem 2-24 if $a_X = 4$ and $\sigma_X = 2$.

2-26 For the gaussian density function of (2.4-1), show that

$$\int_{-\infty}^{\infty} xf_X(x)\, dx = a_X$$

---

† The quantity j is the unit-imaginary; that is, $j = \sqrt{-1}$.

2-27 For the gaussian density function of (2.4-1), show that

$$\int_{-\infty}^{\infty} (x - a_X)^2 f_X(x) \, dx = \sigma_X^2$$

2-28 A production line manufactures 1000-Ω resistors that must satisfy a 10% tolerance.

(a) If resistance is adequately described by a gaussian random variable X for which $a_X = 1000 \ \Omega$ and $\sigma_X = 40 \ \Omega$, what fraction of the resistors is expected to be rejected?

(b) If a machine is not properly adjusted, the product resistances change to the case where $a_X = 1050 \ \Omega$ (5% shift). What fraction is now rejected?

2-29 Cannon shell impact position, as measured along the line of fire from the target point, can be described by a gaussian random variable X. It is found that 15.15% of shells fall 11.2 m or farther from the target in a direction toward the cannon, while 5.05% fall farther than 95.6 m beyond the target. What are a_X and σ_X for X?

2-30 (a) Use the exponential density of (2.5-9) and solve for I_2 defined by

$$I_2 = \int_{-\infty}^{\infty} x^2 f_X(x) \, dx$$

(b) Solve for I_1 defined by

$$I_1 = \int_{-\infty}^{\infty} x f_X(x) \, dx$$

(c) Verify that I_1 and I_2 satisfy the equation $I_2 - I_1^2 = b^2$.

2-31 Verify that the maximum value of $f_X(x)$ for the Rayleigh density function of (2.5-11) occurs at $x = a + \sqrt{b/2}$ and is equal to $\sqrt{2/b} \exp(-\frac{1}{2}) \approx 0.607\sqrt{2/b}$. This value of x is called the *mode* of the random variable. (In general, a random variable may have more than one such value—explain.)

2-32 Find the value $x = x_0$ of a Rayleigh random variable for which $P\{X \leq x_0\} = P\{x_0 < X\}$. This value of x is called the *median* of the random variable.

2-33 The lifetime of a system expressed in weeks is a Rayleigh random variable X for which

$$f_X(x) = \begin{cases} (x/200)e^{-x^2/400} & 0 \leq x \\ 0 & x < 0 \end{cases}$$

(a) What is the probability that the system will not last a full week?

(b) What is the probability the system lifetime will exceed one year?

2-34 The *Cauchy*† random variable has the probability density function

$$f_X(x) = \frac{b/\pi}{b^2 + (x - a)^2}$$

† After the French mathematician Augustin Louis Cauchy (1789–1857).

for real numbers $0 < b$ and $-\infty < a < \infty$. Show that the distribution function of X is

$$F_X(x) = \frac{1}{2} + \frac{1}{\pi} \tan^{-1}\left(\frac{x-a}{b}\right)$$

2-35 The *Log-Normal density* function is given by

$$f_X(x) = \begin{cases} \dfrac{\exp\left\{-[\ln\,(x-b)-a_X]^2/2\sigma_X^2\right\}}{\sqrt{2\pi}\,\sigma_X(x-b)} & x \geq b \\ 0 & x < b \end{cases}$$

for real constants $0 < \sigma_X$, $-\infty < a_X < \infty$, and $-\infty < b < \infty$, where $\ln\,(x)$ denotes the natural logarithm of x. Show that the corresponding distribution function is

$$F_X(x) = \begin{cases} F\left[\dfrac{\ln\,(x-b)-a_X}{\sigma_X}\right] & x \geq b \\ 0 & x < b \end{cases}$$

where $F(\cdot)$ is given by (2.4-3).

2-36 A random variable X is known to be Poisson with $b = 4$.

(a) Plot the density and distribution functions for this random variable.

(b) What is the probability of the event $\{0 \leq X \leq 5\}$?

2-37 The number of cars arriving at a certain bank drive-in window during any 10-min period is a Poisson random variable X with $b = 2$. Find:

(a) The probability that more than 3 cars will arrive during any 10-min period.

(b) The probability that no cars will arrive.

2-38 Rework Example 2.6-1 to find $f_X(x\,|\,B_2)$ and $F_X(x\,|\,B_2)$. Sketch the two functions.

*2-39 Extend the analysis of the text, that leads to (2.6-11) and (2.6-12), to the more general event $B = \{a < X \leq b\}$. Specifically, show that now

$$F_X(x\,|\,a < X \leq b) = \begin{cases} 0 & x < a \\ \dfrac{F_X(x) - F_X(a)}{F_X(b) - F_X(a)} & a \leq x < b \\ 1 & b \leq x \end{cases}$$

and

$$f_X(x\,|\,a < X \leq b) = \begin{cases} 0 & x < a \\ \dfrac{f_X(x)}{F_X(b) - F_X(a)} = \dfrac{f_X(x)}{\int_a^b f_X(x)\,dx} & a \leq x < b \\ 0 & b \leq x \end{cases}$$

*2-40 Consider the system having a lifetime defined by the random variable X in Problem 2-33. Given that the system will survive beyond 20 weeks, find the probability that it will survive beyond 26 weeks.

ADDITIONAL PROBLEMS

2-41 A sample space is defined by $S = \{1, 2 \le s \le 3, 4, 5\}$. A random variable is defined by: $X = 2$ for $0 \le s \le 2.5$, $X = 3$ for $2.5 < s < 3.5$, and $X = 5$ for $3.5 \le s \le 6$.

    (a) Is X discrete, continuous, or mixed?

    (b) Give a sct that defines the values X can have.

2-42 A gambler flips a fair coin three times.

    (a) Draw a sample space S for this experiment. A random variable X representing his winnings is defined as follows: He loses \$1 if he gets no heads in three flips; he wins \$1, \$2, and \$3 if he obtains 1, 2, or 3 heads, respectively. Show how elements of S map to values of X.

    (b) What are the probabilities of the various values of X?

2-43 A function $G_X(x) = a[1 + (2/\pi) \sin^{-1} (x/c)] \text{ rect } (x/2c) + (a + b)u(x - c)$ is defined for all $-\infty < x < \infty$, where $c > 0$, b, and a are real constants and rect $(\cdot)$ is defined by $(E$-$2)$. Find any conditions on a, b, and c that will make $G_X(x)$ a valid probability distribution function. Discuss what choices of constants correspond to a continuous, discrete, or mixed random variable.

2-44 (a) Generalize Problem 2-16(a) by finding values of real constants a and b such that

$$G_X(x) = [1 - a \exp (-x/b)]u(x)$$

is a valid distribution function.

    (b) Are there any values of a and b such that $G_X(x)$ corresponds to a mixed random variable X?

2-45 Find a constant $b > 0$ so that the function

$$f_X(x) = \begin{cases} e^{3x}/4 & 0 \le x \le b \\ 0 & \text{elsewhere} \end{cases}$$

is a valid probability density.

2-46 Given the function

$$g_X(x) = 4 \cos (\pi x/2b) \text{ rect } (x/2b)$$

find a value of b so that $g_X(x)$ is a valid probability density.

2-47 A random variable X has the density function

$$f_X(x) = (\tfrac{1}{2})u(x) \exp (-x/2)$$

Define events $A = \{1 < X \le 3\}$, $B = \{X \le 2.5\}$, and $C = A \cap B$. Find the probabilities of events (a) A, (b) B, and (c) C.

★**2-48** Let $\phi(x)$ be a continuous, but otherwise arbitrary real function, and let a and b be real constants. Find $G(a, b)$ defined by

$$G(a, b) = \int_{-\infty}^{\infty} \phi(x)\, \delta(ax + b)\, dx$$

(*Hint*: Use the definition of the impulse function.)

2-49 For real constants $b > 0$, $c > 0$, and any a, find a condition on constant a and a relationship between c and a (for given b) such that the function

$$f_X(x) = \begin{cases} a[1 - (x/b)] & 0 \le x \le c \\ 0 & \text{elsewhere} \end{cases}$$

is a valid probability density.

2-50 A gaussian random variable X has $a_X = 2$, and $\sigma_X = 2$.
  (*a*) Find $P\{X > 1.0\}$.
  (*b*) Find $P\{X \le -1.0\}$.

2-51 In a certain "junior" olympics, javelin throw distances are well approximated by a gaussian distribution for which $a_X = 30$ m and $\sigma_X = 5$ m. In a qualifying round, contestants must throw farther than 26 m to qualify. In the main event the record throw is 42 m.
  (*a*) What is the probability of being disqualified in the qualifying round?
  (*b*) In the main event what is the probability the record will be broken?

2-52 Suppose height to the bottom of clouds is a gaussian random variable X for which $a_X = 4000$ m, and $\sigma_X = 1000$ m. A person bets that cloud height tomorrow will fall in the set $A = \{1000 \text{ m} < X \le 3300 \text{ m}\}$ while a second person bets that height will be satisfied by $B = \{2000 \text{ m} < X \le 4200 \text{ m}\}$. A third person bets they are both correct. Find the probabilities that each person will win the bet.

2-53 Let X be a Rayleigh random variable with $a = 0$. Find the probability that X will have values larger than its mode (see Problem 2-31).

2-54 A certain large city averages three murders per week and their occurrences follow a Poisson distribution.
  (*a*) What is the probability that there will be five or more murders in a given week?
  (*b*) On the average, how many weeks a year can this city expect to have no murders?
  (*c*) How many weeks per year (average) can the city expect the number of murders per week to equal or exceed the average number per week?

2-55 A certain military radar is set up at a remote site with no repair facilities. If the radar is known to have a *mean-time-between-failures* (MTBF) of 200 h find the probability that the radar is still in operation one week later when picked up for maintenance and repairs.

2-56 If the radar of Problem 2-55 is permanently located at the remote site, find the probability that it will be operational as a function of time since its set up.

2-57 A computer undergoes down-time if a certain critical component fails. This component is known to fail at an average rate of once per four weeks. No significant down-time occurs if replacement components are on hand because repair can be made rapidly. There are three components on hand and ordered replacements are not due for six weeks.

(*a*) What is the probability of significant down-time occurring before the ordered components arrive?

(*b*) If the shipment is delayed two weeks what is the probability of significant down-time occurring before the shipment arrives?

*★**2-58** Assume the lifetime of a laboratory research animal is defined by a Rayleigh density with $a = 0$ and $b = 30$ weeks in (2.5-11) and (2.5-12). If for some clinical reasons it is known that the animal will live *at most* 20 weeks, what is the probability it will live 10 weeks or less?

*★**2-59** Suppose the depth of water, measured in meters, behind a dam is described by an exponential random variable having a density

$$f_X(x) = (1/13.5) \exp(-x/13.5)$$

There is an emergency overflow at the top of the dam that prevents the depth from exceeding 40.6 m. There is a pipe placed 32.0 m below the overflow (ignore the pipe's finite diameter) that feeds water to a hydroelectric generator.

(*a*) What is the probability that water is wasted through emergency overflow?

(*b*) Given that water is not wasted in overflow, what is the probability the generator will have water to drive it?

(*c*) What is the probability that water will be too low to produce powe ?

*★**2-60** In Problem 2-59 find and sketch the distribution and density functions of water depth given that water will be deep enough to generate power but no water is wasted by emergency overflow. Also sketch for comparisons the distribution and density of water depth without any conditions?

*★**2-61** In Example 2.6-2 a parachuter is an "expert" if he hits the bull's eye. If he falls outside the bull's eye but within a circle of 25-m radius he is called "qualified" for competition. Given that a parachuter is not an expert but hits the target what is the probability of being "qualified?"

THREE

OPERATIONS ON
ONE RANDOM VARIABLE—EXPECTATION

3.0 INTRODUCTION

The random variable was introduced in Chapter 2 as a means of providing a systematic definition of events defined on a sample space. Specifically, it formed a mathematical model for describing characteristics of some real, physical world random phenomenon. In this chapter we extend our work to include some important *operations* that may be performed on a random variable. Most of these operations are based on a single concept—expectation.

3.1 EXPECTATION

Expectation is the name given to the process of averaging when a random variable is involved. For a random variable X, we use the notation $E[X]$, which may be read "the mathematical *expectation* of X," "the *expected* value of X," "the *mean* value of X," or "the *statistical average* of X." Occasionally we also use the notation $\bar{X}$ which is read the same way as $E[X]$; that is, $\bar{X} = E[X]$.†

Nearly everyone is familiar with averaging procedures. An example that serves to tie a familiar problem to the new concept of expectation may be the easiest way to proceed.

---

† Up to this point in this book an overbar represented the complement of a set or event. Henceforth, unless specifically stated otherwise, the overbar will always represent a mean value.

Example 3.1-1 Ninety people are randomly selected and the fractional dollar value of coins in their pockets is counted. If the count goes above a dollar, the dollar value is discarded and only the portion from 0¢ to 99¢ is accepted. It is found that 8, 12, 28, 22, 15, and 5 people had 18¢, 45¢, 64¢, 72¢, 77¢, and 95¢ in their pockets, respectively.

Our everyday experiences indicate that the average of these values is

$$\text{Average } \$ = 0.18\left(\frac{8}{90}\right) + 0.45\left(\frac{12}{90}\right) + 0.64\left(\frac{28}{90}\right) + 0.72\left(\frac{22}{90}\right)$$

$$+ 0.77\left(\frac{15}{90}\right) + 0.95\left(\frac{5}{90}\right)$$

$$\approx \$0.632$$

Expected Value of a Random Variable

The everyday averaging procedure used in the above example carries over directly to random variables. In fact, if X is the discrete random variable "fractional dollar value of pocket coins," it has 100 discrete values x_i that occur with probabilities $P(x_i)$, and its expected value $E[X]$ is found in the same way as in the example:

$$E[X] = \sum_{i=1}^{100} x_i P(x_i) \tag{3.1-1}$$

The values x_i identify with the fractional dollar values in the example, while $P(x_i)$ is identified with the ratio of the number of people for the given dollar value to the total number of people. If a large number of people had been used in the "sample" of the example, all fractional dollar values would have shown up and the ratios would have approached $P(x_i)$. Thus, the average in the example would have become more like (3.1-1) for many more than 90 people.

In general, the expected value of any random variable X is defined by

$$E[X] = \bar{X} = \int_{-\infty}^{\infty} x f_X(x) \, dx \tag{3.1-2}$$

If X happens to be discrete with N possible values x_i having probabilities $P(x_i)$ of occurrence, then

$$f_X(x) = \sum_{i=1}^{N} P(x_i)\delta(x - x_i) \tag{3.1-3}$$

from (2.3-5). Upon substitution of (3.1-3) into (3.1-2), we have

$$E[X] = \sum_{i=1}^{N} x_i P(x_i) \qquad \text{discrete random variable} \tag{3.1-4}$$

Hence, (3.1-1) is a special case of (3.1-4) when $N = 100$. For some discrete random variables, N may be infinite in (3.1-3) and (3.1-4).

Example 3.1-2 We determine the mean value of the continuous, exponentially distributed random variable for which (2.5-9) applies:

$$f_X(x) = \begin{cases} \dfrac{1}{b} e^{-(x-a)/b} & x > a \\ 0 & x < a \end{cases}$$

From (3.1-2) and an integral from Appendix C:

$$E[X] = \int_a^\infty \frac{x}{b} e^{-(x-a)/b}\, dx = \frac{e^{a/b}}{b} \int_a^\infty x e^{-x/b}\, dx = a + b$$

If a random variable's density is symmetrical about a line $x = a$, then $E[X] = a$; that is,

$$E[X] = a \qquad \text{if} \qquad f_X(x + a) = f_X(-x + a) \tag{3.1-5}$$

Expected Value of a Function of a Random Variable

As will be evident in the next section, many useful parameters relating to a random variable X can be derived by finding the expected value of a real function $g(\cdot)$ of X. It can be shown (Papoulis, 1965, p. 142) that this expected value is given by

$$E[g(X)] = \int_{-\infty}^\infty g(x) f_X(x)\, dx \tag{3.1-6}$$

If X is a discrete random variable, (3.1-3) applies and (3.1-6) reduces to

$$E[g(X)] = \sum_{i=1}^N g(x_i) P(x_i) \qquad \text{discrete random variable} \tag{3.1-7}$$

where N may be infinite for some random variables.

Example 3.1-3 It is known that a particular random voltage can be represented as a Rayleigh random variable V having a density function given by (2.5-11) with $a = 0$ and $b = 5$. The voltage is applied to a device that generates a voltage $Y = g(V) = V^2$ that is equal, numerically, to the power in V (in a 1-Ω resistor). We find the average power in V by means of (3.1-6):

$$\text{Power in } V = E[g(V)] = E[V^2] = \int_0^\infty \frac{2v^3}{5} e^{-v^2/5}\, dv$$

By letting $\xi = v^2/5$, $d\xi = 2v\, dv/5$, we obtain

$$\text{Power in } V = 5 \int_0^\infty \xi e^{-\xi}\, d\xi = 5 \text{ W}$$

after using (C-46).

⋆Conditional Expected Value

If, in (3.1-2), $f_X(x)$ is replaced by the conditional density $f_X(x \mid B)$, where B is any event defined on the sample space, we have the *conditional expected value* of X, denoted $E[X \mid B]$:

$$E[X \mid B] = \int_{-\infty}^{\infty} x f_X(x \mid B) \, dx \tag{3.1-8}$$

One way to define event B, as shown in Chapter 2, is to let it depend on the random variable X by defining

$$B = \{X \le b\} \qquad -\infty < b < \infty \tag{3.1-9}$$

We showed there that

$$f_X(x \mid X \le b) = \begin{cases} \dfrac{f_X(x)}{\int_{-\infty}^{b} f_X(x) \, dx} & x < b \\ 0 & x \ge b \end{cases} \tag{3.1-10}$$

Thus, by substituting (3.1-10) into (3.1-8):

$$E[X \mid X \le b] = \frac{\int_{-\infty}^{b} x f_X(x) \, dx}{\int_{-\infty}^{b} f_X(x) \, dx} \tag{3.1-11}$$

which is the mean value of X when X is constrained to the set $\{X \le b\}$.

3.2 MOMENTS

An immediate application of the expected value of a function $g(\cdot)$ of a random variable X is in calculating moments. Two types of moments are of interest, those about the origin and those about the mean.

Moments About the Origin

The function

$$g(X) = X^n \qquad n = 0, 1, 2, \ldots \tag{3.2-1}$$

when used in (3.1-6) gives the moments about the origin of the random variable X. Denote the nth moment by m_n. Then,

$$m_n = E[X^n] = \int_{-\infty}^{\infty} x^n f_X(x) \, dx \tag{3.2-2}$$

Clearly $m_0 = 1$, the area of the function $f_X(x)$, while $m_1 = \bar{X}$, the expected value of X.

Central Moments

Moments about the mean value of X are called *central moments* and are given the symbol μ_n. They are defined as the expected value of the function

$$g(X) = (X - \bar{X})^n \qquad n = 0, 1, 2, \ldots \tag{3.2-3}$$

which is

$$\mu_n = E[(X - \bar{X})^n] = \int_{-\infty}^{\infty} (x - \bar{X})^n f_X(x)\, dx \tag{3.2-4}$$

The moment $\mu_0 = 1$, the area of $f_X(x)$, while $\mu_1 = 0$. (Why?)

Variance and Skew

The second central moment μ_2 is so important we shall give it the name *variance* and the special notation σ_X^2. Thus, variance is given by†

$$\sigma_X^2 = \mu_2 = E[(X - \bar{X})^2] = \int_{-\infty}^{\infty} (x - \bar{X})^2 f_X(x)\, dx \tag{3.2-5}$$

The positive square root σ_X of variance is called the *standard deviation* of X; it is a measure of the spread in the function $f_X(x)$ about the mean.

Variance can be found from a knowledge of first and second moments. By expanding (3.2-5), we have‡

$$\sigma_X^2 = E[X^2 - 2\bar{X}X + \bar{X}^2] = E[X^2] - 2\bar{X}E[X] + \bar{X}^2$$
$$= E[X^2] - \bar{X}^2 = m_2 - m_1^2 \tag{3.2-6}$$

Example 3.2-1 Let X have the exponential density function given in Example 3.1-2. By substitution into (3.2-5), the variance of X is

$$\sigma_X^2 = \int_a^{\infty} (x - \bar{X})^2 \frac{1}{b} e^{-(x-a)/b}\, dx$$

By making the change of variable $\xi = x - \bar{X}$ we obtain

$$\sigma_X^2 = \frac{e^{-(\bar{X}-a)/b}}{b} \int_{a-\bar{X}}^{\infty} \xi^2 e^{-\xi/b}\, d\xi = (a + b - \bar{X})^2 + b^2$$

---

† The subscript indicates that σ_X^2 is the variance of a random variable X. For a random variable Y its variance would be σ_Y^2.

‡ We use the fact that the expected value of a sum of functions of X equals the sum of expected values of individual functions, as the reader can readily verify as an exercise.

after using an integral from Appendix C. However, from Example 3.1-2, $\bar{X} = E[X] = (a + b)$, so

$$\sigma_X^2 = b^2$$

The reader may wish to verify this result by finding the second moment $E[X^2]$ and using (3.2-6).

The third central moment $\mu_3 = E[(X - \bar{X})^3]$ is a measure of the asymmetry of $f_X(x)$ about $x = \bar{X} = m_1$. It will be called the *skew* of the density function. If a density is symmetric about $x = \bar{X}$, it has zero skew. In fact, for this case $\mu_n = 0$ for all odd values of n. (Why?) The normalized third central moment μ_3/σ_X^3 is known as the *skewness* of the density function, or, alternatively, as the *coefficient of skewness*.

Example 3.2-2 We continue Example 3.2-1 and compute the skew and coefficient of skewness for the exponential density. From (3.2-4) with $n = 3$ we have

$$\mu_3 = E[(X - \bar{X})^3] = E[X^3 - 3\bar{X}X^2 + 3\bar{X}^2 X - \bar{X}^3]$$
$$= \overline{X^3} - 3\bar{X}\,\overline{X^2} + 2\bar{X}^3 = \overline{X^3} - 3\bar{X}(\sigma_X^2 + \bar{X}^2) + 2\bar{X}^3$$
$$= \overline{X^3} - 3\bar{X}\sigma_X^2 - \bar{X}^3$$

Next, we have

$$\overline{X^3} = \int_a^\infty \frac{x^3}{b} e^{-(x-a)/b}\, dx = a^3 + 3a^2 b + 6ab^2 + 6b^3$$

after using (C-48). On substituting $\bar{X} = a + b$ and $\sigma_X^2 = b^2$ from the earlier example, and reducing the algebra we find

$$\mu_3 = 2b^3$$

$$\frac{\mu_3}{\sigma_X^3} = 2$$

This density has a relatively large coefficient of skewness, as can be seen intuitively from Figure 2.5-3.

★3.3 FUNCTIONS THAT GIVE MOMENTS

Two functions can be defined that allow moments to be calculated for a random variable X. They are the characteristic function and the moment generating function.

*Characteristic Function

The *characteristic function* of a random variable X is defined by

$$\Phi_X(\omega) = E[e^{j\omega X}]$$ (3.3-1)

where $j = \sqrt{-1}$. It is a function of the real number $-\infty < \omega < \infty$. If (3.3-1) is written in terms of the density function, $\Phi_X(\omega)$ is seen to be the *Fourier transform*† (with the sign of ω reversed) of $f_X(x)$:

$$\Phi_X(\omega) = \int_{-\infty}^{\infty} f_X(x)e^{j\omega x}\, dx$$ (3.3-2)

Because of this fact, if $\Phi_X(\omega)$ is known, $f_X(x)$ can be found from the *inverse Fourier transform* (with sign of x reversed)

$$f_X(x) = \frac{1}{2\pi} \int_{-\infty}^{\infty} \Phi_X(\omega)e^{-j\omega x}\, d\omega$$ (3.3-3)

By formal differentiation of (3.3-2) n times with respect to ω and setting $\omega = 0$ in the derivative, we may show that the nth moment of X is given by

$$m_n = (-j)^n \left. \frac{d^n \Phi_X(\omega)}{d\omega^n} \right|_{\omega=0}$$ (3.3-4)

A major advantage of using $\Phi_X(\omega)$ to find moments is that $\Phi_X(\omega)$ always exists (Davenport, 1970, p. 426), so the moments can always be found if $\Phi_X(\omega)$ is known, provided, of course, the derivatives of $\Phi_X(\omega)$ exist.

It can be shown that the maximum magnitude of a characteristic function is unity and occurs at $\omega = 0$; that is,

$$|\Phi_X(\omega)| \le \Phi_X(0) = 1$$ (3.3-5)

(See Problem 3-24.)

Example 3.3-1 Again we consider the random variable with the exponential density of Example 3.1-2 and find its characteristic function and first moment.

---

† Readers unfamiliar with Fourier transforms should interpret $\Phi_X(\omega)$ as simply the expected value of the function $g(X) = \exp(j\omega X)$. Appendix D is included as a review for others wishing to refresh their background in Fourier transform theory.

By substituting the density function into (3.3-2), we get

$$\Phi_X(\omega) = \int_a^\infty \frac{1}{b} e^{-(x-a)/b} e^{j\omega x} \, dx = \frac{e^{a/b}}{b} \int_a^\infty e^{-(1/b - j\omega)x} \, dx$$

Evaluation of the integral follows the use of an integral from Appendix C:

$$\Phi_X(\omega) = \frac{e^{a/b}}{b} \left[\frac{e^{-(1/b - j\omega)x}}{-(1/b - j\omega)} \Big|_a^\infty \right]$$

$$= \frac{e^{j\omega a}}{1 - j\omega b}$$

The derivative of $\Phi_X(\omega)$ is

$$\frac{d\Phi_X(\omega)}{d\omega} = e^{j\omega a} \left[\frac{ja}{1 - j\omega b} + \frac{jb}{(1 - j\omega b)^2} \right]$$

so the first moment becomes

$$m_1 = (-j) \frac{d\Phi_X(\omega)}{d\omega} \Big|_{\omega=0} = a + b,$$

in agreement with m_1 found in Example 3.1-2.

★Moment Generating Function

Another statistical average closely related to the characteristic function is the *moment generating function*, defined by

$$M_X(v) = E[e^{vX}] \tag{3.3-6}$$

where v is a real number $-\infty < v < \infty$. Thus, $M_X(v)$ is given by

$$M_X(v) = \int_{-\infty}^\infty f_X(x) e^{vx} \, dx \tag{3.3-7}$$

The main advantage of the moment generating function derives from its ability to give the moments. Moments are related to $M_X(v)$ by the expression:

$$m_n = \frac{d^n M_X(v)}{dv^n} \Big|_{v=0} \tag{3.3-8}$$

The main disadvantage of the moment generating function, as opposed to the characteristic function, is that it may not exist for all random variables. In fact, $M_X(v)$ exists only if all the moments exist (Davenport and Root, 1958, p. 52).

Example 3.3-2 To illustrate the calculation and use of the moment generating function, let us reconsider the exponential density of the earlier examples. On use of (3.3-7) we have

$$M_X(v) = \int_a^\infty \frac{1}{b} e^{-(x-a)/b} e^{vx} \, dx$$

$$= \frac{e^{a/b}}{b} \int_a^\infty e^{[v-(1/b)]x} \, dx$$

$$= \frac{e^{av}}{1-bv}$$

In evaluating $M_X(v)$ we have used an integral from Appendix C.
By differentiation we have the first moment

$$m_1 = \frac{dM_X(v)}{dv}\bigg|_{v=0}$$

$$= \frac{e^{av}[a(1-bv)+b]}{(1-bv)^2}\bigg|_{v=0} = a+b$$

which, of course, is the same as previously found.

3.4 TRANSFORMATIONS OF A RANDOM VARIABLE

Quite often one may wish to transform (change) one random variable X into a new random variable Y by means of a transformation

$$Y = T(X) \tag{3.4-1}$$

Typically, the density function $f_X(x)$ or distribution function $F_X(x)$ of X is known, and the problem is to determine either the density function $f_Y(y)$ or distribution function $F_Y(y)$ of Y. The problem can be viewed as a "black box" with input X, output Y, and "transfer characteristic" $Y = T(X)$, as illustrated in Figure 3.4-1.

In general, X can be a discrete, continuous, or a mixed random variable. In turn, the transformation T can be linear, nonlinear, segmented, staircase, etc. Clearly, there are many cases to consider in a general study, depending on the form of X and T. In this section we shall consider only three cases: (1) X continuous and T continuous and either monotonically increasing or decreasing with X; (2) X continuous and T continuous but nonmonotonic; (3) X discrete and T

Figure 3.4-1 Transformation of a random variable X to a new random variable Y.

continuous. Note that the transformation in all three cases is assumed continuous. The concepts introduced in these three situations are broad enough that the reader should have no difficulty in extending them to other cases (see Problem 3-32).

Monotonic Transformations of a Continuous Random Variable

A transformation T is called *monotonically increasing* if $T(x_1) < T(x_2)$ for any $x_1 < x_2$. It is *monotonically decreasing* if $T(x_1) > T(x_2)$ for any $x_1 < x_2$.

Consider first the increasing transformation. We assume that T is continuous and differentiable at all values of x for which $f_X(x) \neq 0$. Let Y have a particular value y_0 corresponding to the particular value x_0 of X as shown in Figure 3.4-2a. The two numbers are related by

$$y_0 = T(x_0) \qquad \text{or} \qquad x_0 = T^{-1}(y_0) \qquad (3.4\text{-}2)$$

where T^{-1} represents the inverse of the transformation T. Now the probability of the event $\{Y \leq y_0\}$ must equal the probability of the event $\{X \leq x_0\}$ because of the one-to-one correspondence between X and Y. Thus,

$$F_Y(y_0) = P\{Y \leq y_0\} = P\{X \leq x_0\} = F_X(x_0) \qquad (3.4\text{-}3)$$

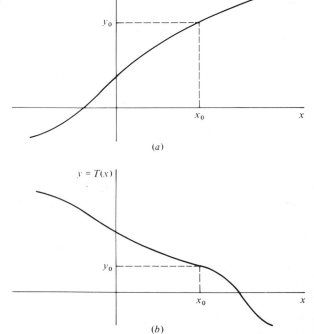

(a)

(b)

Figure 3.4-2 Monotonic transformations: (a) increasing, and (b) decreasing. [*Adapted from Peebles (1976) with permission of publishers Addison-Wesley, Advanced Book Program.*]

or

$$\int_{-\infty}^{y_0} f_Y(y) \, dy = \int_{-\infty}^{x_0 = T^{-1}(y_0)} f_X(x) \, dx \tag{3.4-4}$$

Next, we differentiate both sides of (3.4-4) with respect to y_0 using Leibniz's rule†
to get

$$f_Y(y_0) = f_X[T^{-1}(y_0)] \frac{dT^{-1}(y_0)}{dy_0} \tag{3.4-5}$$

Since this result applies for any y_0, we may now drop the subscript and write

$$f_Y(y) = f_X[T^{-1}(y)] \frac{dT^{-1}(y)}{dy} \tag{3.4-6}$$

or, more compactly,

$$f_Y(y) = f_X(x) \frac{dx}{dy} \tag{3.4-7}$$

In (3.4-7) it is understood that x is a function of y through (3.4-2).

A consideration of Figure 3.4-2b for the decreasing transformation verifies
that

$$F_Y(y_0) = P\{Y \le y_0\} = P\{X \ge x_0\} = 1 - F_X(x_0). \tag{3.4-8}$$

A repetition of the steps leading to (3.4-6) will again produce (3.4-6) except that
the right side is negative. However, since the slope of $T^{-1}(y)$ is also negative, we
conclude that for either type of monotonic transformation

$$f_Y(y) = f_X[T^{-1}(y)] \left| \frac{dT^{-1}(y)}{dy} \right| \tag{3.4-9}$$

or simply

$$f_Y(y) = f_X(x) \left| \frac{dx}{dy} \right| \tag{3.4-10}$$

---

† Leibniz's rule, after the great German mathematician Gottfried Wilhelm von Leibniz (1646–
1716), states that, if $H(x, u)$ is continuous in x and u and

$$G(u) = \int_{\alpha(u)}^{\beta(u)} H(x, u) \, dx$$

then the derivative of the integral with respect to the parameter u is

$$\frac{dG(u)}{du} = H[\beta(u), u] \frac{d\beta(u)}{du} - H[\alpha(u), u] \frac{d\alpha(u)}{du} + \int_{\alpha(u)}^{\beta(u)} \frac{\partial H(x, u)}{\partial u} \, dx$$

Example 3.4-1 If we take T to be the linear transformation $Y = T(X) = aX + b$, where a and b are any real constants, then $X = T^{-1}(Y) = (Y - b)/a$ and $dx/dy = 1/a$. From (3.4-9)

$$f_Y(y) = f_X\left(\frac{y - b}{a}\right)\left|\frac{1}{a}\right|$$

If X is assumed to be gaussian with the density function given by (2.4-1), we get

$$f_Y(y) = \frac{1}{\sqrt{2\pi\sigma_X^2}} e^{-[(y-b)/a - a_X]^2/2\sigma_X^2}\left|\frac{1}{a}\right|$$

$$= \frac{1}{\sqrt{2\pi a^2\sigma_X^2}} e^{-[y - (aa_X + b)]^2/2a^2\sigma_X^2}$$

which is the density function of another gaussian random variable having

$$a_Y = aa_X + b \qquad \text{and} \qquad \sigma_Y^2 = a^2\sigma_X^2$$

Thus, *a linear transformation of a gaussian random variable produces another gaussian random variable*. A linear amplifier having a random voltage X as its input is one example of a linear transformation.

Nonmonotonic Transformations of a Continuous Random Variable

A transformation may not be monotonic in the more general case. Figure 3.4-3 illustrates one such transformation. There may now be more than one interval of values of X that correspond to the event $\{Y \le y_0\}$. For the value of y_0 shown in the figure, the event $\{Y \le y_0\}$ corresponds to the event $\{X \le x_1$ and $x_2 \le X \le x_3\}$. Thus, the probability of the event $\{Y \le y_0\}$ now equals the probability

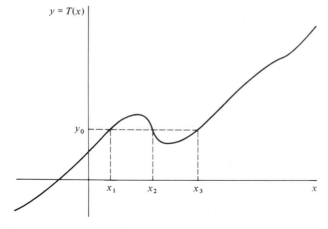

Figure 3.4-3 A nonmonotonic transformation. [*Adapted from Peebles (1976) with permission of publishers Addison-Wesley, Advanced Book Program.*]

of the event $\{x$ values yielding $Y \leq y_0\}$, which we shall write as $\{x \mid Y \leq y_0\}$. In other words

$$F_Y(y_0) = P\{Y \leq y_0\} = P\{x \mid Y \leq y_0\} = \int_{\{x \mid Y \leq y_0\}} f_X(x)\, dx \qquad (3.4\text{-}11)$$

Formally, one may differentiate to obtain the density function of Y:

$$f_Y(y_0) = \frac{d}{dy_0} \int_{\{x \mid Y \leq y_0\}} f_X(x)\, dx \qquad (3.4\text{-}12)$$

Although we shall not give a proof, the density function is also given by (Papoulis, 1965, p. 126)

$$f_Y(y) = \sum_n \frac{f_X(x_n)}{\left| \dfrac{dT(x)}{dx} \right|_{x=x_n}} \qquad (3.4\text{-}13)$$

where the sum is taken so as to include all the roots x_n, $n = 1, 2, \ldots$, which are the real solutions of the equation†

$$y = T(x) \qquad (3.4\text{-}14)$$

We illustrate the above concepts by an example.

Example 3.4-2 We find $f_Y(y)$ for the square-law transformation

$$Y = T(X) = cX^2$$

shown in Figure 3.4-4, where c is a real constant $c > 0$. We shall use both the procedure leading to (3.4-12) and that leading to (3.4-13).

    In the former case, the event $\{Y \leq y\}$ occurs when $\{-\sqrt{y/c} \leq x \leq \sqrt{y/c}\} = \{x \mid Y \leq y\}$, so (3.4-12) becomes

$$f_Y(y) = \frac{d}{dy} \int_{-\sqrt{y/c}}^{\sqrt{y/c}} f_X(x)\, dx \qquad y \geq 0$$

Upon use of Leibniz's rule we obtain

$$f_Y(y) = f_X(\sqrt{y/c}) \frac{d(\sqrt{y/c})}{dy} - f_X(-\sqrt{y/c}) \frac{d(-\sqrt{y/c})}{dy}$$

$$= \frac{f_X(\sqrt{y/c}) + f_X(-\sqrt{y/c})}{2\sqrt{cy}} \qquad y \geq 0$$

---

† If $y = T(x)$ has no real roots for a given value of y, then $f_Y(y) = 0$.

Figure 3.4-4 A square-law transformation. [*Adapted from Peebles (1976) with permission of publishers Addison-Wesley, Advanced Book Program.*]

In the latter case where we use (3.4-13), we have $X = \pm\sqrt{Y/c}$, $Y \geq 0$, so $x_1 = -\sqrt{y/c}$ and $x_2 = \sqrt{y/c}$. Furthermore, $dT(x)/dx = 2cx$ so

$$\left.\frac{dT(x)}{dx}\right|_{x=x_1} = 2cx_1 = -2c\sqrt{\frac{y}{c}} = -2\sqrt{cy}$$

$$\left.\frac{dT(x)}{dx}\right|_{x=x_2} = 2\sqrt{cy}$$

From (3.4-13) we again have

$$f_Y(y) = \frac{f_X(\sqrt{y/c}) + f_X(-\sqrt{y/c})}{2\sqrt{cy}} \qquad y \geq 0$$

Transformation of a Discrete Random Variable

If X is a discrete random variable while $Y = T(X)$ is a continuous transformation, the problem is especially simple. Here

$$f_X(x) = \sum_n P(x_n)\delta(x - x_n) \tag{3.4-15}$$

$$F_X(x) = \sum_n P(x_n)u(x - x_n) \tag{3.4-16}$$

where the sum is taken to include all the possible values $x_n, n = 1, 2, \ldots,$ of X.

If the transformation is monotonic, there is a one-to-one correspondence between X and Y so that a set $\{y_n\}$ corresponds to the set $\{x_n\}$ through the equation $y_n = T(x_n)$. The probability $P(y_n)$ equals $P(x_n)$. Thus,

$$f_Y(y) = \sum_n P(y_n)\delta(y - y_n) \tag{3.4-17}$$

$$F_Y(y) = \sum_n P(y_n)u(y - y_n) \tag{3.4-18}$$

where

$$y_n = T(x_n) \tag{3.4-19}$$

$$P(y_n) = P(x_n) \tag{3.4-20}$$

If T is not monotonic, the above procedure remains valid except there now exists the possibility that more than one value x_n corresponds to a value y_n. In such a case $P(y_n)$ will equal the sum of the probabilities of the various x_n for which $y_n = T(x_n)$.

PROBLEMS

3-1 A discrete random variable X has possible values $x_i = i^2$, $i = 1, 2, 3, 4, 5$, which occur with probabilities 0.4, 0.25, 0.15, 0.1, and 0.1, respectively. Find the mean value $\bar{X} = E[X]$ of X.

3-2 The natural numbers are the possible values of a random variable X; that is, $x_n = n$, $n = 1, 2, \ldots$. These numbers occur with probabilities $P(x_n) = (\frac{1}{2})^n$. Find the expected value of X.

3-3 If the probabilities in Problem 3-2 are $P(x_n) = p^n$, $0 < p < 1$, show that $p = \frac{1}{2}$ is the only value of p that is allowed for the problem as formulated. (*Hint:* Use the fact that $\int_{-\infty}^{\infty} f_X(x)\, dx = 1$ is necessary.)

3-4 Give an example of a random variable where its mean value might not equal any of its possible values.

3-5 Find:

(a) the expected value, and

(b) the variance of the random variable with the triangular density of Figure 2.3-1a if $a = 1/\alpha$.

3-6 Show that the mean value and variance of the random variable having the uniform density function of (2.5-7) are:

$$\bar{X} = E[X] = (a + b)/2$$

and

$$\sigma_X^2 = (b - a)^2/12$$

3-7 A pointer is spun on a fair wheel of chance numbered from 0 to 100 around its circumference.

(a) What is the average value of all possible pointer positions?

(b) What deviation from its average value will pointer position take on the average; that is, what is the pointer's root-mean-squared deviation from its mean? (*Hint:* Use results of Problem 3-6.)

3-8 Find:

(a) the mean value, and

(b) the variance of the random variable X defined by Problems 2-6 and 2-14 of Chapter 2.

*3-9 For the *binomial density* of (2.5-1), show that

$$E[X] = \bar{X} = Np$$

and

$$\sigma_X^2 = Np(1 - p)$$

3-10 (*a*) Let resistance be a random variable in Problem 2-11 of Chapter 2. Find the mean value of resistance.

(*b*) What is the output voltage E_2 if an *average* resistor were used in the circuit?

(*c*) For the resistors specified, what is the mean value of E_2? Does the voltage of part (*b*) equal this value? Explain your results.

3-11 (*a*) Use the symmetry of the density function given by (2.4-1) to justify that the parameter a_X in the *gaussian density* is the mean value of the random variable: $\bar{X} = a_X$.

(*b*) Prove that the parameter σ_X^2 is the variance. (*Hint:* Use an equation from Appendix C.)

3-12 Show that the mean value $E[X]$ and variance σ_X^2 of the Rayleigh random variable, with density given by (2.5-11), are

$$E[X] = a + \sqrt{\pi b/4}$$

and

$$\sigma_X^2 = b(4 - \pi)/4$$

3-13 What is the expected lifetime of the system defined in Problem 2-33 of Chapter 2?

3-14 Find:

(*a*) the mean value, and

(*b*) the variance for a random variable with the *Laplace* density

$$f_X(x) = \frac{1}{2b} e^{-|x - m|/b}$$

where b and m are real constants, $b > 0$ and $-\infty < m < \infty$.

3-15 Determine the mean value of the *Cauchy* random variable in Problem 2-34 of Chapter 2. What can you say about the variance of this random variable?

*3-16 For the *Poisson* random variable defined in (2.5-4) show that:

(*a*) the mean value is b and

(*b*) the variance also equals b.

3-17 (*a*) Use (3.2-2) to find the first three moments m_1, m_2, and m_3 for the exponential density of Example 3.1-2.

(*b*) Find m_1, m_2, and m_3 from the characteristic function found in Example 3.3-1. Verify that they agree with those of part (*a*).

3-18 Find expressions for all the moments about the origin and central moments for the uniform density of (2.5-7).

3-19 Define a function $g(\cdot)$ of a random variable X by

$$g(X) = \begin{cases} 1 & x \geq x_0 \\ 0 & x < x_0 \end{cases}$$

where x_0 is a real number $-\infty < x_0 < \infty$. Show that

$$E[g(X)] = 1 - F_X(x_0)$$

3-20 Show that the second moment of any random variable X about an arbitrary point a is minimum when $a = \bar{X}$; that is, show that $E[(X - a)^2]$ is minimum for $a = \bar{X}$.

3-21 For any discrete random variable X with values x_i having probabilities of occurrence $P(x_i)$, show that the moments of X are

$$m_n = \sum_{i=1}^{N} x_i^n P(x_i)$$

$$\mu_n = \sum_{i=1}^{N} (x_i - \bar{X})^n P(x_i)$$

where N may be infinite for some X.

3-22 Prove that central moments μ_n are related to moments m_k about the origin by

$$\mu_n = \sum_{k=0}^{n} \binom{n}{k} (-\bar{X})^{n-k} m_k$$

3-23 A random variable X has a density function $f_X(x)$ and moments m_n. If the density is shifted higher in x by an amount $\alpha > 0$ to a new origin, show that the moments of the shifted density, denoted m'_n, are related to the moments m_n by

$$m'_n = \sum_{k=0}^{n} \binom{n}{k} \alpha^{n-k} m_k$$

*3-24 Show that any characteristic function $\Phi_X(\omega)$ satisfies

$$|\Phi_X(\omega)| \leq \Phi_X(0) = 1$$

3-25 A random variable X is uniformly distributed on the interval $(-5, 15)$. Another random variable $Y = e^{-X/5}$ is formed. Find $E[Y]$.

3-26 A gaussian voltage random variable X [see (2.4-1)] has a mean value $\bar{X} = a_X = 0$ and variance $\sigma_X^2 = 9$. The voltage X is applied to a square-law, full-wave diode detector with a transfer characteristic $Y = 5X^2$. Find the mean value of the output voltage Y.

*3-27 For the system having a lifetime specified in Problem 2-33 of Chapter 2, determine the expected lifetime of the system given that the system has survived 20 weeks.

*3-28 The characteristic function for a gaussian random variable X, having a mean value of 0, is

$$\Phi_X(\omega) = \exp\left(-\sigma_X^2 \omega^2/2\right)$$

Find all the moments of X using $\Phi_X(\omega)$.

*3-29 Work Problem 3-28 using the moment generating function

$$M_X(v) = \exp\left(\sigma_X^2 v^2/2\right)$$

for the zero-mean gaussian random variable.

*3-30 A discrete random variable X can have $N + 1$ values $x_k = k\Delta$, $k = 0, 1, \ldots,$ N, where $\Delta > 0$ is a real number. Its values occur with equal probability. Show that the characteristic function of X is

$$\Phi_X(\omega) = \frac{1}{N+1} \frac{\sin\left[(N+1)\omega\Delta/2\right]}{\sin\left(\omega\Delta/2\right)} e^{jN\omega\Delta/2}$$

3-31 A random variable X is uniformly distributed on the interval $(-\pi/2, \pi/2)$. X is transformed to the new random variable $Y = T(X) = a \tan(X)$, where $a > 0$. Find the probability density function of Y.

3-32 Work Problem 3-31 if X is uniform on the interval $(-\pi, \pi)$.

3-33 A random variable X undergoes the transformation $Y = a/X$, where a is a real number. Find the density function of Y.

3-34 A random variable X is uniformly distributed on the interval $(-a, a)$. It is transformed to a new variable Y by the transformation $Y = cX^2$ defined in Example 3.4-2. Find and sketch the density function of Y.

3-35 A zero-mean gaussian random variable X is transformed to the random variable Y determined by

$$Y = \begin{cases} cX & X > 0 \\ 0 & X \leq 0 \end{cases}$$

where c is a real constant, $c > 0$. Find and sketch the density function of Y.

3-36 If the transformation of Problem 3-35 is applied to a Rayleigh random variable with $a \geq 0$, what is its effect?

*3-37 A random variable Θ is uniformly distributed on the interval (θ_1, θ_2) where θ_1 and θ_2 are real and satisfy

$$0 \leq \theta_1 < \theta_2 < \pi$$

Find and sketch the probability density function of the transformed random variable $Y = \cos(\Theta)$.

3-38 A random variable X can have values $-4, -1, 2, 3,$ and 4, each with probability $\frac{1}{5}$. Find:
  (a) the density function,
  (b) the mean, and
  (c) the variance of the random variable $Y = 3X^3$.

ADDITIONAL PROBLEMS

3-39 (a) Find the average amount the gambler in Problem 2-42 can expect to win. (b) What is his probability of winning on any given playing of the game?

3-40 The *arcsine* probability density is defined by

$$f_X(x) = \frac{\text{rect } (x/2a)}{\pi\sqrt{a^2 - x^2}}$$

for any real constant $a > 0$. Show that $\bar{X} = 0$ and $\overline{X^2} = a^2/2$ for this density.

*3-41 For the animal described in Problem 2-58 find its expected lifetime given that it will not live beyond 20 weeks.

3-42 Find the expected value of the function $g(X) = X^3$ where X is a random variable defined by the density

$$f_X(x) = (\tfrac{1}{2})u(x) \exp (-x/2)$$

3-43 Continue Problem 3-25 by finding all moments of Y. (*Hint:* Treat Y^n as a function of Y, not as a transformation.)

3-44 Reconsider the production line that manufactures bolts in Problem 2-12.
   (a) What is the average length of bolts that are placed up for sale?
   (b) What is the standard deviation of length of bolts sold?
   (c) What percentage of all bolts sold are expected to have a length within one standard deviation of the average length?
   (d) By what tolerance (as a percentage) does the average length of bolts sold match the nominally desired length of 760 mm?

3-45 A random variable X has a probability density

$$f_X(x) = \begin{cases} (\pi/16) \cos (\pi x/8) & -4 \le x \le 4 \\ 0 & \text{elsewhere} \end{cases}$$

Find: (a) its mean value $\bar{X}$, (b) its second moment $\overline{X^2}$, and (c) its variance.

3-46 A certain meter is designed to measure small dc voltages but makes errors because of noise. The errors are accurately represented as a gaussian random variable with a mean of zero and a standard deviation of 10^{-3} V. When the dc voltage is disconnected it is found that the probability is 0.5 that the meter reading is positive due to noise. With the dc voltage present this probability becomes 0.2514. What is the dc voltage?

3-47 Find the skew and coefficient of skewness for a Rayleigh random variable for which $a = 0$ in (2.5-11).

3-48 A random variable X has the density

$$f_X(x) = \begin{cases} (\tfrac{3}{32})(-x^2 + 8x - 12) & 2 \le x \le 6 \\ 0 & \text{elsewhere} \end{cases}$$

Find the following moments: (a) m_0, (b) m_1, (c) m_2, and (d) μ_2.

3-49 The *chi-square density* with N degrees of freedom is defined by

$$f_X(x) = \frac{x^{(N/2)-1}}{2^{N/2}\Gamma(N/2)} u(x)e^{-x/2}$$

where $\Gamma(\cdot)$ is the gamma function

$$\Gamma(z) = \int_0^\infty \xi^{z-1}e^{-\xi} \, d\xi \qquad \text{real part of } z > 0$$

and $N = 1, 2, \dots$. Show that (a) $\bar{X} = N$, (b) $\overline{X^2} = N(N+2)$, and (c) $\sigma_X^2 = 2N$ for this density.

3-50 For the density of Problem 3-49 find its arbitrary moment $\overline{X^n}$, $n = 0, 1, 2, \dots$.

3-51 A random variable X is called *Weibull*† if its density has the form

$$f_X(x) = abx^{b-1} \exp(-ax^b)u(x)$$

where $a > 0$ and $b > 0$ are real constants. Use the definition of the gamma function of Problem 3-49 to find (a) the mean value, (b) the second moment, and (c) the variance of X.

★3-52 Show that the characteristic function of a random variable having the binomial density of (2.5-1) is

$$\Phi_X(\omega) = [1 - p + pe^{j\omega}]^N$$

★3-53 Show that the characteristic function of a Poisson random variable defined by (2.5-4) is

$$\Phi_X(\omega) = \exp[-b(1 - e^{j\omega})]$$

★3-54 The *Erlang*‡ *random variable* X has a characteristic function

$$\Phi_X(\omega) = \left[\frac{a}{a - j\omega}\right]^N$$

for $a > 0$ and $N = 1, 2, \dots$. Show that $\bar{X} = N/a$, $\overline{X^2} = N(N+1)/a^2$, and $\sigma_X^2 = N/a^2$.

3-55 A random variable X has $\bar{X} = -3$, $\overline{X^2} = 11$, and $\sigma_X^2 = 2$. For a new random variable $Y = 2X - 3$, find (a) $\bar{Y}$, (b) $\overline{Y^2}$, and (c) σ_Y^2.

★3-56 For any real random variable X with mean $\bar{X}$ and variance σ_X^2, *Chebychev's inequality*§ is

$$P\{|X - \bar{X}| \geq \lambda\sigma_X\} \leq 1/\lambda^2$$

where $\lambda > 0$ is a real constant. Prove the inequality. (*Hint:* Define a new random variable $Y = 0$ for $|X - \bar{X}| < \lambda\sigma_X$ and $Y = \lambda^2\sigma_X^2$ for $|X - \bar{X}| > \lambda\sigma_X$, observe that $Y \leq (X - \bar{X})^2$ and find $E[Y]$.)

---

† After Ernst Hjalmar Waloddi Weibull (1887– ·), a Swedish applied physicist.
‡ A. K. Erlang (1878–1929) was a Danish engineer.
§ After the Russian mathematician Pafnuty Lvovich Chebychev (1821–1894).

3-57 A gaussian random variable, for which

$$f_X(x) = (2/\sqrt{\pi}) \exp(-4x^2)$$

is applied to a square-law device to produce a new (output) random variable $Y = X^2/2$. (a) Find the density of Y. (b) Find the moments $m_n = E[Y^n]$, $n = 0, 1, \ldots$. (Hint: Put your answer in terms of the gamma function defined in Problem 3-49.)

3-58 A gaussian random variable, for which $\bar{X} = 0.6$ and $\sigma_X = 0.8$, is transformed to a new random variable by the transformation

$$Y = T(X) = \begin{cases} 4 & 1.0 \le X < \infty \\ 2 & 0 \le X < 1.0 \\ -2 & -1.0 \le X < 0 \\ -4 & -\infty < X < -1.0 \end{cases}$$

(a) Find the density function of Y.

(b) Find the mean and variance of Y.

3-59 Work Problem 3-31 except assume a transformation $Y = T(X) = a \sin(X)$ with $a > 0$.

3-60 Let X be a gaussian random variable with density given by (2.4-1). If X is transformed to a new random variable $Y = b + e^X$, where b is a real constant, show that the density of Y is log-normal as defined in Problem 2-35. This transformation allows log-normal random numbers to be generated from gaussian random numbers by a digital computer.

3-61 A random variable X is uniformly distributed on $(0, 6)$. If X is transformed to a new random variable $Y = 2(X - 3)^2 - 4$, find: (a) the density of Y, (b) $\bar{Y}$, (c) σ_Y^2.

FOUR

MULTIPLE RANDOM VARIABLES

4.0 INTRODUCTION

In Chapters 2 and 3, various aspects of the theory of a single random variable were studied. The random variable was found to be a powerful concept. It enabled many realistic problems to be described in a probabilistic way such that practical measures could be applied to the problem even though it was random. For example, we have seen that shell impact position along the line of fire from a cannon to a target can be described by a random variable (Problem 2-29). From knowledge of the probability distribution or density function of impact position, we can solve for such practical measures as the mean value of impact position, its variance, and skew. These measures are not, however, a complete enough description of the problem in most cases.

Naturally, we may also be interested in how much the impact positions deviate *from* the line of fire in, say, the perpendicular (cross-fire) direction. In other words, we prefer to describe impact position as a point in a plane as opposed to being a point along a line. To handle such situations it is necessary that we extend our theory to include *two* random variables, one for each coordinate axis of the plane in our example. In other problems it may be necessary to extend the theory to include *several* random variables. We accomplish these extensions in this and the next chapter.

Fortunately, many situations of interest in engineering can be handled by the theory of two random variables.† Because of this fact, we emphasize the two-variable case, although the more general theory is also stated in most discussions to follow.

---

† In particular, it will be found in Chapter 6 that such important concepts as autocorrelation, cross-correlation, and covariance functions, which apply to random processes, are based on two random variables.

4.1 VECTOR RANDOM VARIABLES

Suppose two random variables X and Y are defined on a sample space S, where specific values of X and Y are denoted by x and y, respectively. Then any ordered pair of numbers (x, y) may be conveniently considered to be a *random point* in the xy plane. The point may be taken as a specific value of a *vector random variable* or a *random vector*.† Figure 4.1-1 illustrates the mapping involved in going from S to the xy plane.

The plane of all points (x, y) in the ranges of X and Y may be considered a new sample space. It is in reality a vector space where the components of any vector are the values of the random variables X and Y. The new space has been called the *range sample space* (Davenport, 1970) or the *two-dimensional product space*. We shall just call it a *joint sample space* and give it the symbol S_J.

As in the case of one random variable, let us define an event A by

$$A = \{X \le x\} \tag{4.1-1}$$

A similar event B can be defined for Y:

$$B = \{Y \le y\} \tag{4.1-2}$$

Events A and B refer to the sample space S, while events $\{X \le x\}$ and $\{Y \le y\}$ refer to the joint sample space S_J.‡ Figure 4.1-2 illustrates the correspondences

---

† There are some specific conditions that must be satisfied in a complete definition of a random vector (Davenport, 1970, Chapter 5). They are somewhat advanced for our scope and we shall simply assume the validity of our random vectors.

‡ Do not forget that elements s of S form the link between the two events since by writing $\{X \le x\}$ we really refer to the set of those s such that $X(s) \le x$ for some real number x. A similar statement holds for the event $\{Y \le y\}$.

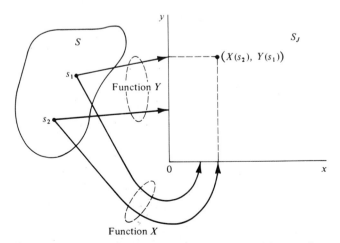

Figure 4.1-1 Mapping from the sample space S to the joint sample space S_J (xy plane).

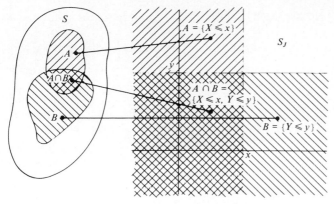

Figure 4.1-2 Comparisons of events in S with those in S_J.

between events in the two spaces. Event A corresponds to all points in S_J for which the X coordinate values are not greater than x. Similarly, event B corresponds to the Y coordinate values in S_J not exceeding y. Of special interest is to observe that the event $A \cap B$ defined on S corresponds to the *joint event* $\{X \le x \text{ and } Y \le y\}$ defined on S_J, which we write $\{X \le x, Y \le y\}$. This joint event is shown crosshatched in Figure 4.1-2.

   In the more general case where N random variables $X_1, X_2, \ldots, X_N$ are defined on a sample space S, we consider them to be components of an N-*dimensional random vector* or N-*dimensional random variable*. The joint sample space S_J is now N-dimensional.

4.2 JOINT DISTRIBUTION AND ITS PROPERTIES

The probabilities of the two events $A = \{X \le x\}$ and $B = \{Y \le y\}$ have already been defined as functions of x and y, respectively, called probability distribution functions:

$$F_X(x) = P\{X \le x\} \qquad (4.2\text{-}1)$$

$$F_Y(y) = P\{Y \le y\} \qquad (4.2\text{-}2)$$

We must introduce a new concept to include the probability of the joint event $\{X \le x, Y \le y\}$.

Joint Distribution Function

We define the probability of the joint event $\{X \le x, Y \le y\}$, which is a function of the numbers x and y, by a *joint probability distribution function* and denote it by the symbol $F_{X,Y}(x, y)$. Hence,

$$F_{X,Y}(x, y) = P\{X \le x, Y \le y\} \qquad (4.2\text{-}3)$$

It should be clear that $P\{X \le x, Y \le y\} = P(A \cap B)$, where the joint event $A \cap B$ is defined on S.

To illustrate joint distribution, we take an example where both random variables X and Y are discrete.

Example 4.2-1 Assume that the joint sample space S_J has only three possible elements: (1, 1), (2, 1), and (3, 3). The probabilities of these elements are assumed to be $P(1, 1) = 0.2$, $P(2, 1) = 0.3$, and $P(3, 3) = 0.5$. We find $F_{X,Y}(x, y)$.

In constructing the joint distribution function, we observe that the event $\{X \le x, Y \le y\}$ has no elements for any $x < 1$ and/or $y < 1$. Only at the point (1, 1) does the function assume a step value. So long as $x \ge 1$ and $y \ge 1$, this probability is maintained so that $F_{X,Y}(x, y)$ has a stair step holding in the region $x \ge 1$ and $y \ge 1$ as shown in Figure 4.2-1a. For larger x and y, the point (2, 1) produces a second stair step of amplitude 0.3 which holds in the region $x \ge 2$ and $y \ge 1$. The second step adds do the first. Finally, a third stair step of amplitude 0.5 is added to the first two when x and y are in the region $x \ge 3$ and $y \ge 3$. The final function is shown in Figure 4.2-1a.

The preceding example can be used to identify the form of the joint distribution function for two general discrete random variables. Let X have N possible values x_n and Y have M possible values y_m, then

$$F_{X,Y}(x, y) = \sum_{n=1}^{N} \sum_{m=1}^{M} P(x_n, y_m)u(x - x_n)u(y - y_m) \tag{4.2-4}$$

where $P(x_n, y_m)$ is the probability of the joint event $\{X = x_n, Y = y_m\}$ and $u(\cdot)$ is the unit-step function. As seen in Example 4.2-1, some couples (x_n, y_m) may have zero probability. In some cases N or M, or both, may be infinite.

If $F_{X,Y}(x, y)$ is plotted for continuous random variables X and Y, the same general behavior as shown in Figure 4.2-1a is obtained except the surface becomes smooth and has no stairstep discontinuities.

For N random variables X_n, $n = 1, 2, \ldots, N$, the generalization of (4.2-3) is direct. The joint distribution function, denoted by $F_{X_1, X_2, \ldots, X_N}(x_1, x_2, \ldots, x_N)$, is defined as the probability of the joint event $\{X_1 \le x_1, X_2 \le x_2, \ldots, X_N \le x_N\}$:

$$F_{X_1, X_2, \ldots, X_N}(x_1, x_2, \ldots, x_N) = P\{X_1 \le x_1, X_2 \le x_2, \ldots, X_N \le x_N\} \tag{4.2-5}$$

For a single random variable X, we found in Chapter 2 that $F_X(x)$ could be expressed in general as the sum of a function of stairstep form (due to the discrete portion of a mixed random variable X) and a function that was continuous (due to the continuous portion of X). Such a simple decomposition of the joint distribution when $N > 1$ is not generally true [Cramér, 1946, Section 8.4]. However,

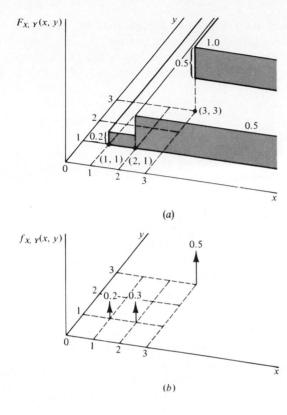

Figure 4.2-1 A joint distribution function (a), and its corresponding joint density function (b), that apply to Examples 4.2-1 and 4.2-2.

it is true that joint density functions in practice often correspond to all random variables being either discrete or continuous. Therefore, we shall limit our consideration in this book almost entirely to these two cases when $N > 1$.

Properties of the Joint Distribution

A joint distribution function for two random variables X and Y has several properties that follow readily from its definition. We list them:

(1) $F_{X,Y}(-\infty, -\infty) = 0 \qquad F_{X,Y}(-\infty, y) = 0 \qquad F_{X,Y}(x, -\infty) = 0$ (4.2-6a)

(2) $F_{X,Y}(\infty, \infty) = 1$ (4.2-6b)

(3) $0 \leq F_{X,Y}(x, y) \leq 1$ (4.2-6c)

(4) $F_{X,Y}(x, y)$ is a nondecreasing function of both x and y (4.2-6d)

(5) $F_{X,Y}(x_2, y_2) + F_{X,Y}(x_1, y_1) - F_{X,Y}(x_1, y_2) - F_{X,Y}(x_2, y_1)$

$\qquad = P\{x_1 < X \leq x_2, y_1 < Y \leq y_2\} \geq 0$ (4.2-6e)

(6) $F_{X,Y}(x, \infty) = F_X(x) \qquad F_{X,Y}(\infty, y) = F_Y(y)$ (4.2-6f)

The first five of these properties are just the two-dimensional extensions of the properties of one random variable given in (2.2-2). Properties 1, 2, and 5 may be used as tests to determine whether some function can be a valid distribution function for two random variables X and Y (Papoulis, 1965, p. 169). Property 6 deserves a few special comments.

Marginal Distribution Functions

Property 6 above states that the distribution function of one random variable can be obtained by setting the value of the other variable to infinity in $F_{X,Y}(x, y)$. The functions $F_X(x)$ or $F_Y(y)$ obtained in this manner are called *marginal distribution functions*.

To justify property 6, it is easiest to return to the basic events A and B, defined by $A = \{X \leq x\}$ and $B = \{Y \leq y\}$, and observe that $F_{X,Y}(x, y) = P\{X \leq x, Y \leq y\} = P(A \cap B)$. Now if we set y to ∞, this is equivalent to making B the certain event; that is, $B = \{Y \leq \infty\} = S$. Furthermore, since $A \cap B = A \cap S = A$, then we have $F_{X,Y}(x, \infty) = P(A \cap S) = P(A) = P\{X \leq x\} = F_X(x)$. A similar proof can be stated for obtaining $F_Y(y)$.

Example 4.2-2 We find explicit expressions for $F_{X,Y}(x, y)$, and the marginal distributions $F_X(x)$ and $F_Y(y)$ for the joint sample space of Example 4.2-1.

The joint distribution derives from (4.2-4) if we recognize that only three probabilities are nonzero:

$$F_{X,Y}(x, y) = P(1, 1)u(x - 1)u(y - 1)$$
$$+ P(2, 1)u(x - 2)u(y - 1)$$
$$+ P(3, 3)u(x - 3)u(y - 3)$$

where $P(1, 1) = 0.2$, $P(2, 1) = 0.3$, and $P(3, 3) = 0.5$. If we set $y = \infty$:

$$F_X(x) = F_{X,Y}(x, \infty)$$
$$= P(1, 1)u(x - 1) + P(2, 1)u(x - 2) + P(3, 3)u(x - 3)$$
$$= 0.2u(x - 1) + 0.3u(x - 2) + 0.5u(x - 3)$$

If we set $x = \infty$:

$$F_Y(y) = F_{X,Y}(\infty, y)$$
$$= 0.2u(y - 1) + 0.3u(y - 1) + 0.5u(y - 3)$$
$$= 0.5u(y - 1) + 0.5u(y - 3)$$

Plots of these marginal distributions are shown in Figure 4.2-2.

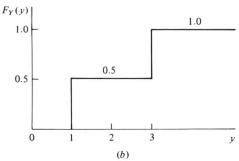

Figure 4.2-2 Marginal distributions applicable to Figure 4.2-1 and Example 4.2-2: (a) $F_X(x)$ and (b) $F_Y(y)$.

From an N-dimensional joint distribution function we may obtain a k-*dimensional marginal distribution function*, for any selected group of k of the N random variables, by setting the values of the other $N - k$ random variables to infinity. Here k can be any integer $1, 2, 3, \ldots, N - 1$.

4.3 JOINT DENSITY AND ITS PROPERTIES

In this section the concept of a probability density function is extended to include multiple random variables.

Joint Density Function

For two random variables X and Y, the *joint probability density function*, denoted $f_{X, Y}(x, y)$, is defined by the second derivative of the joint distribution function wherever it exists:

$$f_{X, Y}(x, y) = \frac{\partial^2 F_{X, Y}(x, y)}{\partial x \, \partial y} \tag{4.3-1}$$

We shall refer often to $f_{X, Y}(x, y)$ as the *joint density function*.

If X and Y are discrete random variables, $F_{X, Y}(x, y)$ will possess step discontinuities (see Example 4.2-1 and Figure 4.2-1). Derivatives at these discontinuities are normally undefined. However, by admitting impulse functions (see Appendix A), we are able to define $f_{X,Y}(x, y)$ at these points. Therefore, the joint density function may be found for any two discrete random variables by substitution of (4.2-4) into (4.3-1):

$$f_{X,Y}(x, y) = \sum_{n=1}^{N} \sum_{m=1}^{M} P(x_n, y_m) \, \delta(x - x_n) \, \delta(y - y_m) \qquad (4.3-2)$$

An example of the joint density function of two discrete random variables is shown in Figure 4.2-1b.

When N random variables $X_1, X_2, \ldots, X_N$ are involved, the joint density function becomes the N-fold partial derivative of the N-dimensional distribution function:

$$f_{X_1, X_2, \ldots, X_N}(x_1, x_2, \ldots, x_N) = \frac{\partial^N F_{X_1, X_2, \ldots, X_N}(x_1, x_2, \ldots, x_N)}{\partial x_1 \, \partial x_2 \cdots \partial x_N} \qquad (4.3-3)$$

By direct integration this result is equivalent to

$$F_{X_1, X_2, \ldots, X_N}(x_1, x_2, \ldots, x_N)$$

$$= \int_{-\infty}^{x_N} \cdots \int_{-\infty}^{x_2} \int_{-\infty}^{x_1} f_{X_1, X_2, \ldots, X_N}(\xi_1, \xi_2, \ldots, \xi_N) \, d\xi_1 \, d\xi_2 \cdots d\xi_N \qquad (4.3-4)$$

Properties of the Joint Density

Several properties of a joint density function may be listed that derive from its definition (4.3-1) and the properties (4.2-6) of the joint distribution function:

(1) $f_{X,Y}(x, y) \geq 0$ \qquad (4.3-5a)

(2) $\displaystyle\int_{-\infty}^{\infty} \int_{-\infty}^{\infty} f_{X,Y}(x, y) \, dx \, dy = 1$ \qquad (4.3-5b)

(3) $F_{X,Y}(x, y) = \displaystyle\int_{-\infty}^{y} \int_{-\infty}^{x} f_{X,Y}(\xi_1, \xi_2) \, d\xi_1 \, d\xi_2$ \qquad (4.3-5c)

(4) $F_X(x) = \displaystyle\int_{-\infty}^{x} \int_{-\infty}^{\infty} f_{X,Y}(\xi_1, \xi_2) \, d\xi_2 \, d\xi_1$ \qquad (4.3-5d)

$F_Y(y) = \displaystyle\int_{-\infty}^{y} \int_{-\infty}^{\infty} f_{X,Y}(\xi_1, \xi_2) \, d\xi_1 \, d\xi_2$ \qquad (4.3-5e)

(5) $P\{x_1 < X \leq x_2, y_1 < Y \leq y_2\} = \displaystyle\int_{y_1}^{y_2} \int_{x_1}^{x_2} f_{X,Y}(x, y) \, dx \, dy$ \qquad (4.3-5f)

(6) $f_X(x) = \displaystyle\int_{-\infty}^{\infty} f_{X,Y}(x, y) \, dy$ \qquad (4.3-5g)

$f_Y(y) = \displaystyle\int_{-\infty}^{\infty} f_{X,Y}(x, y) \, dx$ \qquad (4.3-5h)

Properties 1 and 2 may be used as sufficient tests to determine if some function can be a valid density function. Both tests must be satisfied (Papoulis, 1965, p. 169).

The first five of these properties are readily verified from earlier work and the reader should go through the necessary logic as an exercise. Property 6 introduces a new concept.

Marginal Density Functions

The functions $f_X(x)$ and $f_Y(y)$ of property 6 are called *marginal probability density functions* or just *marginal density functions*. They are the density functions of the single variables X and Y and are defined as the derivatives of the marginal distribution functions:

$$f_X(x) = \frac{dF_X(x)}{dx} \tag{4.3-6}$$

$$f_Y(y) = \frac{dF_Y(y)}{dy} \tag{4.3-7}$$

By substituting (4.3-5d) and (4.3-5e) into (4.3-6) and (4.3-7), respectively, we are able to verify the equations of property 6.

We shall illustrate the calculation of marginal density functions from a given joint density function with an example.

Example 4.3-1 We find $f_X(x)$ and $f_Y(y)$ when the joint density function is given by (Clarke and Disney, 1970, p. 108):

$$f_{X, Y}(x, y) = u(x)u(y)xe^{-x(y+1)}$$

From (4.3-5g) and the above equation:

$$f_X(x) = \int_0^\infty u(x)xe^{-x(y+1)}\, dy = u(x)xe^{-x} \int_0^\infty e^{-xy}\, dy$$

$$= u(x)xe^{-x}(1/x) = u(x)e^{-x}$$

after using an integral from Appendix C.

From (4.3-5h):

$$f_Y(y) = \int_0^\infty u(y)xe^{-x(y+1)}\, dx = \frac{u(y)}{(y+1)^2}$$

after using another integral from Appendix C.

For N random variables $X_1, X_2, \ldots, X_N$, the *k-dimensional marginal density function* is defined as the k-fold partial derivative of the k-dimensional marginal distribution function. It can also be found from the joint density function by integrating out all variables except the k variables of interest $X_1, X_2, \ldots, X_k$:

$$f_{X_1, X_2, \ldots, X_k}(x_1, x_2, \ldots, x_k)$$

$$= \int_{-\infty}^{\infty} \cdots \int_{-\infty}^{\infty} f_{X_1, X_2, \ldots, X_N}(x_1, x_2, \ldots, x_N) \, dx_{k+1} \, dx_{k+2} \cdots dx_N \quad (4.3\text{-}8)$$

4.4 CONDITIONAL DISTRIBUTION AND DENSITY

In Section 2.6, the conditional distribution function of a random variable X, given some event B, was defined as

$$F_X(x \mid B) = P\{X \le x \mid B\} = \frac{P\{X \le x \cap B\}}{P(B)} \quad (4.4\text{-}1)$$

for any event B with nonzero probability. The corresponding conditional density function was defined through the derivative

$$f_X(x \mid B) = \frac{dF_X(x \mid B)}{dx} \quad (4.4\text{-}2)$$

In this section these two functions are extended to include a second random variable through suitable definitions of event B.

Conditional Distribution and Density—Point Conditioning

Often in practical problems we are interested in the distribution function of one random variable X conditioned by the fact that a second random variable Y has some specific value y. This is called *point conditioning* and we can handle such problems by defining event B by

$$B = \{y - \Delta y < Y \le y + \Delta y\} \quad (4.4\text{-}3)$$

where Δy is a small quantity that we eventually let approach 0. For this event, (4.4-1) can be written

$$F_X(x \mid y - \Delta y < Y \le y + \Delta y) = \frac{\int_{y-\Delta y}^{y+\Delta y} \int_{-\infty}^{x} f_{X, Y}(\xi_1, \xi_2) \, d\xi_1 \, d\xi_2}{\int_{y-\Delta y}^{y+\Delta y} f_Y(\xi) \, d\xi} \quad (4.4\text{-}4)$$

where we have used (4.3-5f) and (2.3-6d).

Consider two cases of (4.4-4). In the first case, assume X and Y are both discrete random variables with values x_i, $i = 1, 2, \ldots, N$, and y_j, $j = 1, 2, \ldots, M$, respectively, while the probabilities of these values are denoted $P(x_i)$ and $P(y_j)$, respectively. The probability of the joint occurrence of x_i and y_j is denoted $P(x_i, y_j)$. Thus,

$$f_Y(y) = \sum_{j=1}^{M} P(y_j)\, \delta(y - y_j) \tag{4.4-5}$$

$$f_{X,Y}(x, y) = \sum_{i=1}^{N} \sum_{j=1}^{M} P(x_i, y_j)\, \delta(x - x_i)\, \delta(y - y_j) \tag{4.4-6}$$

Now suppose that the specific value of y of interest is y_k. With substitution of (4.4-5) and (4.4-6) into (4.4-4) and allowing $\Delta y \to 0$, we obtain

$$F_X(x \mid Y = y_k) = \sum_{i=1}^{N} \frac{P(x_i, y_k)}{P(y_k)}\, u(x - x_i) \tag{4.4-7}$$

After differentiation we have

$$f_X(x \mid Y = y_k) = \sum_{i=1}^{N} \frac{P(x_i, y_k)}{P(y_k)}\, \delta(x - x_i) \tag{4.4-8}$$

Example 4.4-1 To illustrate the use of (4.4-8) assume a joint density function as given in Figure 4.4-1a. Here $P(x_1, y_1) = \frac{2}{15}$, $P(x_2, y_1) = \frac{3}{15}$, etc. Since $P(y_3) = (\frac{4}{15}) + (\frac{5}{15}) = \frac{9}{15}$, use of (4.4-8) will give $f_X(x \mid Y = y_3)$ as shown in Figure 4.4-1b.

The second case of (4.4-4) that is of interest corresponds to X and Y both continuous random variables. As $\Delta y \to 0$ the denominator in (4.4-4) becomes 0. However, we can still show that the conditional density $f_X(x \mid Y = y)$ may exist. If Δy is very small, (4.4-4) can be written as

$$F_X(x \mid y - \Delta y < Y \le y + \Delta y) = \frac{\int_{-\infty}^{x} f_{X,Y}(\xi_1, y)\, d\xi_1\, 2\Delta y}{f_Y(y) 2\Delta y} \tag{4.4-9}$$

and, in the limit as $\Delta y \to 0$

$$F_X(x \mid Y = y) = \frac{\int_{-\infty}^{x} f_{X,Y}(\xi, y)\, d\xi}{f_Y(y)} \tag{4.4-10}$$

for every y such that $f_Y(y) \ne 0$. After differentiation of both sides of (4.4-10) with respect to x:

$$f_X(x \mid Y = y) = \frac{f_{X,Y}(x, y)}{f_Y(y)} \tag{4.4-11}$$

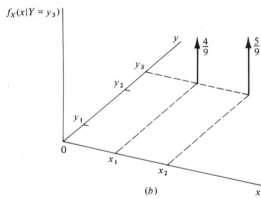

Figure 4.4-1 A joint density function (a) and a conditional density function (b) applicable to Example 4.4-1.

When there is no confusion as to meaning, we shall often write (4.4-11) as

$$f_X(x \mid y) = \frac{f_{X,Y}(x, y)}{f_Y(y)} \qquad (4.4\text{-}12)$$

It can also be shown that

$$f_Y(y \mid x) = \frac{f_{X,Y}(x, y)}{f_X(x)} \qquad (4.4\text{-}13)$$

Example 4.4-2 We find $f_Y(y \mid x)$ for the density functions defined in Example 4.3-1. Since

$$f_{X,Y}(x, y) = u(x)u(y)xe^{-x(y+1)}$$

and

$$f_X(x) = u(x)e^{-x}$$

are nonzero only for $0 < y$ and $0 < x$, $f_Y(y \mid x)$ is nonzero only for $0 < y$ and $0 < x$. It is

$$f_Y(y \mid x) = u(x)u(y)xe^{-xy}$$

from (4.4-13).

⋆Conditional Distribution and Density—Interval Conditioning

It is sometimes convenient to define event B in (4.4-1) and (4.4-2) in terms of a random variable Y by

$$B = \{y_a < Y \le y_b\} \tag{4.4-14}$$

where y_a and y_b are real numbers and we assume $P(B) = P\{y_a < Y \le y_b\} \ne 0$. With this definition it is readily shown that (4.4-1) and (4.4-2) become

$$F_X(x \mid y_a < Y \le y_b) = \frac{F_{X,Y}(x, y_b) - F_{X,Y}(x, y_a)}{F_Y(y_b) - F_Y(y_a)}$$

$$= \frac{\int_{y_a}^{y_b} \int_{-\infty}^{x} f_{X,Y}(\xi, y) \, d\xi \, dy}{\int_{y_a}^{y_b} \int_{-\infty}^{\infty} f_{X,Y}(x, y) \, dx \, dy} \tag{4.4-15}$$

and

$$f_X(x \mid y_a < Y \le y_b) = \frac{\int_{y_a}^{y_b} f_{X,Y}(x, y) \, dy}{\int_{y_a}^{y_b} \int_{-\infty}^{\infty} f_{X,Y}(x, y) \, dx \, dy} \tag{4.4-16}$$

These last two expressions hold for X and Y either continuous or discrete random variables. In the discrete case, the joint density is given by (4.3-2). The resulting distribution and density will be defined, however, only for y_a and y_b such that the denominators of (4.4-15) and (4.4-16) are nonzero. This requirement is satisfied so long as the interval $y_a < y \le y_b$ spans at least one possible value of Y having a nonzero probability of occurrence.

An example will serve to illustrate the application of (4.4-16) when X and Y are continuous random variables.

Example 4.4-3 We use (4.4-16) to find $f_X(x \mid Y \le y)$ for the joint density function of Example 4.3-1. Since we have here defined $B = \{Y \le y\}$, then $y_a = -\infty$ and $y_b = y$. Furthermore, since $f_{X,Y}(x, y)$ is nonzero only for $0 < x$ and $0 < y$, we need only consider this region of x and y in finding the conditional density function. The denominator of (4.4-16) can be written as $\int_{-\infty}^{y} f_Y(\xi) \, d\xi$. By using results from Example 4.3-1:

$$\int_{-\infty}^{y} f_Y(\xi) \, d\xi = \int_{-\infty}^{y} \frac{u(\xi) \, d\xi}{(\xi + 1)^2} = \int_{0}^{y} \frac{d\xi}{(\xi + 1)^2} = \frac{y}{y + 1} \qquad y > 0$$

Figure 4.4-2 Conditional probability density functions applicable to Example 4.4-3.

and zero for $y < 0$, after using an integral from Appendix C. The numerator of (4.4-16) becomes

$$\int_{-\infty}^{y} f_{X,Y}(x, \xi)\, d\xi = \int_{0}^{y} u(x)xe^{-x(\xi+1)}\, d\xi$$

$$= u(x)xe^{-x}\int_{0}^{y} e^{-x\xi}\, d\xi$$

$$= u(x)e^{-x}(1 - e^{-xy}) \qquad y > 0$$

and zero for $y < 0$, after using another integral from Appendix C. Thus

$$f_X(x \mid Y \le y) = u(x)u(y)\left(\frac{y+1}{y}\right)e^{-x}(1 - e^{-xy})$$

This function is plotted in Figure 4.4-2 for several values of y.

4.5 STATISTICAL INDEPENDENCE

It will be recalled from (1.5-3) that two events A and B are statistically independent if (and only if)

$$P(A \cap B) = P(A)P(B) \tag{4.5-1}$$

This condition can be used to apply to two random variables X and Y by defining the events $A = \{X \le x\}$ and $B = \{Y \le y\}$ for two real numbers x and y. Thus, X and Y are said to be *statistically independent random variables* if (and only if)

$$P\{X \le x, Y \le y\} = P\{X \le x\}P\{Y \le y\} \qquad (4.5\text{-}2)$$

From this expression and the definitions of distribution functions, it follows that

$$F_{X,Y}(x, y) = F_X(x)F_Y(y) \qquad (4.5\text{-}3)$$

if X and Y are independent. From the definitions of density functions, (4.5-3) gives

$$f_{X,Y}(x, y) = f_X(x)f_Y(y) \qquad (4.5\text{-}4)$$

by differentiation, if X and Y are independent. Either (4.5-3) or (4.5-4) may serve as a sufficient definition of, or test for, independence of two random variables.

The form of the conditional distribution function for independent events is found by use of (4.4-1) with $B = \{Y \le y\}$:

$$F_X(x \mid Y \le y) = \frac{P\{X \le x, Y \le y\}}{P\{Y \le y\}} = \frac{F_{X,Y}(x, y)}{F_Y(y)} \qquad (4.5\text{-}5)$$

By substituting (4.5-3) into (4.5-5), we have

$$F_X(x \mid Y \le y) = F_X(x) \qquad (4.5\text{-}6)$$

In other words, the conditional distribution ceases to be conditional and simply equals the marginal distribution for independent random variables. It can also be shown that

$$F_Y(y \mid X \le x) = F_Y(y) \qquad (4.5\text{-}7)$$

Conditional density function forms, for independent X and Y, are found by differentiation of (4.5-6) and (4.5-7):

$$f_X(x \mid Y \le y) = f_X(x) \qquad (4.5\text{-}8)$$
$$f_Y(y \mid X \le x) = f_Y(y) \qquad (4.5\text{-}9)$$

Example 4.5-1 For the densities of Example 4.3-1:

$$f_{X,Y}(x, y) = u(x)u(y)xe^{-x(y+1)}$$

$$f_X(x)f_Y(y) = u(x)u(y)\frac{e^{-x}}{(y+1)^2} \ne f_{X,Y}(x, y)$$

Therefore the random variables X and Y are not independent.

In the more general study of the statistical independence of N random variables $X_1, X_2, \ldots, X_N$, we define events A_i by

$$A_i = \{X_i \le x_i\} \qquad i = 1, 2, \ldots, N \tag{4.5-10}$$

where the x_i are real numbers. With these definitions, the random variables X_i are said to be statistically independent if (1.5-6) is satisfied.

It can be shown that if $X_1, X_2, \ldots, X_N$ are statistically independent then any group of these random variables is independent of any other group. Furthermore, a function of any group is independent of any function of any other group of the random variables. For example, with $N = 4$ random variables: X_4 is independent of $X_3 + X_2 + X_1$; X_3 is independent of $X_2 + X_1$, etc. (see Papoulis, 1965, p. 238).

4.6 DISTRIBUTION AND DENSITY OF A SUM OF RANDOM VARIABLES

The problem of finding the distribution and density functions for a sum of *statistically independent* random variables is considered in this section.

Sum of Two Random Variables

Let W be a random variable equal to the sum of two independent random variables X and Y:

$$W = X + Y \tag{4.6-1}$$

This is a very practical problem because X might represent a random signal voltage and Y could represent random noise at some instant in time. The sum W would represent a signal-plus-noise voltage available to some receiver.

The probability distribution function we seek is defined by

$$F_W(w) = P\{W \le w\} = P\{X + Y \le w\} \tag{4.6-2}$$

Figure 4.6-1 illustrates the region in the xy plane where $x + y \le w$. Now from (4.3-5f), the probability corresponding to an elemental area $dx\,dy$ in the xy plane located at the point (x, y) is $f_{X,Y}(x, y)\,dx\,dy$. If we sum all such probabilities over the region where $x + y \le w$ we will obtain $F_W(w)$. Thus

$$F_W(w) = \int_{-\infty}^{\infty} \int_{x=-\infty}^{w-y} f_{X,Y}(x, y)\,dx\,dy \tag{4.6-3}$$

and, after using (4.5-4):

$$F_W(w) = \int_{-\infty}^{\infty} f_Y(y) \int_{x=-\infty}^{w-y} f_X(x)\,dx\,dy \tag{4.6-4}$$

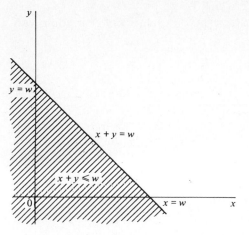

Figure 4.6-1 Region in xy plane where $x + y \leq w$.

By differentiating (4.6-4), using Leibniz's rule, we get the desired density function

$$f_W(w) = \int_{-\infty}^{\infty} f_Y(y) f_X(w - y) \, dy \tag{4.6-5}$$

This expression is recognized as a convolution integral. Consequently, we have shown that *the density function of the sum of two statistically independent random variables is the convolution of their individual density functions.*

Example 4.6-1 We use (4.6-5) to find the density of $W = X + Y$ where the densities of X and Y are assumed to be

$$f_X(x) = \frac{1}{a} \left[u(x) - u(x - a) \right]$$

$$f_Y(y) = \frac{1}{b} \left[u(y) - u(y - b) \right]$$

with $0 < a < b$, as shown in Figure 4.6-2a and b. Now because $0 < X$ and $0 < Y$, we only need examine the case $W = X + Y > 0$. From (4.6-5) we write

$$f_W(w) = \int_{-\infty}^{\infty} \frac{1}{ab} \left[u(y) - u(y - b) \right] \left[u(w - y) - u(w - y - a) \right] dy$$

$$= \frac{1}{ab} \int_{0}^{\infty} \left[1 - u(y - b) \right] \left[u(w - y) - u(w - y - a) \right] dy$$

$$= \frac{1}{ab} \left[\int_{0}^{\infty} u(w - y) \, dy - \int_{0}^{\infty} u(w - y - a) \, dy \right.$$

$$\left. - \int_{0}^{\infty} u(y - b) u(w - y) \, dy + \int_{0}^{\infty} u(y - b) u(w - y - a) \, dy \right]$$

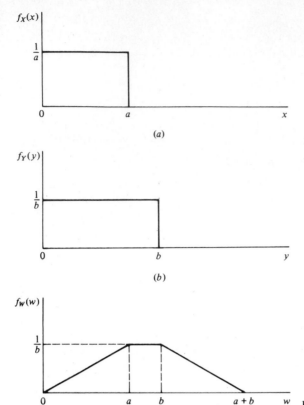

Figure 4.6-2 Two density functions (a) and (b) and their convolution (c).

All these integrands are unity; the values of the integrals are determined by the unit-step functions through their control over limits of integration. After straightforward evaluation we get

$$f_W(w) = \begin{cases} w/ab & 0 \le w < a \\ 1/b & a \le w < b \\ (a + b - w)/ab & b \le w < a + b \\ 0 & w \ge a + b \end{cases}$$

which is sketched in Figure 4.6-2c.

★Sum of Several Random Variables

When the sum Y of N independent random variables $X_1, X_2, \ldots, X_N$ is to be considered, we may extend the above analysis for two random variables. Let $Y_1 = X_1 + X_2$. Then we know from the preceding work that $f_{Y_1}(y_1) =$

$f_{X_2}(x_2) * f_{X_1}(x_1)$.† Next, we know that X_3 will be independent of $Y_1 = X_1 + X_2$ because X_3 is independent of both X_1 and X_2. Thus, by applying (4.6-5) to the two variables X_3 and Y_1 to find the density function of $Y_2 = X_3 + Y_1$, we get

$$f_{Y_2 = X_1 + X_2 + X_3}(y_2) = f_{X_3}(x_3) * f_{Y_1 = X_1 + X_2}(y_1)$$

$$= f_{X_3}(x_3) * f_{X_2}(x_2) * f_{X_1}(x_1) \qquad (4.6\text{-}6)$$

By continuing the process we find that the density function of $Y = X_1 + X_2 + \cdots + X_N$ is the $(N-1)$-fold convolution of the N individual density functions:

$$f_Y(y) = f_{X_N}(x_N) * f_{X_{N-1}}(x_{N-1}) * \cdots * f_{X_1}(x_1) \qquad (4.6\text{-}7)$$

The distribution function of Y is found from the integral of $f_Y(y)$ using (2.3-6c).

*4.7 CENTRAL LIMIT THEOREM

Broadly defined, the *central limit theorem* says that the probability distribution function of the sum of a large number of random variables approaches a gaussian distribution. Although the theorem is known to apply to some cases of statistically *dependent* random variables (Cramér, 1946, p. 219), most applications, and the largest body of knowledge, are directed toward statistically independent random variables. Thus, in all succeeding discussions we assume statistically independent random variables.

*Unequal Distributions

Let $\bar{X}_i$ and $\sigma_{X_i}^2$ be the means and variances, respectively, of N random variables X_i, $i = 1, 2, \ldots, N$, which may have arbitrary probability densities. The central limit theorem states that the sum $Y_N = X_1 + X_2 + \cdots + X_N$, which has mean $\bar{Y}_N = \bar{X}_1 + \bar{X}_2 + \cdots + \bar{X}_N$ and variance $\sigma_{Y_N}^2 = \sigma_{X_1}^2 + \sigma_{X_2}^2 + \cdots + \sigma_{X_N}^2$, has a probability distribution that asymptotically approaches gaussian as $N \to \infty$. Necessary conditions for the theorem's validity are difficult to state, but sufficient conditions are known to be (Cramér, 1946; Thomas, 1969)

$$\sigma_{X_i}^2 > B_1 > 0 \qquad i = 1, 2, \ldots, N \qquad (4.7\text{-}1a)$$

$$E[\,|X_i - \bar{X}_i|^3\,] < B_2 \qquad i = 1, 2, \ldots, N \qquad (4.7\text{-}1b)$$

where B_1 and B_2 are positive numbers. These conditions guarantee that no one random variable in the sum dominates.

The reader should observe that the central limit theorem guarantees only that the *distribution* of the sum of random variables becomes gaussian. It does not follow that the probability *density* is always gaussian. For continuous

---

† The asterisk denotes convolution.

random variables there is usually no problem, but certain conditions imposed on the individual random variables (Cramér, 1946; Papoulis, 1965 and 1984) will guarantee that the density is gaussian.

For discrete random variables X_i the sum Y_N will also be discrete so its density will contain impulses and is, therefore, not gaussian, even though the distribution approaches gaussian. When the possible discrete values of each random variable are kb, $k = 0$, ± 1, ± 2, ..., with b a constant,† the envelope of the impulses in the density of the sum will be gaussian (with mean $\bar{Y}_N$ and variance $\sigma_{Y_N}^2$). This case is discussed in some detail by Papoulis (1965).

The practical usefulness of the central limit theorem does not reside so much in the exactness of the gaussian distribution for $N \to \infty$ because the variance of Y_N becomes infinite from (4.7-1a). Usefulness derives more from the fact that Y_N for *finite* N may have a distribution that is closely approximated as gaussian. The approximation can be quite accurate, even for relatively small values of N, in the central region of the gaussian curve near the mean. However, the approximation can be very inaccurate in the tail regions away from the mean, even for large values of N (Davenport, 1970; Melsa and Sage, 1973). Of course, the approximation is made more accurate by increasing N.

*Equal Distributions

If all of the statistically independent random variables being summed are continuous and have the same distribution function, and therefore the same density, the proof of the central limit theorem is relatively straightforward and is next developed.

Because the sum $Y_N = X_1 + X_2 + \cdots + X_N$ has an infinite variance as $N \to \infty$, we shall work with the zero-mean, unit-variance random variable

$$W_N = (Y_N - \bar{Y}_N)/\sigma_{Y_N} = \sum_{i=1}^{N} (X_i - \bar{X}_i) \bigg/ \left[\sum_{i=1}^{N} \sigma_{X_i}^2 \right]^{1/2}$$

$$= \frac{1}{\sqrt{N}\,\sigma_X} \sum_{i=1}^{N} (X_i - \bar{X}) \tag{4.7-2}$$

instead. Here we define $\bar{X}$ and σ_X^2 by

$$\bar{X}_i = \bar{X} \qquad \text{all } i \tag{4.7-3}$$

$$\sigma_{X_i}^2 = \sigma_X^2 \qquad \text{all } i \tag{4.7-4}$$

since all the X_i have the same distribution.

The theorem's proof consists of showing that the characteristic function of W_N is that of a zero-mean, unit-variance gaussian random variable, which is

$$\Phi_{W_N}(\omega) = \exp\left(-\omega^2/2\right) \tag{4.7-5}$$

---

† These are called *lattice-type* discrete random variables (Papoulis, 1965).

from Problem 3-28. If this is proved the density of W_N must be gaussian from (3.3-3) and the fact that Fourier transforms are unique. The characteristic function of W_N is

$$\Phi_{W_N}(\omega) = E[e^{j\omega W_N}] = E\left[\exp\left\{\frac{j\omega}{\sqrt{N}\,\sigma_X}\sum_{i=1}^{N}(X_i - \bar{X})\right\}\right]$$

$$= \left\langle E\left\{\exp\left[\frac{j\omega}{\sqrt{N}\,\sigma_X}(X_i - \bar{X})\right]\right\}\right\rangle^N \qquad (4.7\text{-}6)$$

The last step in (4.7-6) follows from the independence and equal distribution of the X_i. Next, the exponential in (4.7-6) is expanded in a Taylor polynomial with a remainder term R_N/N:

$$E\left\{\exp\left[\frac{j\omega}{\sqrt{N}\,\sigma_X}(X_i - \bar{X})\right]\right\}$$

$$= E\left\{1 + \left(\frac{j\omega}{\sqrt{N}\,\sigma_X}\right)(X_i - \bar{X}) + \left(\frac{j\omega}{\sqrt{N}\,\sigma_X}\right)^2\frac{(X_i - \bar{X})^2}{2} + \frac{R_N}{N}\right\}$$

$$= 1 - (\omega^2/2N) + E[R_N]/N \qquad (4.7\text{-}7)$$

where $E[R_N]$ approaches zero as $N \to \infty$ (Davenport, 1970, p. 442). On substitution of (4.7-7) into (4.7-6) and forming the natural logarithm, we have

$$\ln\left[\Phi_{W_N}(\omega)\right] = N\ln\left\{1 - (\omega^2/2N) + E[R_N]/N\right\} \qquad (4.7\text{-}8)$$

Since

$$\ln(1-z) = -\left[z + \frac{z^2}{2} + \frac{z^3}{3} + \cdots\right] \qquad |z| < 1 \qquad (4.7\text{-}9)$$

we identify z with $(\omega^2/2N) - E[R_N]/N$ and write (4.7-8) as

$$\ln\left[\Phi_{W_N}(\omega)\right] = -(\omega^2/2) + E[R_N] - \frac{N}{2}\left[\frac{\omega^2}{2N} - \frac{E[R_N]}{N}\right]^2 + \cdots \qquad (4.7\text{-}10)$$

so

$$\lim_{N\to\infty}\left\{\ln\left[\Phi_{W_N}(\omega)\right]\right\} = \ln\left\{\lim_{N\to\infty}\Phi_{W_N}(\omega)\right\} = -\omega^2/2 \qquad (4.7\text{-}11)$$

Finally, we have

$$\lim_{N\to\infty}\Phi_{W_N}(\omega) = e^{-\omega^2/2} \qquad (4.7\text{-}12)$$

which was to be shown.

We illustrate the use of the central limit theorem through an example.

Example 4.7-1 Consider the sum of just two independent uniformly distributed random variables X_1 and X_2 having the same density

$$f_X(x) = \frac{1}{a} [u(x) - u(x - a)]$$

where $a > 0$ is a constant. The means and variances of X_1 and X_2 are $\bar{X} = a/2$ and $\sigma_X^2 = a^2/12$, respectively. The density of the sum $W = X_1 + X_2$ is available from Example 4.6-1 (with $b = a$):

$$f_W(w) = \frac{1}{a} \operatorname{tri} \left(\frac{w}{a} \right)$$

where the function tri $(\cdot)$ is defined in (E-4). The gaussian approximation to W has variance $\sigma_W^2 = 2\sigma_X^2 = a^2/6$ and mean $\bar{W} = 2(a/2) = a$:

$$\text{Approximation to } f_W(w) = \frac{e^{-(w-a)^2/(a^2/3)}}{\sqrt{\pi(a^2/3)}}$$

Figure 4.7-1 illustrates $f_W(w)$ and its gaussian approximation. Even for the case of only two random variables being summed the gaussian approximation is a fairly good one. For other densities the approximation may be very poor (see Problem 4-63).

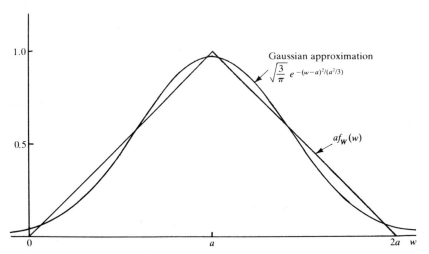

Figure 4.7-1 The triangular density function of Example 4.7-1 and its gaussian approximation.

PROBLEMS

4-1 Two events A and B defined on a sample space S are related to a joint sample space through random variables X and Y and are defined by $A = \{X \leq x\}$ and $B = \{y_1 < Y \leq y_2\}$. Make a sketch of the two sample spaces showing areas corresponding to both events and the event $A \cap B = \{X \leq x, y_1 < Y \leq y_2\}$.

4-2 Work Problem 4-1 for the two events $A = \{x_1 < X \leq x_2\}$ and $B = \{y_1 < Y \leq y_2\}$.

4-3 Work Problem 4-1 for the two events $A = \{x_1 < X \leq x_2$ or $x_3 < X \leq x_4\}$ and $B = \{y_1 < Y \leq y_2\}$.

4-4 Three events A, B, and C satisfy $C \subset B \subset A$ and are defined by $A = \{X \leq x_a, \ Y \leq y_a\}$, $B = \{X \leq x_b, \ Y \leq y_b\}$, and $C = \{X \leq x_c, \ Y \leq y_c\}$ for two random variables X and Y.

(a) Sketch the two sample spaces S and S_J and show the regions corresponding to the three events.

(b) What region corresponds to the event $A \cap B \cap C$?

4-5 A joint sample space for two random variables X and Y has four elements $(1, 1)$, $(2, 2)$, $(3, 3)$, and $(4, 4)$. Probabilities of these elements are 0.1, 0.35, 0.05, and 0.5 respectively.

(a) Determine through logic and sketch the distribution function $F_{X, Y}(x, y)$.

(b) Find the probability of the event $\{X \leq 2.5, Y \leq 6\}$.

(c) Find the probability of the event $\{X \leq 3\}$.

4-6 Write a mathematical equation for $F_{X, Y}(x, y)$ of Problem 4-5.

4-7 The joint distribution function for two random variables X and Y is

$$F_{X, Y}(x, y) = u(x)u(y)[1 - e^{-ax} - e^{-ay} + e^{-a(x+y)}]$$

where $u(\cdot)$ is the unit-step function and $a > 0$. Sketch $F_{X, Y}(x, y)$.

4-8 By use of the joint distribution function in Problem 4-7, and assuming $a = 0.5$ in each case, find the probabilities:

(a) $P\{X \leq 1, Y \leq 2\}$ (b) $P\{0.5 < X < 1.5\}$

(c) $P\{-1.5 < X \leq 2, 1 < Y \leq 3\}$.

4-9 Find and sketch the marginal distribution functions for the joint distribution function of Problem 4-5.

4-10 Find and sketch the marginal distribution functions for the joint distribution function of Problem 4-7.

4-11 Given the function

$$G_{X, Y}(x, y) = u(x)u(y)[1 - e^{-(x+y)}]$$

Show that this function satisfies the first four properties of (4.2-6) but fails the fifth one. The function is therefore not a valid joint probability distribution function.

4-12 Random variables X and Y are components of a two-dimensional random vector and have a joint distribution

$$F_{X,Y}(x, y) = \begin{cases} 0 & x < 0 \quad \text{or} \quad y < 0 \\ xy & 0 \le x < 1 \quad \text{and} \quad 0 \le y < 1 \\ x & 0 \le x < 1 \quad \text{and} \quad 1 \le y \\ y & 1 \le x \quad \text{and} \quad 0 \le y < 1 \\ 1 & 1 \le x \quad \text{and} \quad 1 \le y \end{cases}$$

(a) Sketch $F_{X,Y}(x, y)$.

(b) Find and sketch the marginal distribution functions $F_X(x)$ and $F_Y(y)$.

4-13 Show that the function

$$G_{X,Y}(x, y) = \begin{cases} 0 & x < y \\ 1 & x \ge y \end{cases}$$

cannot be a valid joint distribution function. [*Hint:* Use (4.2-6e).]

4-14 A fair coin is tossed twice. Define random variables by: X = "number of heads on the first toss" and Y = "number of heads on the second toss" (note that X and Y can have only the values 0 or 1).

(a) Find and sketch the joint density function of X and Y.

(b) Find and sketch the joint distribution function.

4-15 A joint probability density function is

$$f_{X,Y}(x, y) = \begin{cases} 1/ab & 0 < x < a \quad \text{and} \quad 0 < y < b \\ 0 & \text{elsewhere} \end{cases}$$

Find and sketch $F_{X,Y}(x, y)$.

4-16 If $a < b$ in Problem 4-15, find:

(a) $P\{X + Y \le 3a/4\}$ (b) $P\{Y \le 2bX/a\}$.

4-17 Find the joint distribution function applicable to Example 4.3-1.

4-18 Sketch the joint density function $f_{X,Y}(x, y)$ applicable to Problem 4-5. Write an equation for $f_{X,Y}(x, y)$.

4-19 Determine the joint density and both marginal density functions for Problem 4-7.

4-20 Find and sketch the joint density function for the distribution function in Problem 4-12.

4-21 (a) Find a constant b (in terms of a) so that the function

$$f_{X,Y}(x, y) = \begin{cases} be^{-(x+y)} & 0 < x < a \quad \text{and} \quad 0 < y < \infty \\ 0 & \text{elsewhere} \end{cases}$$

is a valid joint density function.

(b) Find an expression for the joint distribution function.

4-22 (*a*) By use of the joint density function of Problem 4-21, find the marginal density functions.

(*b*) What is $P\{0.5a < X \le 0.75a\}$ in terms of a and b?

4-23 Determine a constant b such that each of the following are valid joint density functions:

(*a*) $f_{X,Y}(x, y) = \begin{cases} 3xy & 0 < x < 1 \quad \text{and} \quad 0 < y < b \\ 0 & \text{elsewhere} \end{cases}$

(*b*) $f_{X,Y}(x, y) = \begin{cases} bx(1 - y) & 0 < x < 0.5 \quad \text{and} \quad 0 < y < 1 \\ 0 & \text{elsewhere} \end{cases}$

(*c*) $f_{X,Y}(x, y) = \begin{cases} b(x^2 + 4y^2) & 0 \le |x| < 1 \quad \text{and} \quad 0 \le y < 2 \\ 0 & \text{elsewhere} \end{cases}$

*4-24 Given the function

$$f_{X,Y}(x, y) = \begin{cases} (x^2 + y^2)/8\pi & x^2 + y^2 < b \\ 0 & \text{elsewhere} \end{cases}$$

(*a*) Find a constant b so that this is a valid joint density function.

(*b*) Find $P\{0.5b < X^2 + Y^2 \le 0.8b\}$. (*Hint:* Use polar coordinates in both parts.)

*4-25 On a firing range the coordinates of bullet strikes relative to the target bull's-eye are random variables X and Y having a joint density given by

$$f_{X,Y}(x, y) = \frac{e^{-(x^2 + y^2)/2\sigma^2}}{2\pi\sigma^2}$$

Here σ^2 is a constant related to the accuracy of manufacturing a gun's barrel. What value of σ^2 will allow 80% of all bullets to fall inside a circle of diameter 6 cm? (*Hint:* Use polar coordinates.)

4-26 Given the function

$$f_{X,Y}(x, y) = \begin{cases} b(x + y)^2 & -2 < x < 2 \quad \text{and} \quad -3 < y < 3 \\ 0 & \text{elsewhere} \end{cases}$$

(*a*) Find the constant b such that this is a valid joint density function.

(*b*) Determine the marginal density functions $f_X(x)$ and $f_Y(y)$.

4-27 Find the conditional density functions $f_X(x \mid y_1)$, $f_X(x \mid y_2)$, $f_Y(y \mid x_1)$, and $f_Y(y \mid x_2)$ for the joint density defined in Example 4.4-1.

4-28 Find the conditional density function $f_X(x \mid y)$ applicable to Example 4.4-2.

4-29 By using the results of Example 4.4-2, calculate the probability of the event $\{Y \le 2 \mid X = 1\}$.

4-30 Random variables X and Y are *jointly gaussian and normalized* if

$$f_{X,Y}(x, y) = \frac{1}{2\pi\sqrt{1 - \rho^2}} \exp\left[-\frac{x^2 - 2\rho xy + y^2}{2(1 - \rho^2)} \right] \qquad \text{where} \qquad -1 \le \rho \le 1$$

(a) Show that the marginal density functions are

$$f_X(x) = \frac{1}{\sqrt{2\pi}} \exp\left(-x^2/2\right) \qquad f_Y(y) = \frac{1}{\sqrt{2\pi}} \exp\left(-y^2/2\right)$$

(*Hint:* Complete the square and use the fact that the area under a gaussian density is unity.)

(b) Are X and Y statistically independent?

4-31 By use of the joint density of Problem 4-30, show that

$$f_X(x \mid Y = y) = \frac{1}{\sqrt{2\pi(1 - \rho^2)}} \exp\left[-\frac{(x - \rho y)^2}{2(1 - \rho^2)} \right]$$

4-32 Given the joint distribution function

$$F_{X,Y}(x, y) = u(x)u(y)[1 - e^{-ax} - e^{-ay} + e^{-a(x+y)}]$$

find:

(a) The conditional density functions $f_X(x \mid Y = y)$ and $f_Y(y \mid X = x)$.

(b) Are the random variables X and Y statistically independent?

4-33 For two independent random variables X and Y show that

$$P\{Y \le X\} = \int_{-\infty}^{\infty} F_Y(x) f_X(x) \, dx$$

or

$$P\{Y \le X\} = 1 - \int_{-\infty}^{\infty} F_X(y) f_Y(y) \, dy$$

4-34 Two random variables X and Y have a joint probability density function

$$f_{X,Y}(x, y) = \begin{cases} \dfrac{5}{16} x^2 y & 0 < y < x < 2 \\ 0 & \text{elsewhere} \end{cases}$$

(a) Find the marginal density functions of X and Y.

(b) Are X and Y statistically independent?

4-35 Show, by use of (4.4-13), that the area under $f_Y(y \mid x)$ is unity.

★4-36 Two random variables R and Θ have the joint density function

$$f_{R,\Theta}(r, \theta) = \frac{u(r)[u(\theta) - u(\theta - 2\pi)]r}{2\pi} e^{-r^2/2}$$

(a) Find $P\{0 < R \le 1, 0 < \Theta \le \pi/2\}$.

(b) Find $f_R(r \mid \Theta = \pi)$.

(c) Find $f_R(r \mid \Theta \leq \pi)$ and compare to the result found in part (b), and explain the comparison.

4-37 Random variables X and Y have respective density functions

$$f_X(x) = \frac{1}{a} [u(x) - u(x - a)]$$

$$f_Y(y) = bu(y)e^{-by}$$

where $a > 0$ and $b > 0$. Find and sketch the density function of $W = X + Y$ if X and Y are statistically independent.

4-38 Random variables X and Y have respective density functions

$$f_X(x) = 0.1\delta(x - 1) + 0.2\delta(x - 2) + 0.4\delta(x - 3) + 0.3\delta(x - 4)$$

$$f_Y(y) = 0.4\delta(y - 5) + 0.5\delta(y - 6) + 0.1\delta(y - 7)$$

Find and sketch the density function of $W = X + Y$ if X and Y are independent.

4-39 Find and sketch the density function of $W = X + Y$, where the random variable X is that of Problem 4-37 with $a = 5$ and Y is that of Problem 4-38. Assume X and Y are independent.

4-40 Find the density function of $W = X + Y$, where the random variable X is that of Problem 4-38 and Y is that of Problem 4-37. Assume X and Y are independent. Sketch the density function for $b = 1$ and $b = 4$.

*4-41 Three statistically independent random variables X_1, X_2, and X_3 all have the same density function

$$f_{X_i}(x_i) = \frac{1}{a} [u(x_i) - u(x_i - a)] \qquad i = 1, 2, 3$$

Find and sketch the density function of $Y = X_1 + X_2 + X_3$ if $a > 0$ is constant.

ADDITIONAL PROBLEMS

4-42 In a gambling game two fair dice are tossed and the sum of the numbers that show up determines who wins among two players. Random variables X and Y represent the winnings of the first and second numbered players, respectively. The first wins \$3 if the sum is 4, 5, or 6, and loses \$2 if the sum is 11 or 12; he neither wins nor loses for all other sums. The second player wins \$2 for a sum of 8 or more, loses \$3 for a sum of 5 or less, and neither wins nor loses for other sums.

(a) Draw sample spaces S and S_J and show how elements of S map to elements of S_J.

(b) Find the probabilities of all joint outcomes possible in S_J.

4-43 Discrete random variables X and Y have a joint distribution function

$$F_{X,Y}(x, y) = 0.10u(x + 4)u(y - 1) + 0.15u(x + 3)u(y + 5)$$
$$+ 0.17u(x + 1)u(y - 3) + 0.05u(x)u(y - 1)$$
$$+ 0.18u(x - 2)u(y + 2) + 0.23u(x - 3)u(y - 4)$$
$$+ 0.12u(x - 4)u(y + 3)$$

Find: (a) the marginal distributions $F_X(x)$ and $F_Y(y)$ and sketch the two functions, (b) $\bar{X}$ and $\bar{Y}$, and (c) the probability $P\{-1 < X \le 4, -3 < Y \le 3\}$.

4-44 Random variables X and Y have the joint distribution

$$F_{X,Y}(x, y) = \begin{cases} \dfrac{5}{4}\left(\dfrac{x + e^{-(x+1)y^2}}{x + 1} - e^{-y^2}\right)u(y) & 0 \le x < 4 \\ 0 & x < 0 \text{ or } y < 0 \\ 1 + \dfrac{1}{4}e^{-5y^2} - \dfrac{5}{4}e^{-y^2} & 4 \le x \text{ and any } y \ge 0 \end{cases}$$

Find: (a) The marginal distribution functions of X and Y, and (b) the probability $P\{3 < X \le 5, 1 < Y \le 2\}$.

4-45 Find the joint distribution function of the random variables having the joint density of Problem 4-48.

4-46 Find a value of the constant b so that the function

$$f_{X,Y}(x, y) = bxy^2 \exp{(-2xy)}u(x - 2)u(y - 1)$$

is a valid joint probability density.

4-47 The locations of hits of darts thrown at a round dartboard of radius r are determined by a vector random variable with components X and Y. The joint density of X and Y is uniform, that is,

$$f_{X,Y}(x, y) = \begin{cases} 1/\pi r^2 & x^2 + y^2 < r^2 \\ 0 & \text{elsewhere} \end{cases}$$

Find the densities of X and Y.

4-48 Two random variables X and Y have a joint density

$$f_{X,Y}(x, y) = {}^{10}\!/_{4}[u(x) - u(x - 4)]u(y)y^3 \exp{[-(x + 1)y^2]}$$

Find the marginal densities and distributions of X and Y.

4-49 Find the marginal densities of X and Y using the joint density

$$f_{X,Y}(x, y) = 2u(x)u(y) \exp\left[-\left(4y + \dfrac{x}{2}\right)\right]$$

4-50 Random variables X and Y have the joint density of Problem 4-49. Find the probability that the values of Y are not greater than twice the values of X for $x \le 3$.

4-51 Find the conditional densities $f_X(x \mid Y = y)$ and $f_Y(y \mid X = x)$ applicable to the joint density of Problem 4-47.

4-52 For the joint density of Problem 4-48 determine the conditional densities $f_X(x \mid Y = y)$ and $f_Y(y \mid X = x)$.

4-53 The time it takes a person to drive to work is a random variable Y. Because of traffic driving time depends on the (random) time of departure, denoted X, which occurs in an interval of duration T_0 that begins at 7:30 A.M. each day. There is a minimum driving time T_1 required, regardless of the time of departure. The joint density of X and Y is known to be

$$f_{X,Y}(x, y) = c(y - T_1)^3 u(y - T_1)[u(x) - u(x - T_0)] \exp\left[-(y - T_1)(x + 1)\right]$$

where

$$c = (1 + T_0)^3 / 2[(1 + T_0)^3 - 1]$$

(a) Find the average driving time that results when it is given that departure occurs at 7:30 A.M. Evaluate your result for $T_0 = 1$ h.

(b) Repeat part (a) given that departure time is at 7:30 A.M. plus T_0.

(c) What is the average time of departure if $T_0 = 1$ h? (*Hint:* Note that point conditioning applies.)

*__**4-54**__ Start with the expressions

$$F_Y(y \mid B) = P\{Y \le y \mid B\} = \frac{P\{Y \le y \cap B\}}{P(B)}$$

$$f_Y(y \mid B) = \frac{dF_Y(y \mid B)}{dy}$$

which are analogous to (4.4-1) and (4.4-2), and derive $F_Y(y \mid x_a < X \le x_b)$ and $f_Y(y \mid x_a < X \le x_b)$ which are analogous to (4.4-15) and (4.4-16).

*__**4-55**__ Extend the procedures of the text that lead to (4.4-16) to show that the joint distribution and density of random variables X and Y, conditional on the event $B = \{y_a < Y \le y_b\}$, are

$$F_{X,Y}(x, y \mid y_a < Y \le y_b) = \begin{cases} 0 & y \le y_a \\ \dfrac{F_{X,Y}(x, y) - F_{X,Y}(x, y_a)}{F_Y(y_b) - F_Y(y_a)} & y_a < y \le y_b \\ \dfrac{F_{X,Y}(x, y_b) - F_{X,Y}(x, y_a)}{F_Y(y_b) - F_Y(y_a)} & y_b < y \end{cases}$$

and

$$f_{X,Y}(x, y \mid y_a < Y \le y_b) = \begin{cases} 0 & y \le y_a \quad \text{and} \quad y > y_b \\ \dfrac{f_{X,Y}(x, y)}{F_Y(y_b) - F_Y(y_a)} & y_a < y \le y_b \end{cases}$$

4-56 Determine if random variables X and Y of Problem 4-53 are statistically independent.

4-57 Determine if X and Y of Problem 4-49 are statistically independent.

4-58 The joint density of four random variables X_i, $i = 1, 2, 3$, and 4, is

$$f_{X_1, X_2, X_3, X_4}(x_1, x_2, x_3, x_4) = \prod_{i=1}^{4} \exp\left(-2|x_i|\right)$$

Find densities (a) $f_{X_1, X_2, X_3}(x_1, x_2, x_3|x_4)$ (b) $f_{X_1, X_2}(x_1, x_2|x_3, x_4)$, and (c) $f_{X_1}(x_1|x_2, x_3, x_4)$.

4-59 If the difference $W = X - Y$ is formed instead of the sum in (4.6-1), develop the probability density of W. Compare the result with (4.6-5). Is the density still a convolution of the densities of X and Y? Discuss.

4-60 Statistically independent random variables X and Y have respective densities

$$f_X(x) = [u(x + 12) - u(x - 12)][1 - |x/12|]/12$$

$$f_Y(y) = (1/4)u(y) \exp\left(-y/4\right)$$

Find the probabilities of the events:

(a) $\{Y \le 8 - (2|X|/3)\}$, and (b) $\{Y \le 8 + (2|X|/3)\}$.

Compare the two results.

4-61 Statistically independent random variables X and Y have respective densities

$$f_X(x) = 5u(x) \exp\left(-5x\right)$$

$$f_Y(y) = 2u(y) \exp\left(-2y\right)$$

Find the density of the sum $W = X + Y$.

★4-62 N statistically independent random variables X_i, $i = 1, 2, \ldots, N$, all have the same density

$$f_{X_i}(x_i) = au(x_i) \exp\left(-ax_i\right)$$

where $a > 0$ is a constant. Find an expression for the density of the sum $W = X_1 + X_2 + \cdots + X_N$ for any N.

★4-63 Find the exact probability density for the sum of two statistically independent random variables each having the density

$$f_X(x) = 3[u(x + a) - u(x - a)]x^2/2a^3$$

where $a > 0$ is a constant. Plot the density along with the gaussian approximation (to the density of the sum) that has variance $2\sigma_X^2$ and mean $2\bar{X}$. Is the approximation a good one?

★4-64 Work Problem 4-63 except assume

$$f_X(x) = (1/2) \cos\left(x\right) \operatorname{rect}\left(x/\pi\right).$$

FIVE

OPERATIONS ON
MULTIPLE RANDOM VARIABLES

5.0 INTRODUCTION

After establishing some of the basic theory of several random variables in the previous chapter, it is appropriate to now extend the operations described in Chapter 3 to include multiple random variables. This chapter is dedicated to these extensions. Mainly, the concept of expectation is enlarged to include two or more random variables. Other operations involving moments, characteristic functions, and transformations are all special applications of expectation.

5.1 EXPECTED VALUE OF A
FUNCTION OF RANDOM VARIABLES

When more than a single random variable is involved, expectation must be taken with respect to all the variables involved. For example, if $g(X, Y)$ is some function of two random variables X and Y the expected value of $g(\cdot, \cdot)$ is given by

$$\bar{g} = E[g(X, Y)] = \int_{-\infty}^{\infty} \int_{-\infty}^{\infty} g(x, y) f_{X, Y}(x, y) \, dx \, dy \qquad (5.1\text{-}1)$$

This expression is the two-variable extension of (3.1-6).

For N random variables $X_1, X_2, \ldots, X_N$ and some function of these variables, denoted $g(X_1, \ldots, X_N)$, the expected value of the function becomes

$$\bar{g} = E[g(X_1, \ldots, X_N)]$$

$$= \int_{-\infty}^{\infty} \cdots \int_{-\infty}^{\infty} g(x_1, \ldots, x_N) f_{X_1, \ldots, X_N}(x_1, \ldots, x_N) \, dx_1 \cdots dx_N \qquad (5.1\text{-}2)$$

Thus, expectation in general involves an N-fold integration when N random variables are involved.

We illustrate the application of (5.1-2) with an example that will develop an important point.

Example 5.1-1 We shall find the mean (expected) value of a sum of N weighted random variables. If we let

$$g(X_1, \ldots, X_N) = \sum_{i=1}^{N} \alpha_i X_i$$

where the "weights" are the constants α_i, the mean value of the weighted sum becomes

$$E[g(X_1, \ldots, X_N)] = E\left[\sum_{i=1}^{N} \alpha_i X_i\right]$$

$$= \sum_{i=1}^{N} \int_{-\infty}^{\infty} \cdots \int_{-\infty}^{\infty} \alpha_i x_i \, f_{X_1, \ldots, X_N}(x_1, \ldots, x_N) \, dx_1 \cdots dx_N$$

from (5.1-2). After using (4.3-8), the terms in the sum all reduce to the form

$$\int_{-\infty}^{\infty} \alpha_i x_i \, f_{X_i}(x_i) \, dx_i = E[\alpha_i X_i] = \alpha_i E[X_i]$$

so

$$E\left[\sum_{i=1}^{N} \alpha_i X_i\right] = \sum_{i=1}^{N} \alpha_i E[X_i]$$

which says that *the mean value of a weighted sum of random variables equals the weighted sum of mean values.*

The above extensions (5.1-1) and (5.1-2) of expectation do not invalidate any of our single random variable results. For example, let

$$g(X_1, \ldots, X_N) = g(X_1) \qquad (5.1\text{-}3)$$

and substitute into (5.1-2). After integrating with respect to all random variables except X_1, (5.1-2) becomes

$$\bar{g} = E[g(X_1)] = \int_{-\infty}^{\infty} g(x_1) f_{X_1}(x_1)\, dx_1 \tag{5.1-4}$$

which is the same as previously given in (3.1-6) for one random variable. Some reflection on the reader's part will verify that (5.1-4) also validates such earlier topics as moments, central moments, characteristic function, etc., for a single random variable.

Joint Moments About the Origin

One important application of (5.1-1) is in defining *joint moments* about the origin. They are denoted by m_{nk} and are defined by

$$m_{nk} = E[X^n Y^k] = \int_{-\infty}^{\infty} \int_{-\infty}^{\infty} x^n y^k f_{X,Y}(x, y)\, dx\, dy \tag{5.1-5}$$

for the case of two random variables X and Y. Clearly $m_{n0} = E[X^n]$ are the moments m_n of X, while $m_{0k} = E[Y^k]$ are the moments of Y. The sum $n + k$ is called the *order* of the moments. Thus m_{02}, m_{20}, and m_{11} are all second-order moments of X and Y. The first-order moments $m_{01} = E[Y] = \bar{Y}$ and $m_{10} = E[X] = \bar{X}$ are the expected values of Y and X, respectively, and are the coordinates of the "center of gravity" of the function $f_{X,Y}(x, y)$.

The second-order moment $m_{11} = E[XY]$ is called the *correlation* of X and Y. It is so important to later work that we give it the symbol R_{XY}. Hence,

$$R_{XY} = m_{11} = E[XY] = \int_{-\infty}^{\infty} \int_{-\infty}^{\infty} xy f_{X,Y}(x, y)\, dx\, dy \tag{5.1-6}$$

If correlation can be written in the form

$$R_{XY} = E[X]E[Y] \tag{5.1-7}$$

then X and Y are said to be *uncorrelated*. Statistical independence of X and Y is sufficient to guarantee they are uncorrelated, as is readily proven by (5.1-6) using (4.5-4). The converse of this statement, that is, that X and Y are independent if X and Y are uncorrelated, is *not* necessarily true in general.†

If

$$R_{XY} = 0 \tag{5.1-8}$$

for two random variables X and Y, they are called *orthogonal*.

A simple example is next developed that illustrates the important new topic of correlation.

---

† Uncorrelated *gaussian* random variables are, however, known to also be independent (see Section 5.3).

Example 5.1-2 Let X be a random variable that has a mean value $\bar{X} = E[X] = 3$ and variance $\sigma_X^2 = 2$. From (3.2-6) we easily determine the second moment of X about the origin: $E[X^2] = m_{20} = \sigma_X^2 + \bar{X}^2 = 11$.

Next, let another random variable Y be defined by

$$Y = -6X + 22$$

The mean value of Y is $\bar{Y} = E[Y] = E[-6X + 22] = -6\bar{X} + 22 = 4$. The correlation of X and Y is found from (5.1-6)

$$R_{XY} = m_{11} = E[XY] = E[-6X^2 + 22X] = -6E[X^2] + 22\bar{X}$$
$$= -6(11) + 22(3) = 0$$

Since $R_{XY} = 0$, X and Y are orthogonal from (5.1-8). On the other hand, $R_{XY} \neq E[X]E[Y] = 12$, so X and Y are *not* uncorrelated [see (5.1-7)].

We note that two random variables can be orthogonal even though correlated when one, Y, is related to the other, X, by the linear function $Y = aX + b$. It can be shown that X and Y are always correlated if $|a| \neq 0$, regardless of the value of b (see Problem 5-9). They are uncorrelated if $a = 0$, but this is not a case of much practical interest. Orthogonality can likewise be shown to occur when a and b are related by $b = -aE[X^2]/E[X]$ whenever $E[X] \neq 0$. If $E[X] = 0$, X and Y cannot be orthogonal for any value of a except $a = 0$, a noninteresting problem. The reader may wish to verify these statements as an exercise.

For N random variables $X_1, X_2, \ldots, X_N$, the $(n_1 + n_2 + \cdots + n_N)$-order moments are defined by

$$m_{n_1 n_2 \cdots n_N} = E[X_1^{n_1} X_2^{n_2} \cdots X_N^{n_N}]$$
$$= \int_{-\infty}^{\infty} \cdots \int_{-\infty}^{\infty} X_1^{n_1} \cdots X_N^{n_N} f_{X_1, \ldots, X_N}(x_1, \ldots, x_N) \, dx_1 \cdots dx_N \quad (5.1\text{-}9)$$

where $n_1, n_2, \ldots, n_N$ are all integers $= 0, 1, 2, \ldots$.

Joint Central Moments

Another important application of (5.1-1) is in defining *joint central moments*. For two random variables X and Y, these moments, denoted by μ_{nk}, are given by

$$\mu_{nk} = E[(X - \bar{X})^n (Y - \bar{Y})^k]$$
$$= \int_{-\infty}^{\infty} \int_{-\infty}^{\infty} (x - \bar{X})^n (y - \bar{Y})^k f_{X,Y}(x, y) \, dx \, dy \quad (5.1\text{-}10)$$

The second-order central moments

$$\mu_{20} = E[(X - \bar{X})^2] = \sigma_X^2 \tag{5.1-11}$$

$$\mu_{02} = E[(Y - \bar{Y})^2] = \sigma_Y^2 \tag{5.1-12}$$

are just the variances of X and Y.

The second-order joint moment μ_{11} is very important. It is called the *covariance* of X and Y and is given the symbol C_{XY}. Hence

$$C_{XY} = \mu_{11} = E[(X - \bar{X})(Y - \bar{Y})]$$

$$= \int_{-\infty}^{\infty} \int_{-\infty}^{\infty} (x - \bar{X})(y - \bar{Y}) f_{X,Y}(x, y) \, dx \, dy \tag{5.1-13}$$

By direct expansion of the product $(x - \bar{X})(y - \bar{Y})$, this integral reduces to the form

$$C_{XY} = R_{XY} - \bar{X}\bar{Y} = R_{XY} - E[X]E[Y] \tag{5.1-14}$$

when (5.1-6) is used. If X and Y are either independent or uncorrelated, then (5.1-7) applies and (5.1-14) shows their covariance is zero:

$$C_{XY} = 0 \qquad X \text{ and } Y \text{ independent or uncorrelated} \tag{5.1-15}$$

If X and Y are orthogonal random variables, then

$$C_{XY} = -E[X]E[Y] \qquad X \text{ and } Y \text{ orthogonal} \tag{5.1-16}$$

from use of (5.1-8) with (5.1-14). Clearly, $C_{XY} = 0$ if either X or Y also has zero mean value.

The normalized second-order moment

$$\rho = \mu_{11}/\sqrt{\mu_{20}\mu_{02}} = C_{XY}/\sigma_X \sigma_Y \tag{5.1-17a}$$

given by

$$\rho = E\left[\frac{(X - \bar{X})}{\sigma_X} \frac{(Y - \bar{Y})}{\sigma_Y}\right] \tag{5.1-17b}$$

is known as the *correlation coefficient* of X and Y. It can be shown (see Problem 5-10) that

$$-1 \leq \rho \leq 1 \tag{5.1-18}$$

For N random variables $X_1, X_2, \ldots, X_N$ the $(n_1 + n_2 + \cdots + n_N)$-order joint central moment is defined by

$$\mu_{n_1 n_2 \cdots n_N} = E[(X_1 - \bar{X}_1)^{n_1}(X_2 - \bar{X}_2)^{n_2} \cdots (X_N - \bar{X}_N)^{n_N}]$$

$$= \int_{-\infty}^{\infty} \cdots \int_{-\infty}^{\infty} (x_1 - \bar{X}_1)^{n_1} \cdots$$

$$(x_N - \bar{X}_N)^{n_N} f_{X_1, \ldots, X_N}(x_1, \ldots, x_N) \, dx_1 \cdots dx_N \tag{5.1-19}$$

An example is next developed that involves the use of covariances.

Example 5.1-3 Again let X be a weighted sum of N random variables X_i; that is, let

$$X = \sum_{i=1}^{N} \alpha_i X_i$$

where the α_i are real weighting constants. The variance of X will be found. From Example 5.1-1,

$$E[X] = \sum_{i=1}^{N} \alpha_i E[X_i] = \sum_{i=1}^{N} \alpha_i \bar{X}_i = \bar{X}$$

so we have

$$X - \bar{X} = \sum_{i=1}^{N} \alpha_i(X_i - \bar{X}_i)$$

and

$$\sigma_X^2 = E[(X - \bar{X})^2] = E\left[\sum_{i=1}^{N} \alpha_i(X_i - \bar{X}_i) \sum_{j=1}^{N} \alpha_j(X_j - \bar{X}_j)\right]$$

$$= \sum_{i=1}^{N} \sum_{j=1}^{N} \alpha_i \alpha_j E[(X_i - \bar{X}_i)(X_j - \bar{X}_j)] = \sum_{i=1}^{N} \sum_{j=1}^{N} \alpha_i \alpha_j C_{X_i X_j}$$

Thus, the variance of a weighted sum of N random variables X_i (weights α_i) equals the weighted sum of all their covariances $C_{X_i X_j}$ (weights $\alpha_i \alpha_j$). For the special case of uncorrelated random variables, where

$$C_{X_i X_j} = \begin{cases} 0 & i \neq j \\ \sigma_{X_i}^2 & i = j \end{cases}$$

is true, we get

$$\sigma_X^2 = \sum_{i=1}^{N} \alpha_i^2 \sigma_{X_i}^2$$

In words: *the variance of a weighted sum of uncorrelated random variables (weights α_i) equals the weighted sum of the variances of the random variables (weights α_i^2).*

★5.2 JOINT CHARACTERISTIC FUNCTIONS

The *joint characteristic function* of two random variables X and Y is defined by

$$\Phi_{X,Y}(\omega_1, \omega_2) = E[e^{j\omega_1 X + j\omega_2 Y}] \tag{5.2-1}$$

where ω_1 and ω_2 are real numbers. An equivalent form is

$$\Phi_{X,Y}(\omega_1, \omega_2) = \int_{-\infty}^{\infty} \int_{-\infty}^{\infty} f_{X,Y}(x, y) e^{j\omega_1 x + j\omega_2 y} \, dx \, dy \tag{5.2-2}$$

This expression is recognized as the two-dimensional Fourier transform (with signs of ω_1 and ω_2 reversed) of the joint density junction. From the inverse Fourier transform we also have

$$f_{X,Y}(x, y) = \frac{1}{(2\pi)^2} \int_{-\infty}^{\infty} \int_{-\infty}^{\infty} \Phi_{X,Y}(\omega_1, \omega_2) e^{-j\omega_1 x - j\omega_2 y} \, d\omega_1 \, d\omega_2 \qquad (5.2\text{-}3)$$

By setting either $\omega_2 = 0$ or $\omega_1 = 0$ in (5.2-2), the characteristic functions of X or Y are obtained. They are called *marginal characteristic functions*:

$$\Phi_X(\omega_1) = \Phi_{X,Y}(\omega_1, 0) \qquad (5.2\text{-}4)$$

$$\Phi_Y(\omega_2) = \Phi_{X,Y}(0, \omega_2) \qquad (5.2\text{-}5)$$

Joint moments m_{nk} can be found from the joint characteristic function as follows:

$$m_{nk} = (-j)^{n+k} \left. \frac{\partial^{n+k} \Phi_{X,Y}(\omega_1, \omega_2)}{\partial \omega_1^n \, \partial \omega_2^k} \right|_{\omega_1 = 0, \, \omega_2 = 0} \qquad (5.2\text{-}6)$$

This expression is the two-dimensional extension of (3.3-4).

Example 5.2-1 Two random variables X and Y have the joint characteristic function

$$\Phi_{X,Y}(\omega_1, \omega_2) = \exp\left(-2\omega_1^2 - 8\omega_2^2\right)$$

We show that X and Y are both zero-mean random variables and that they are uncorrelated.

The means derive from (5.2-6):

$$\bar{X} = E[X] = m_{10} = -j \left. \frac{\partial \Phi_{X,Y}(\omega_1, \omega_2)}{\partial \omega_1} \right|_{\omega_1 = 0, \, \omega_2 = 0}$$

$$= -j(-4\omega_1) \exp\left(-2\omega_1^2 - 8\omega_2^2\right) \Big|_{\omega_1 = 0, \, \omega_2 = 0} = 0$$

$$\bar{Y} = E[Y] = m_{01} = -j(-16\omega_2) \exp\left(-2\omega_1^2 - 8\omega_2^2\right) \Big|_{\omega_1 = 0, \, \omega_2 = 0} = 0$$

Also from (5.2-6):

$$R_{XY} = E[XY] = m_{11} = (-j)^2 \left. \frac{\partial^2}{\partial \omega_1 \, \partial \omega_2} \left[\exp\left(-2\omega_1^2 - 8\omega_2^2\right)\right] \right|_{\omega_1 = 0, \, \omega_2 = 0}$$

$$= -(-4\omega_1)(-16\omega_2) \exp\left(-2\omega_1^2 - 8\omega_2^2\right) \Big|_{\omega_1 = 0, \, \omega_2 = 0} = 0$$

Since means are zero, $C_{XY} = R_{XY}$ from (5.1-14). Therefore, $C_{XY} = 0$ and X and Y are uncorrelated.

The joint characteristic function for N random variables $X_1, X_2, \ldots, X_N$ is defined by

$$\Phi_{X_1, \ldots, X_N}(\omega_1, \ldots, \omega_N) = E[e^{j\omega_1 X_1 + \cdots + j\omega_N X_N}] \tag{5.2-7}$$

Joint moments are obtained from

$$m_{n_1 n_2 \cdots n_N} = (-j)^R \frac{\partial^R \Phi_{X_1, \ldots, X_N}(\omega_1, \ldots, \omega_N)}{\partial \omega_1^{n_1} \partial \omega_2^{n_2} \cdots \partial \omega_N^{n_N}} \bigg|_{\text{all } \omega_i = 0} \tag{5.2-8}$$

where

$$R = n_1 + n_2 + \cdots + n_N \tag{5.2-9}$$

5.3 JOINTLY GAUSSIAN RANDOM VARIABLES

Gaussian random variables are very important because they show up in nearly every area of science and engineering. In this section, the case of two gaussian random variables is first examined. The more advanced case of N random variables is then introduced.

Two Random Variables

Two random variables X and Y are said to be *jointly gaussian* if their joint density function is of the form

$$f_{X, Y}(x, y) = \frac{1}{2\pi\sigma_X \sigma_Y \sqrt{1 - \rho^2}}$$

$$\cdot \exp \left\{ \frac{-1}{2(1 - \rho^2)} \left[\frac{(x - \bar{X})^2}{\sigma_X^2} - \frac{2\rho(x - \bar{X})(y - \bar{Y})}{\sigma_X \sigma_Y} + \frac{(y - \bar{Y})^2}{\sigma_Y^2} \right] \right\} \tag{5.3-1}$$

which is sometimes called the *bivariate gaussian density*. Here

$$\bar{X} = E[X] \tag{5.3-2}$$

$$\bar{Y} = E[Y] \tag{5.3-3}$$

$$\sigma_X^2 = E[(X - \bar{X})^2] \tag{5.3-4}$$

$$\sigma_Y^2 = E[(Y - \bar{Y})^2] \tag{5.3-5}$$

$$\rho = E[(X - \bar{X})(Y - \bar{Y})]/\sigma_X \sigma_Y \tag{5.3-6}$$

Figure 5.3-1a illustrates the appearance of the joint gaussian density function (5.3-1). Its maximum is located at the point $(\bar{X}, \bar{Y})$. The maximum value is obtained from

$$f_{X, Y}(x, y) \leq f_{X, Y}(\bar{X}, \bar{Y}) = \frac{1}{2\pi\sigma_X \sigma_Y \sqrt{1 - \rho^2}} \tag{5.3-7}$$

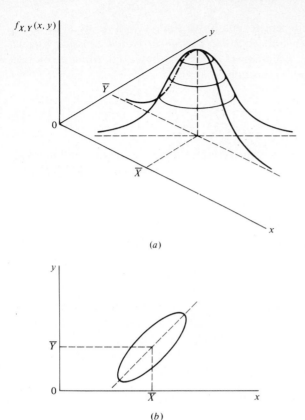

(a)

(b)

Figure 5.3-1 Sketch of the joint density function of two gaussian random variables.

The locus of constant values of $f_{X,Y}(x, y)$ will be an ellipse† as shown in Figure 5.3-1b. This is equivalent to saying that the line of intersection formed by slicing the function $f_{X,Y}(x, y)$ with a plane parallel to the xy plane is an ellipse.

Observe that if $\rho = 0$, corresponding to uncorrelated X and Y, (5.3-1) can be written as

$$f_{X,Y}(x, y) = f_X(x)f_Y(y) \tag{5.3-8}$$

where $f_X(x)$ and $f_Y(y)$ are the marginal density functions of X and Y given by

$$f_X(x) = \frac{1}{\sqrt{2\pi\sigma_X^2}} \exp\left[-\frac{(x - \bar{X})^2}{2\sigma_X^2}\right] \tag{5.3-9}$$

$$f_Y(y) = \frac{1}{\sqrt{2\pi\sigma_Y^2}} \exp\left[-\frac{(y - \bar{Y})^2}{2\sigma_Y^2}\right] \tag{5.3-10}$$

---

† When $\sigma_X = \sigma_Y$ and $\rho = 0$ the ellipse degenerates into a circle; when $\rho = +1$ or -1 the ellipses degenerate into axes rotated by angles $\pi/4$ and $-\pi/4$ respectively that pass through the point $(\bar{X}, \bar{Y})$.

Now the form of (5.3-8) is sufficient to guarantee that X and Y are statistically independent. Therefore we conclude that *any two uncorrelated gaussian random variables are also statistically independent*. It results that a coordinate rotation (linear transformation of X and Y) through an angle

$$\theta = \frac{1}{2} \tan^{-1} \left[\frac{2\rho \sigma_X \sigma_Y}{\sigma_X^2 - \sigma_Y^2} \right] \tag{5.3-11}$$

is sufficient to convert correlated random variables X and Y, having variances σ_X^2 and σ_Y^2, respectively, correlation coefficient ρ, and the joint density of (5.3-1), into two statistically independent gaussian random variables.†

By direct application of (4.4-12) and (4.4-13), the conditional density functions $f_X(x \mid Y = y)$ and $f_Y(y \mid X = x)$ can be found from the above expressions (see Problem 5-29).

Example 5.3-1 We show by example that (5.3-11) applies to arbitrary as well as gaussian random variables. Consider random variables Y_1 and Y_2 related to arbitrary random variables X and Y by the coordinate rotation

$$Y_1 = X \cos (\theta) + Y \sin (\theta)$$
$$Y_2 = -X \sin (\theta) + Y \cos (\theta)$$

If $\bar{X}$ and $\bar{Y}$ are the means of X and Y, respectively, the means of Y_1 and Y_2 are clearly $\bar{Y}_1 = \bar{X} \cos (\theta) + \bar{Y} \sin (\theta)$ and $\bar{Y}_2 = -\bar{X} \sin (\theta) + \bar{Y} \cos (\theta)$, respectively. The covariance of Y_1 and Y_2 is

$$\begin{aligned}
C_{Y_1 Y_2} &= E[(Y_1 - \bar{Y}_1)(Y_2 - \bar{Y}_2)] \\
&= E[\{(X - \bar{X}) \cos (\theta) + (Y - \bar{Y}) \sin (\theta)\} \\
&\quad \cdot \{-(X - \bar{X}) \sin (\theta) + (Y - \bar{Y}) \cos (\theta)\}] \\
&= (\sigma_Y^2 - \sigma_X^2) \sin (\theta) \cos (\theta) + C_{XY}[\cos^2 (\theta) - \sin^2 (\theta)] \\
&= (\sigma_Y^2 - \sigma_X^2)(\tfrac{1}{2}) \sin (2\theta) + C_{XY} \cos (2\theta)
\end{aligned}$$

Here $C_{XY} = E[(X - \bar{X})(Y - \bar{Y})] = \rho \sigma_X \sigma_Y$. If we require Y_1 and Y_2 to be uncorrelated, we must have $C_{Y_1 Y_2} = 0$. By equating the above equation to zero we obtain (5.3-11). Thus, (5.3-11) applies to arbitrary as well as gaussian random variables.

---

† Wozencraft and Jacobs (1965), p. 155.

⋆*N* **Random Variables**

N random variables X_1, X_2, ..., X_N are called *jointly gaussian* if their joint density function can be written as†

$$f_{X_1, ..., X_N}(x_1, ..., x_N) = \frac{||C_X]^{-1}|^{1/2}}{(2\pi)^{N/2}} \exp\left\{-\frac{[x - \bar{X}]'[C_X]^{-1}[x - \bar{X}]}{2}\right\}$$

(5.3-12)

where we define matrices

$$[x - \bar{X}] = \begin{bmatrix} x_1 - \bar{X}_1 \\ x_2 - \bar{X}_2 \\ \vdots \\ x_N - \bar{X}_N \end{bmatrix}$$

(5.3-13)

and

$$[C_X] = \begin{bmatrix} C_{11} & C_{12} & \cdots & C_{1N} \\ C_{21} & C_{22} & \cdots & C_{2N} \\ \vdots & \vdots & & \vdots \\ C_{N1} & C_{N2} & \cdots & C_{NN} \end{bmatrix}$$

(5.3-14)

We use the notation $[\cdot]'$ for the matrix transpose, $[\cdot]^{-1}$ for the matrix inverse, and $||\cdot||$ for the determinant. Elements of $[C_X]$, called the *covariance matrix* of the *N* random variables, are given by

$$C_{ij} = E[(X_i - \bar{X}_i)(X_j - \bar{X}_j)] = \begin{cases} \sigma_{X_i}^2 & i = j \\ C_{X_i X_j} & i \neq j \end{cases}$$

(5.3-15)

The density (5.3-12) is often called the *N-variate gaussian density* function.

For the special case where $N = 2$, the covariance matrix becomes

$$[C_X] = \begin{bmatrix} \sigma_{X_1}^2 & \rho\sigma_{X_1}\sigma_{X_2} \\ \rho\sigma_{X_1}\sigma_{X_2} & \sigma_{X_2}^2 \end{bmatrix}$$

(5.3-16)

so

$$[C_X]^{-1} = \frac{1}{(1 - \rho^2)} \begin{bmatrix} 1/\sigma_{X_1}^2 & -\rho/\sigma_{X_1}\sigma_{X_2} \\ -\rho/\sigma_{X_1}\sigma_{X_2} & 1/\sigma_{X_2}^2 \end{bmatrix}$$

(5.3-17)

$$||C_X]^{-1}| = 1/\sigma_{X_1}^2\sigma_{X_2}^2(1 - \rho^2)$$

(5.3-18)

On substitution of (5.3-17) and (5.3-18) into (5.3-12), and letting $X_1 = X$ and $X_2 = Y$, it is easy to verify that the bivariate density of (5.3-1) results.

---

† We denote a matrix symbolically by use of heavy brackets $[\cdot]$.

⋆Some Properties of Gaussian Random Variables

We state without proof some of the properties exhibited by N jointly gaussian random variables $X_1, \ldots, X_N$.

1. Gaussian random variables are completely defined through only their first- and second-order moments; that is, by their means, variances, and covariances. This fact is readily apparent since only these quantities are needed to completely determine (5.3-12).
2. If the random variables are uncorrelated, they are also statistically independent. This property was given earlier for two variables.
3. Random variables produced by a linear transformation of $X_1, \ldots, X_N$ will also be gaussian, as proven in Section 5.5.
4. Any k-dimensional (k-variate) marginal density function obtained from the N-dimensional density function (5.3-12) by integrating out $N - k$ random variables will be gaussian. If the variables are ordered so that $X_1, \ldots, X_k$ occur in the marginal density and $X_{k+1}, \ldots, X_N$ are integrated out, then the covariance matrix of $X_1, \ldots, X_k$ is equal to the leading $k \times k$ submatrix of the covariance matrix of $X_1, \ldots, X_N$ (Wilks, 1962, p. 168).
5. The conditional density $f_{X_1, \ldots, X_k}(x_1, \ldots, x_k \mid X_{k+1} = x_{k+1}, \ldots, X_N = x_N)$ is gaussian (Papoulis, 1965, p. 257). This holds for any $k < N$.

⋆5.4 TRANSFORMATIONS OF MULTIPLE RANDOM VARIABLES

The function g in either (5.1-1) or (5.1-2) can be considered a transformation involving more than one random variable. By defining a new variable $Y = g(X_1, X_2, \ldots, X_N)$, we see that (5.1-2) is the expected value of Y. In calculating expected values it was not necessary to determine the density function of the new random variable Y. It may be, however, that the density function of Y is required in some practical problems, and its determination is briefly considered in this section.

In fact, one may be more generally interested in finding the joint density function for a set of new random variables

$$Y_i = T_i(X_1, X_2, \ldots, X_N) \qquad i = 1, 2, \ldots, N \qquad (5.4-1)$$

defined by functional transformations T_i. Now all the possible cases described in Chapter 3 for one random variable carry over to the N-dimensional problem. That is, the X_i can be continuous, discrete, or mixed, while the functions T_i can be linear, nonlinear, continuous, segmented, etc. Because so many cases are possible, many of them being beyond our scope, we shall discuss only one representative problem.

We shall assume that the new random variables Y_i, given by (5.4-1), are produced by single-valued continuous functions T_i having continuous partial derivatives everywhere. It is further assumed that a set of inverse continuous functions T_j^{-1} exists such that the old variables may be expressed as single-valued continuous functions of the new variables:

$$X_j = T_j^{-1}(Y_1, Y_2, \ldots, Y_N) \qquad j = 1, 2, \ldots, N \qquad (5.4\text{-}2)$$

These assumptions mean that a point in the joint sample space of the X_i maps into only one point in the space of the new variables Y_j.

Let R_X be a closed region of points in the space of the X_i and R_Y be the corresponding region of mapped points in the space of the Y_j, then the probability that a point falls in R_X will equal the probability that its mapped point falls in R_Y. These probabilities, in terms of joint densities, are given by

$$\int_{R_X} \cdots \int f_{X_1, \ldots, X_N}(x_1, \ldots, x_N)\, dx_1 \cdots dx_N$$

$$= \int_{R_Y} \cdots \int f_{Y_1, \ldots, Y_N}(y_1, \ldots, y_N)\, dy_1 \cdots dy_N \qquad (5.4\text{-}3)$$

This equation may be solved for $f_{Y_1, \ldots, Y_N}(y_1, \ldots, y_N)$ by treating it as simply a multiple integral involving a change of variables.

By working on the left side of (5.4-3) we change the variables x_i to new variables y_j by means of the variable changes (5.4-2). The integrand is changed by direct functional substitution. The limits change from the region R_X to the region R_Y. Finally, the differential hypervolume $dx_1 \cdots dx_N$ will change to the value $|J|\, dy_1 \cdots dy_N$ (Spiegel, 1963, p. 182), where $|J|$ is the magnitude of the jacobian† J of the transformations. The jacobian is the determinant of a matrix of derivatives defined by

$$J = \begin{vmatrix} \dfrac{\partial T_1^{-1}}{\partial Y_1} & \cdots & \dfrac{\partial T_1^{-1}}{\partial Y_N} \\ \vdots & & \vdots \\ \dfrac{\partial T_N^{-1}}{\partial Y_1} & \cdots & \dfrac{\partial T_N^{-1}}{\partial Y_N} \end{vmatrix} \qquad (5.4\text{-}4)$$

Thus, the left side of (5.4-3) becomes

$$\int_{R_X} \cdots \int f_{X_1, \ldots, X_N}(x_1, \ldots, x_N)\, dx_1 \cdots dx_N$$

$$= \int_{R_Y} \cdots \int f_{X_1, \ldots, X_N}(x_1 = T_1^{-1}, \ldots, x_N = T_N^{-1})|J|\, dy_1 \cdots dy_N \qquad (5.4\text{-}5)$$

---

† After the German mathematician Karl Gustav Jakob Jacobi (1804–1851).

Since this result must equal the right side of (5.4-3), we conclude that

$$f_{Y_1, \ldots, Y_N}(y_1, \ldots, y_N) = f_{X_1, \ldots, X_N}(x_1 = T_1^{-1}, \ldots, x_N = T_N^{-1})|J| \quad (5.4\text{-}6)$$

When $N = 1$, (5.4-6) reduces to (3.4-9) previously derived for a single random variable.

The solution (5.4-6) for the joint density of the new variables Y_j is illustrated here with an example.

Example 5.4-1 Let the transformations be linear and given by

$$Y_1 = T_1(X_1, X_2) = aX_1 + bX_2$$

$$Y_2 = T_2(X_1, X_2) = cX_1 + dX_2$$

where a, b, c, and d are real constants. The inverse functions are easy to obtain by solving these two equations for the two variables X_1 and X_2:

$$X_1 = T_1^{-1}(Y_1, Y_2) = (dY_1 - bY_2)/(ad - bc)$$

$$X_2 = T_2^{-1}(Y_1, Y_2) = (-cY_1 + aY_2)/(ad - bc)$$

where we shall assume $(ad - bc) \neq 0$. From (5.4-4):

$$J = \begin{vmatrix} d/(ad - bc) & -b/(ad - bc) \\ -c/(ad - bc) & a/(ad - bc) \end{vmatrix} = \frac{1}{(ad - bc)}$$

Finally, from (5.4-6),

$$f_{Y_1, Y_2}(y_1, y_2) = \frac{f_{X_1, X_2}\left(\dfrac{dy_1 - by_2}{ad - bc}, \dfrac{-cy_1 + ay_2}{ad - bc}\right)}{|ad - bc|}$$

★5.5 LINEAR TRANSFORMATION OF GAUSSIAN RANDOM VARIABLES

Equation (5.4-6) can be readily applied to the problem of linearly transforming a set of gaussian random variables $X_1, X_2, \ldots, X_N$ for which the joint density of (5.3-12) applies. The new variables $Y_1, Y_2, \ldots, Y_N$ are

$$Y_1 = a_{11}X_1 + a_{12}X_2 + \cdots + a_{1N}X_N$$

$$Y_2 = a_{21}X_1 + a_{22}X_2 + \cdots + a_{2N}X_N \quad (5.5\text{-}1)$$

$$\vdots$$

$$Y_N = a_{N1}X_1 + a_{N2}X_2 + \cdots + a_{NN}X_N$$

where the coefficients a_{ij}, i and $j = 1, 2, \ldots, N$, are real numbers. Now if we define the following matrices:

$$[T] = \begin{bmatrix} a_{11} & a_{12} & \cdots & a_{1N} \\ a_{21} & a_{22} & \cdots & a_{2N} \\ \vdots & \vdots & & \vdots \\ a_{N1} & a_{N2} & \cdots & a_{NN} \end{bmatrix} \tag{5.5-2}$$

$$[Y] = \begin{bmatrix} Y_1 \\ \vdots \\ Y_N \end{bmatrix} \quad [\bar{Y}] = \begin{bmatrix} \bar{Y}_1 \\ \vdots \\ \bar{Y}_N \end{bmatrix} \quad [X] = \begin{bmatrix} X_1 \\ \vdots \\ X_N \end{bmatrix} \quad [\bar{X}] = \begin{bmatrix} \bar{X}_1 \\ \vdots \\ \bar{X}_N \end{bmatrix}$$

$$\tag{5.5-3}$$

then it is clear from (5.5-1) that

$$[Y] = [T][X] \qquad [Y - \bar{Y}] = [T][X - \bar{X}] \tag{5.5-4}$$

$$[X] = [T]^{-1}[Y] \qquad [X - \bar{X}] = [T]^{-1}[Y - \bar{Y}] \tag{5.5-5}$$

so long as $[T]$ is nonsingular. Thus,

$$X_i = T_i^{-1}(Y_1, \ldots, Y_N) = a^{i1}Y_1 + a^{i2}Y_2 + \cdots + a^{iN}Y_N \tag{5.5-6}$$

$$\frac{\partial X_i}{\partial Y_j} = \frac{\partial T_i^{-1}}{\partial Y_j} = a^{ij} \tag{5.5-7}$$

$$X_i - \bar{X}_i = a^{i1}(Y_1 - \bar{Y}_1) + \cdots + a^{iN}(Y_N - \bar{Y}_N) \tag{5.5-8}$$

from (5.5-5). Here a^{ij} represents the ijth element of $[T]^{-1}$.

The density function of the new variables $Y_1, \ldots, Y_N$ is found by solving the right side of (5.4-6) in two steps. The first step is to determine $|J|$. By using (5.5-7) with (5.4-4) we find that J equals the determinant of the matrix $[T]^{-1}$. Hence,†

$$|J| = \|[T]^{-1}\| = \frac{1}{\|[T]\|} \tag{5.5-9}$$

The second step in solving (5.4-6) proceeds by using (5.5-8) to obtain

$$C_{X_iX_j} = E[(X_i - \bar{X}_i)(X_j - \bar{X}_j)] = \sum_{k=1}^{N} a^{ik} \sum_{m=1}^{N} a^{jm} E[(Y_k - \bar{Y}_k)(Y_m - \bar{Y}_m)]$$

$$= \sum_{k=1}^{N} a^{ik} \sum_{m=1}^{N} a^{jm} C_{Y_kY_m} \tag{5.5-10}$$

Since $C_{X_iX_j}$ is the ijth element in the covariance matrix $[C_X]$ of (5.3-12) and $C_{Y_kY_m}$

---

† We represent the magnitude of the determinant of a matrix by $\|[\cdot]\|$.

is the kmth element in the covariance matrix of the new variables Y_i, which we denote $[C_Y]$, (5.5-10) can be written in the form

$$[C_X] = [T]^{-1}[C_Y]([T]')^{-1} \tag{5.5-11}$$

Here $[T]'$ represents the transpose of $[T]$. The inverse of (5.5-11) is

$$[C_X]^{-1} = [T]'[C_Y]^{-1}[T] \tag{5.5-12}$$

which has a determinant

$$|[C_X]^{-1}| = |[C_Y]^{-1}| \, |[T]|^2 \tag{5.5-13}$$

On substitution of (5.5-13) and (5.5-12) into (5.3-12):

$$f_{X_1, \dots, X_N}(x_1 = T_1^{-1}, \dots, x_N = T_N^{-1})$$

$$= \frac{|[T]| \, |[C_Y]^{-1}|^{1/2}}{(2\pi)^{N/2}} \exp\left\{ -\frac{[x - \bar{X}]'[T]'[C_Y]^{-1}[T][x - \bar{X}]}{2} \right\} \tag{5.5-14}$$

Finally, (5.5-14) and (5.5-9) are substituted into (5.4-6), and (5.5-4) is used to obtain

$$f_{Y_1, \dots, Y_N}(y_1, \dots, y_N) = \frac{|[C_Y]^{-1}|^{1/2}}{(2\pi)^{N/2}} \exp\left\{ -\frac{[y - \bar{Y}]'[C_Y]^{-1}[y - \bar{Y}]}{2} \right\} \tag{5.5-15}$$

This result shows that the new random variables $Y_1, Y_2, \dots, Y_N$ are jointly gaussian because (5.5-15) is of the required form.

In summary, (5.5-15) shows that a linear transformation of gaussian random variables produces gaussian random variables. The new variables have mean values

$$\bar{Y}_j = \sum_{k=1}^{N} a_{jk} \bar{X}_k \tag{5.5-16}$$

from (5.5-1) and covariances given by the elements of the covariance matrix

$$[C_Y] = [T][C_X][T]' \tag{5.5-17}$$

as found from (5.5-11).

Example 5.5-1 Two gaussian random variables X_1 and X_2 have zero means and variances $\sigma_{X_1}^2 = 4$ and $\sigma_{X_2}^2 = 9$. Their covariance $C_{X_1 X_2}$ equals 3. If X_1 and X_2 are linearly transformed to new variables Y_1 and Y_2 according to

$$Y_1 = X_1 - 2X_2$$

$$Y_2 = 3X_1 + 4X_2$$

we use the above results to find the means, variances, and covariance of Y_1 and Y_2.

Here

$$[T] = \begin{bmatrix} 1 & -2 \\ 3 & 4 \end{bmatrix} \quad \text{and} \quad [C_X] = \begin{bmatrix} 4 & 3 \\ 3 & 9 \end{bmatrix}$$

Since X_1 and X_2 are zero-mean and gaussian, Y_1 and Y_2 will also be zero-mean and gaussian, thus $\bar{Y}_1 = 0$ and $\bar{Y}_2 = 0$. From (5.5-17):

$$[C_Y] = [T][C_X][T]^t = \begin{bmatrix} 1 & -2 \\ 3 & 4 \end{bmatrix} \begin{bmatrix} 4 & 3 \\ 3 & 9 \end{bmatrix} \begin{bmatrix} 1 & 3 \\ -2 & 4 \end{bmatrix} = \begin{bmatrix} 28 & -66 \\ -66 & 252 \end{bmatrix}$$

Thus, $\sigma_{Y_1}^2 = 28$, $\sigma_{Y_2}^2 = 252$, and $C_{Y_1 Y_2} = -66$.

---

*5.6 COMPLEX RANDOM VARIABLES

A *complex random variable* Z can be defined in terms of real random variables X and Y by

$$Z = X + jY \tag{5.6-1}$$

where $j = \sqrt{-1}$. In considering expected values involving Z, the joint density of X and Y must be used. For instance, if $g(\cdot)$ is some function (real or complex) of Z, the expected value of $g(Z)$ is obtained from

$$E[g(Z)] = \int_{-\infty}^{\infty} \int_{-\infty}^{\infty} g(z) f_{X, Y}(x, y) \, dx \, dy \tag{5.6-2}$$

Various important quantities such as the mean value and variance are obtained through application of (5.6-2). The mean value of Z is

$$\bar{Z} = E[Z] = E[X] + jE[Y] = \bar{X} + j\bar{Y} \tag{5.6-3}$$

The variance σ_Z^2 of Z is defined as the mean value of the function $g(Z) = |Z - E[Z]|^2$; that is,

$$\sigma_Z^2 = E[|Z - E[Z]|^2] \tag{5.6-4}$$

Equation (5.6-2) can be extended to include functions of two random variables

$$Z_m = X_m + jY_m \tag{5.6-5}$$

and

$$Z_n = X_n + jY_n \tag{5.6-6}$$

$n \neq m$, if expectation is taken with respect to four random variables X_m, Y_m, X_n, Y_n through their joint density function $f_{X_m, Y_m, X_n, Y_n}(x_m, y_m, x_n, y_n)$. If this density satisfies

$$f_{X_m, Y_m, X_n, Y_n}(x_m, y_m, x_n, y_n) = f_{X_m, Y_m}(x_m, y_m) f_{X_n, Y_n}(x_n, y_n) \tag{5.6-7}$$

then Z_m and Z_n are called *statistically independent*. The extension to N random variables is straightforward.

The *correlation* and *covariance* of Z_m and Z_n are defined by

$$R_{Z_m Z_n} = E[Z_m^* Z_n] \qquad n \neq m \qquad (5.6\text{-}8)$$

and

$$C_{Z_m Z_n} = E[\{Z_m - E[Z_m]\}^* \{Z_n - E[Z_n]\}] \qquad n \neq m \qquad (5.6\text{-}9)$$

respectively, where the superscripted asterisk* represents the complex conjugate. If the covariance is 0, Z_m and Z_n are said to be *uncorrelated random variables*. By setting (5.6-9) to 0, we find that

$$R_{Z_m Z_n} = E[Z_m^*]E[Z_n] \qquad m \neq n \qquad (5.6\text{-}10)$$

for uncorrelated random variables. Statistical independence is sufficient to guarantee that Z_m and Z_n are uncorrelated.

Finally, we note that two complex random variables are called *orthogonal* if their correlation, given by (5.6-8), equals 0.

PROBLEMS

5-1 Random variables X and Y have the joint density

$$f_{X,Y}(x, y) = \begin{cases} \dfrac{1}{24} & 0 < x < 6 \quad \text{and} \quad 0 < y < 4 \\ 0 & \text{elsewhere} \end{cases}$$

What is the expected value of the function $g(X, Y) = (XY)^2$?

5-2 Extend Problem 5-1 by finding the expected value of $g(X_1, X_2, X_3, X_4) = X_1^{n_1} X_2^{n_2} X_3^{n_3} X_4^{n_4}$, where $n_1, n_2, n_3,$ and n_4 are integers ≥ 0 and

$$f_{X_1, X_2, X_3, X_4}(x_1, x_2, x_3, x_4) = \begin{cases} \dfrac{1}{abcd} & 0 < x_1 < a \text{ and } 0 < x_2 < b \text{ and } 0 < x_3 < c \\ & \text{and } 0 < x_4 < d \\ 0 & \text{elsewhere} \end{cases}$$

5-3 The density function of two random variables X and Y is

$$f_{X,Y}(x, y) = u(x)u(y)16e^{-4(x+y)}$$

Find the mean value of the function

$$g(X, Y) = \begin{cases} 5 & 0 < X \leq \dfrac{1}{2} \quad \text{and} \quad 0 < Y \leq \dfrac{1}{2} \\ -1 & \dfrac{1}{2} < X \quad \text{and/or} \quad \dfrac{1}{2} < Y \\ 0 & \text{all other } X \text{ and } Y \end{cases}$$

5-4 For the random variables in Problem 5-3, find the mean value of the function

$$g(X, Y) = e^{-2(X^2 + Y^2)}$$

5-5 Three statistically independent random variables X_1, X_2, and X_3 have mean values $\bar{X}_1 = 3$, $\bar{X}_2 = 6$, and $\bar{X}_3 = -2$. Find the mean values of the following functions:

(a) $g(X_1, X_2, X_3) = X_1 + 3X_2 + 4X_3$
(b) $g(X_1, X_2, X_3) = X_1 X_2 X_3$
(c) $g(X_1, X_2, X_3) = -2X_1 X_2 - 3X_1 X_3 + 4X_2 X_3$
(d) $g(X_1, X_2, X_3) = X_1 + X_2 + X_3$

5-6 Find the mean value of the function

$$g(X, Y) = X^2 + Y^2$$

where X and Y are random variables defined by the density function

$$f_{X, Y}(x, y) = \frac{e^{-(x^2 + y^2)/2\sigma^2}}{2\pi\sigma^2}$$

with σ^2 a constant.

5-7 Two statistically independent random variables X and Y have mean values $\bar{X} = E[X] = 2$ and $\bar{Y} = E[Y] = 4$. They have second moments $\overline{X^2} = E[X^2] = 8$ and $\overline{Y^2} = E[Y^2] = 25$. Find:

(a) the mean value   (b) the second moment and
(c) the variance of the random variable $W = 3X - Y$.

5-8 Two random variables X and Y have means $\bar{X} = 1$ and $\bar{Y} = 2$, variances $\sigma_X^2 = 4$ and $\sigma_Y^2 = 1$, and a correlation coefficient $\rho_{XY} = 0.4$. New random variables W and V are defined by

$$V = -X + 2Y \qquad W = X + 3Y$$

Find:

(a) the means   (b) the variances   (c) the correlation and
(d) the correlation coefficient ρ_{VW} of V and W.

5-9 Two random variables X and Y are related by the expression

$$Y = aX + b$$

where a and b are any real numbers.

(a) Show that their correlation coefficient is

$$\rho = \begin{cases} 1 & \text{if } a > 0 \text{ for any } b \\ -1 & \text{if } a < 0 \text{ for any } b \end{cases}$$

(b) Show that their covariance is

$$C_{XY} = a\sigma_X^2$$

where σ_X^2 is the variance of X.

*5-10 Show that the correlation coefficient satisfies the expression

$$|\rho| = \frac{|\mu_{11}|}{\sqrt{\mu_{02}\,\mu_{20}}} \leq 1$$

5-11 Find all the second-order moments and central moments for the density function given in Problem 5-3.

5-12 Random variables X and Y have the joint density function

$$f_{X,Y}(x, y) = \begin{cases} (x + y)^2/40 & -1 < x < 1 \quad \text{and} \quad -3 < y < 3 \\ 0 & \text{elsewhere} \end{cases}$$

(a) Find all the second-order moments of X and Y.
(b) What are the variances of X and Y?
(c) What is the correlation coefficient?

5-13 Find all the third-order moments by using (5.1-5) for X and Y defined in Problem 5-12.

5-14 For discrete random variables X and Y, show that:
(a) Joint moments are

$$m_{nk} = \sum_{i=1}^{N} \sum_{j=1}^{M} P(x_i, y_j) x_i^n y_j^k$$

(b) Joint central moments are

$$\mu_{nk} = \sum_{i=1}^{N} \sum_{j=1}^{M} P(x_i, y_j)(x_i - \bar{X})^n (y_j - \bar{Y})^k$$

where $P(x_i, y_j) = P\{X = x_i, Y = y_j\}$, X has N possible values x_i, and Y has M possible values y_j.

5-15 For two random variables X and Y:

$$f_{X,Y}(x, y) = 0.15\delta(x + 1)\delta(y) + 0.1\delta(x)\delta(y) + 0.1\delta(x)\delta(y - 2) + 0.4\delta(x - 1)\delta(y + 2)$$
$$+ 0.2\delta(x - 1)\delta(y - 1) + 0.05\delta(x - 1)\delta(y - 3)$$

Find: (a) the correlation, (b) the covariance, and (c) the correlation coefficient of X and Y.
(d) Are X and Y either uncorrelated or orthogonal?

5-16 Discrete random variables X and Y have the joint density

$$f_{X,Y}(x, y) = 0.4\delta(x + \alpha)\delta(y - 2) + 0.3\delta(x - \alpha)\delta(y - 2)$$
$$+ 0.1\delta(x - \alpha)\delta(y - \alpha) + 0.2\delta(x - 1)\delta(y - 1)$$

Determine the value of α, if any, that minimizes the correlation between X and Y and find the minimum correlation. Are X and Y orthogonal?

5-17 For two discrete random variables X and Y:

$$f_{X, Y}(x, y) = 0.3\delta(x - \alpha)\delta(y - \alpha) + 0.5\delta(x + \alpha)\delta(y - 4) + 0.2\delta(x + 2)\delta(y + 2)$$

Determine the value of α, if any, that minimizes the covariance of X and Y. Find the minimum covariance. Are X and Y uncorrelated?

5-18 The density function

$$f_{X, Y}(x, y) = \begin{cases} \dfrac{xy}{9} & 0 < x < 2 \quad \text{and} \quad 0 < y < 3 \\ 0 & \text{elsewhere} \end{cases}$$

applies to two random variables X and Y.
    (a) Show, by use of (5.1-6) and (5.1-7), that X and Y are uncorrelated.
    (b) Show that X and Y are also statistically independent.

5-19 Two random variables X and Y have the density function

$$f_{X, Y}(x, y) = \begin{cases} \dfrac{2}{43}(x + 0.5y)^2 & 0 < x < 2 \quad \text{and} \quad 0 < y < 3 \\ 0 & \text{elsewhere} \end{cases}$$

    (a) Find all the first- and second-order moments.
    (b) Find the covariance.
    (c) Are X and Y uncorrelated?

5-20 Define random variables V and W by

$$V = X + aY$$
$$W = X - aY$$

where a is a real number and X and Y are random variables. Determine a in terms of moments of X and Y such that V and W are orthogonal.

★5-21 If X and Y in Problems 5-20 are gaussian, show that W and V are statistically independent if $a^2 = \sigma_X^2/\sigma_Y^2$, where σ_X^2 and σ_Y^2 are the variances of X and Y, respectively.

5-22 Three uncorrelated random variables X_1, X_2, and X_3 have means $\bar{X}_1 = 1$, $\bar{X}_2 = -3$, and $\bar{X}_3 = 1.5$ and second moments $E[X_1^2] = 2.5$, $E[X_2^2] = 11$, and $E[X_3^2] = 3.5$. Let $Y = X_1 - 2X_2 + 3X_3$ be a new random variable and find:
    (a) the mean value,   (b) the variance of Y.

5-23 Given $W = (aX + 3Y)^2$ where X and Y are zero-mean random variables with variances $\sigma_X^2 = 4$ and $\sigma_Y^2 = 16$. Their correlation coefficient is $\rho = -0.5$.
    (a) Find a value for the parameter a that minimizes the mean value of W.
    (b) Find the minimum mean value.

★5-24 Find the joint characteristic function for X and Y defined in Problem 5-3.

*5-25 Show that the joint characteristic function of N independent random variables X_i, having characteristic functions $\Phi_{X_i}(\omega_i)$ is

$$\Phi_{X_1, \ldots, X_N}(\omega_1, \ldots, \omega_N) = \prod_{i=1}^{N} \Phi_{X_i}(\omega_i)$$

*5-26 For N random variables, show that

$$|\Phi_{X_1, \ldots, X_N}(\omega_1, \ldots, \omega_N)| \leq \Phi_{X_1, \ldots, X_N}(0, \ldots, 0) = 1$$

*5-27 For two zero-mean gaussian random variables X and Y, show that their joint characteristic function is

$$\Phi_{X, Y}(\omega_1, \omega_2) = \exp \left\{ -\frac{1}{2}[\sigma_X^2 \omega_1^2 + 2\rho\sigma_X \sigma_Y \omega_1\omega_2 + \sigma_Y^2 \omega_2^2] \right\}$$

*5-28 Zero-mean gaussian random variables X and Y have variances $\sigma_X^2 = 3$ and $\sigma_Y^2 = 4$, respectively, and a correlation coefficient $\rho = -\frac{1}{4}$.

(a) Write an expression for the joint density function.

(b) Show that a rotation of coordinates through the angle given by (5.3-11) will produce new statistically independent random variables.

*5-29 Find the conditional density functions $f_X(x \mid Y = y)$ and $f_Y(y \mid X = x)$ applicable to two gaussian random variables X and Y defined by (5.3-1) and show that they are also gaussian.

*5-30 Zero-mean gaussian random variables X_1, X_2, and X_3 having a covariance matrix

$$[C_X] = \begin{bmatrix} 4 & 2.05 & 1.05 \\ 2.05 & 4 & 2.05 \\ 1.05 & 2.05 & 4 \end{bmatrix}$$

are transformed to new variables

$$Y_1 = 5X_1 + 2X_2 - X_3$$
$$Y_2 = -X_1 + 3X_2 + X_3$$
$$Y_3 = 2X_1 - X_2 + 2X_3$$

(a) Find the covariance matrix of Y_1, Y_2, and Y_3.

(b) Write an expression for the joint density function of Y_1, Y_2, and Y_3.

*5-31 A complex random variable Z is defined by

$$Z = \cos(X) + j \sin(Y)$$

where X and Y are independent real random variables uniformly distributed from $-\pi$ to π.

(a) Find the mean value of Z.

(b) Find the variance of Z.

ADDITIONAL PROBLEMS

5-32 Two random variables have a uniform density on a circular region defined by

$$f_{X, Y}(x, y) = \begin{cases} 1/\pi r^2 & x^2 + y^2 \leq r^2 \\ 0 & \text{elsewhere} \end{cases}$$

Find the mean value of the function $g(X, Y) = X^2 + Y^2$.

*__5-33__ Define the conditional expected value of a function $g(X, Y)$ of random variables X and Y as

$$E[g(X, Y)|B] = \int_{-\infty}^{\infty} \int_{-\infty}^{\infty} g(x, y) f_{X, Y}(x, y|B) \, dx \, dy$$

(a) If event B is defined as $B = \{y_a < Y \leq y_b\}$, where $y_a < y_b$ are constants, evaluate $E[g(X, Y)|B]$. (*Hint:* Use results of Problem 4-55.)

(b) If B is defined by $B = \{Y = y\}$ what does the conditional expected value of part (a) become?

5-34 For random variables X and Y having $\bar{X} = 1$, $\bar{Y} = 2$, $\sigma_X^2 = 6$, $\sigma_Y^2 = 9$, and $\rho = -\frac{2}{3}$, find (a) the covariance of X and Y, (b) the correlation of X and Y, and (c) the moments m_{20} and m_{02}.

5-35 $\bar{X} = \frac{1}{2}$, $\overline{X^2} = \frac{5}{2}$, $\bar{Y} = 2$, $\overline{Y^2} = \frac{19}{2}$, and $C_{XY} = -1/2\sqrt{3}$ for random variables X and Y.

(a) Find σ_X^2, σ_Y^2, R_{XY}, and ρ.

(b) What is the mean value of the random variable $W = (X + 3Y)^2 + 2X + 3$?

5-36 Let X and Y be statistically independent random variables with $\bar{X} = \frac{3}{4}$, $\overline{X^2} = 4$, $\bar{Y} = 1$, and $\overline{Y^2} = 5$. For a random variable $W = X - 2Y + 1$ find (a) R_{XY}, (b) R_{XW}, (c) R_{YW}, and (d) C_{XY}. (e) Are X and Y uncorrelated?

5-37 Statistically independent random variables X and Y have moments $m_{10} = 2$, $m_{20} = 14$, $m_{02} = 12$, and $m_{11} = -6$. Find the moment μ_{22}.

5-38 A joint density is given as

$$f_{X, Y}(x, y) = \begin{cases} x(y + 1.5) & 0 < x < 1 \quad \text{and} \quad 0 < y < 1 \\ 0 & \text{elsewhere} \end{cases}$$

Find all the joint moments m_{nk}, n and $k = 0, 1, \ldots$.

5-39 Find all the joint central moments μ_{nk}, n and $k = 0, 1, \ldots$, for the density of Problem 5-38.

*__5-40__ Find the joint characteristic function for random variables X and Y defined by

$$f_{X, Y}(x, y) = (1/2\pi) \text{ rect } (x/\pi) \text{ rect } [(x + y)/\pi] \cos (x + y)$$

Use the result to find the marginal characteristic functions of X and Y.

★5-41 Random variables X_1 and X_2 have the joint characteristic function

$$\Phi_{X_1, X_2}(\omega_1, \omega_2) = [(1 - j2\omega_1)(1 - j2\omega_2)]^{-N/2}$$

where $N > 0$ is an integer.
  (a) Find the correlation and moments m_{20} and m_{02}.
  (b) Determine the means of X_1 and X_2.
  (c) What is the correlation coefficient?

★5-42 The joint probability density of two discrete random variables X and Y consists of impulses located at all lattice points (mb, nd), where $m = 0, 1, \ldots, M$ and $n = 1, 2, \ldots, N$ with $b > 0$ and $d > 0$ being constants. All possible points are equally probable. Determine the joint characteristic function.

★5-43 Let X_k, $k = 1, 2, \ldots, K$, be statistically independent Poisson random variables, each with its own variance b_k (Problem 3-16). Show that the sum $X = X_1 + X_2 + \cdots + X_K$ is a Poisson random variable. (*Hint:* Use results of Problems 5-25 and 3-53.)

5-44 Assume $\sigma_X = \sigma_Y = \sigma$ in (5.3-1) and show that the locus of the maximum of the joint density is a line passing through the point $(\bar{X}, \bar{Y})$ with slope $\pi/4$ (or $-\pi/4$) when $\rho = 1$ (or -1).

5-45 Two gaussian random variables X and Y have variances $\sigma_X^2 = 9$ and $\sigma_Y^2 = 4$, respectively, and correlation coefficient ρ. It is known that a coordinate rotation by an angle $-\pi/8$ results in new random variables Y_1 and Y_2 that are uncorrelated. What is ρ?

★5-46 Let X and Y be jointly gaussian random variables where $\sigma_X^2 = \sigma_Y^2$ and $\rho = -1$. Find a transformation matrix such that new random variables Y_1 and Y_2 are statistically independent.

★5-47 Random variables X and Y having the joint density

$$f_{X, Y}(x, y) = (8/3)u(x - 2)u(y - 1)xy^2 \exp(4 - 2xy)$$

undergo a transformation

$$[T] = \begin{bmatrix} 1 & 1 \\ 1 & -1 \end{bmatrix}$$

to generate new random variables Y_1 and Y_2.
  (a) Find the joint density of Y_1 and Y_2.
  (b) Show what points in the $y_1 y_2$ plane correspond to a nonzero value of the new density.

★5-48 Equation (5.4-5) can sometimes be used to find the density of a single function of several random variables if *auxiliary random variables* are used. Apply the idea to finding the density function of $Z = aXY$, where a is a constant and X and Y are random variables, by defining the auxiliary variable $W = X$.

★5-49 Apply the method of Problem 5-48 to finding the density function of $Z = bY/X$, with b a constant, when using the auxiliary variable $W = X$.

*5-50 Two gaussian random variables X_1 and X_2 are defined by the mean and covariance matrices

$$[\bar{X}] = \begin{bmatrix} 2 \\ -1 \end{bmatrix} \quad [C_X] = \begin{bmatrix} 5 & -2/\sqrt{5} \\ -2/\sqrt{5} & 4 \end{bmatrix}$$

Two new random variables Y_1 and Y_2 are formed using the transformation

$$[T] = \begin{bmatrix} 1 & \frac{1}{2} \\ \frac{1}{2} & 1 \end{bmatrix}$$

Find matrices (a) $[\bar{Y}]$ and (b) $[C_Y]$. (c) Also find the correlation coefficient of Y_1 and Y_2.

*5-51 Complex random variables Z_1 and Z_2 have zero means. The correlation of the real parts of Z_1 and Z_2 is 4, while the correlation of the imaginary parts is 6. The real part of Z_1 and the imaginary part of Z_2 are statistically independent as a pair, as are the imaginary part of Z_1 and the real part of Z_2.

    (a) What is the correlation of Z_1 and Z_2?

    (b) Are Z_1 and Z_2 statistically independent?

SIX

RANDOM PROCESSES

6.0 INTRODUCTION

In the real world of engineering and science, it is necessary that we be able to deal with time waveforms. Indeed, we frequently encounter *random* time waveforms in practical systems. More often than not, a *desired* signal in some system is random. For example, the bit stream in a binary communication system is a random message because each bit in the stream occurs randomly. On the other hand, a desired signal is often accompanied by an *undesired* random waveform, noise. The noise interferes with the message and ultimately limits the performance of the system. Thus, any hope we have of determining the performance of systems with random waveforms hinges on our ability to describe and deal with such waveforms. In this chapter we introduce concepts that allow the description of random waveforms in a probabilistic sense.

6.1 THE RANDOM PROCESS CONCEPT

The concept of a random process is based on enlarging the random variable concept to include time. Since a random variable X is, by its definition, a function of the possible outcomes s of an experiment, it now becomes a function of both s and time. In other words, we assign, according to some rule, a time function

$$x(t, s) \tag{6.1-1}$$

to every outcome s. The family of all such functions, denoted $X(t, s)$, is called a *random process*. As with random variables where x was denoted as a specific value of the random variable X, we shall often use the convenient short-form notation $x(t)$ to represent a specific waveform of a random process denoted by $X(t)$.

Clearly, a random process $X(t, s)$ represents a family or *ensemble* of time functions when t and s are variables. Figure 6.1-1 illustrates a few members of an ensemble. Each member time function is called a *sample function, ensemble member*, or sometimes a *realization* of the process. Thus, a random process also represents a *single* time function when t is a variable and s is fixed at a specific value (outcome).

A random process also represents a random variable when t is fixed and s is a variable. For example, the random variable $X(t_1, s) = X(t_1)$ is obtained from the process when time is "frozen" at the value t_1. We often use the notation X_1 to denote the random variable associated with the process $X(t)$ at time t_1. X_1 corresponds to a vertical "slice" through the ensemble at time t_1, as illustrated in Figure 6.1-1. The statistical properties of $X_1 = X(t_1)$ describe the statistical properties of the random process at time t_1. The expected value of X_1 is called the *ensemble average* as well as the expected or mean value of the random process (at time t_1). Since t_1 may have various values, the mean value of a process may not be constant; in general, it may be a function of time. We easily visualize any

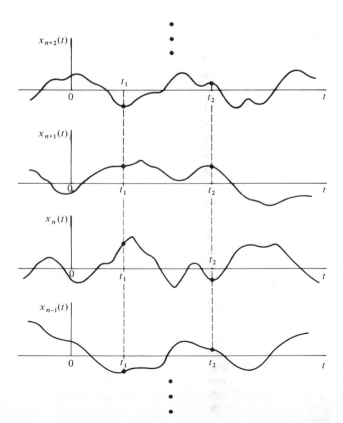

Figure 6.1-1 A continuous random process. [*Reproduced from Peebles (1976) with permission of publishers Addison-Wesley, Advanced Book Program.*]

number of random variables X_i derived from a random process $X(t)$ at times t_i, $i = 1, 2, \ldots$:

$$X_i = X(t_i, s) = X(t_i) \qquad (6.1\text{-}2)$$

A random process can also represent a mere number when t and s are both fixed.

Classification of Processes

It is convenient to classify random processes according to the characteristics of t and the random variable $X = X(t)$ at time t. We shall consider only four cases based on t and X having values in the ranges $-\infty < t < \infty$ and $-\infty < x < \infty$.†

---

† Other cases can be defined based on a definition of random processes on a finite time interval (see for example: Rosenblatt (1974), p. 91; Prabhu (1965), p. 1; Miller (1974), p. 31; Parzen (1962), p. 7; Dubes (1968), p. 320; Ross (1972), p. 56). Other recent texts on random processes are Helstrom (1984), and Gray and Davisson (1986).

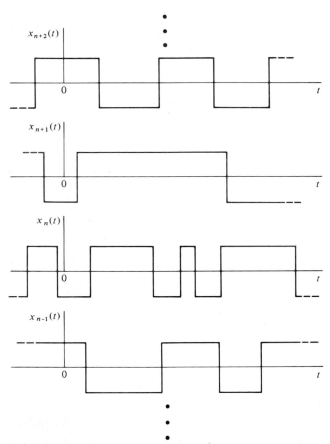

Figure 6.1-2 A discrete random process formed by heavily limiting the waveforms of Figure 6.1-1. [*Reproduced from Peebles (1976) with permission of publishers Addison-Wesley, Advanced Book Program.*]

If X is continuous and t can have any of a continuum of values, then $X(t)$ is called a *continuous random process*. Figure 6.1-1 is an illustration of this class of process. Thermal noise generated by any realizable network is a practical example of a waveform that is modeled as a sample function of a continuous random process. In this example, the network is the outcome in the underlying random experiment of selecting a network. (The presumption is that many networks are available from which to choose; this may not be the case in the real world, but it should not prevent us from imagining a production line producing any number of similar networks.) Each network establishes a sample function, and all sample functions form the process.†

A second class of random process, called a *discrete random process*, corresponds to the random variable X having only discrete values while t is continuous. Figure 6.1-2 illustrates such a process derived by heavily limiting the sample functions shown in Figure 6.1-1. The sample functions have only two dis-

---

† Note that finding the mean value of the process at any time t is equivalent to finding the average voltage that would be produced by all the various networks at time t.

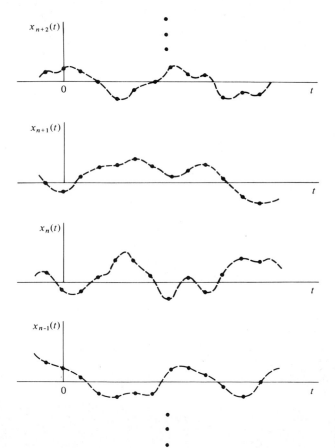

Figure 6.1-3 A continuous random sequence formed by sampling the waveforms of Figure 6.1-1. [*Reproduced from Peebles (1976), with permission of publishers Addison-Wesley, Advanced Book Program.*]

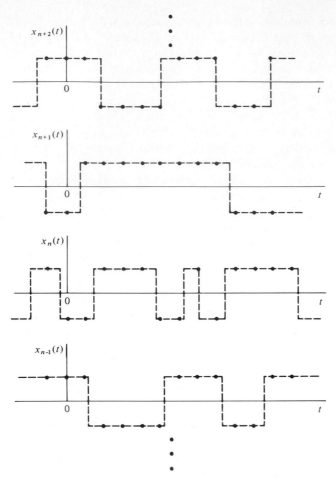

Figure 6.1-4 A discrete random sequence formed by sampling the waveforms of Figure 6.1-2. [*Adapted from Peebles (1976) with permission of publishers Addison-Wesley, Advanced Book Program.*]

crete values: the positive level is generated whenever a sample function in Figure 6.1-1 is positive and the negative level occurs for other times.

A random process for which X is continuous but time has only discrete values is called a *continuous random sequence* (Thomas, 1969, p. 80). Such a sequence can be formed by periodically sampling the ensemble members of Figure 6.1-1. The result is illustrated in Figure 6.1-3.

A fourth class of random process, called a *discrete random sequence,* corresponds to both time and the random variable being discrete. Figure 6.1-4 illustrates a discrete random sequence developed by sampling the sample functions of Figure 6.1-2.

In this text we are concerned almost entirely with discrete and continuous random processes.

Deterministic and Nondeterministic Processes

In addition to the classes described above, a random process can be described by the form of its sample functions. If future values of any sample function cannot be predicted exactly from observed past values, the process is called *nondeterministic.* The process of Figure 6.1-1 is one example.

A process is called *deterministic* if future values of any sample function can be predicted from past values. An example is the random process defined by

$$X(t) = A \cos (\omega_0 t + \Theta) \tag{6.1-3}$$

Here A, Θ, or ω_0 (or all) may be random variables. Any one sample function corresponds to (6.1-3) with particular values of these random variables. Therefore, knowledge of the sample function prior to any time instant automatically allows prediction of the sample function's future values because its form is known.

6.2 STATIONARITY AND INDEPENDENCE

As previously stated, a random process becomes a random variable when time is fixed at some particular value. The random variable will possess statistical properties, such as a mean value, moments, variance, etc., that are related to its density function. If *two* random variables are obtained from the process for two time instants, they will have statistical properties (means, variances, joint moments, etc.) related to their joint density function. More generally, N random variables will possess statistical properties related to their N-dimensional joint density function.

Broadly speaking, a random process is said to be *stationary* if all its statistical properties do not change with time. Other processes are called *nonstationary*. These statements are not intended as definitions of stationarity but are meant to convey only a general meaning. More concrete definitions follow. Indeed, there are several "levels" of stationarity, all of which depend on the density functions of the random variables of the process.

Distribution and Density Functions

To define stationarity, we must first define distribution and density functions as they apply to a random process $X(t)$. For a particular time t_1, the distribution function associated with the random variable $X_1 = X(t_1)$ will be denoted $F_X(x_1; t_1)$. It is defined as†

$$F_X(x_1; t_1) = P\{X(t_1) \leq x_1\} \tag{6.2-1}$$

† $F_X(x_1; t_1)$ is known as the *first-order distribution function* of the process $X(t)$.

for any real number x_1. This is the same definition used all along for the distribution function of one random variable. Only the notation has been altered to reflect the fact that it is possibly now a function of time choice t_1.

For two random variables $X_1 = X(t_1)$ and $X_2 = X(t_2)$, the *second-order joint distribution function* is the two-dimensional extension of (6.2-1):

$$F_X(x_1, x_2; t_1, t_2) = P\{X(t_1) \le x_1, X(t_2) \le x_2\} \tag{6.2-2}$$

In a similar manner, for N random variables $X_i = X(t_i)$, $i = 1, 2, \ldots, N$, the *Nth-order joint distribution function* is

$$F_X(x_1, \ldots, x_N; t_1, \ldots, t_N) = P\{X(t_1) \le x_1, \ldots, X(t_N) \le x_N\} \tag{6.2-3}$$

Joint density functions of interest are found from appropriate derivatives of the above three relationships:†

$$f_X(x_1; t_1) = dF_X(x_1; t_1)/dx_1 \tag{6.2-4}$$

$$f_X(x_1, x_2; t_1, t_2) = \partial^2 F_X(x_1, x_2; t_1, t_2)/(\partial x_1 \, \partial x_2) \tag{6.2-5}$$

$$f_X(x_1, \ldots, x_N; t_1, \ldots, t_N) = \partial^N F_X(x_1, \ldots, x_N; t_1, \ldots, t_N)/(\partial x_1 \cdots \partial x_N) \tag{6.2-6}$$

Statistical Independence

Two processes $X(t)$ and $Y(t)$ are *statistically independent* if the random variable group $X(t_1), X(t_2), \ldots, X(t_N)$ is independent of the group $Y(t_1'), Y(t_2'), \ldots, Y(t_M')$ for any choice of times $t_1, t_2, \ldots, t_N, t_1', t_2', \ldots, t_M'$. Independence requires that the joint density be factorable by groups:

$$f_{X, Y}(x_1, \ldots, x_N, y_1, \ldots, y_M; t_1, \ldots, t_N, t_1', \ldots, t_M')$$
$$= f_X(x_1, \ldots, x_N; t_1, \ldots, t_N) f_Y(y_1, \ldots, y_M; t_1', \ldots, t_M') \tag{6.2-7}$$

First-Order Stationary Processes

A random process is called *stationary to order one* if its first-order density function does not change with a shift in time origin. In other words

$$f_X(x_1; t_1) = f_X(x_1; t_1 + \Delta) \tag{6.2-8}$$

must be true for any t_1 and any real number Δ if $X(t)$ is to be a first-order stationary process.

Consequences of (6.2-8) are that $f_X(x_1; t_1)$ is independent of t_1 and the process mean value $E[X(t)]$ is a constant:

$$E[X(t)] = \bar{X} = \text{constant} \tag{6.2-9}$$

---

† Analogous to distribution functions, these are called *first-*, *second-*, and *Nth-order density functions*, respectively.

To prove (6.2-9), we find mean values of the random variables $X_1 = X(t_1)$ and $X_2 = X(t_2)$. For X_1:

$$E[X_1] = E[X(t_1)] = \int_{-\infty}^{\infty} x_1 f_X(x_1; t_1) \, dx_1 \qquad (6.2\text{-}10)$$

For X_2:

$$E[X_2] = E[X(t_2)] = \int_{-\infty}^{\infty} x_1 f_X(x_1; t_2) \, dx_1 \dagger \qquad (6.2\text{-}11)$$

Now by letting $t_2 = t_1 + \Delta$ in (6.2-11), substituting (6.2-8), and using (6.2-10), we get

$$E[X(t_1 + \Delta)] = E[X(t_1)] \qquad (6.2\text{-}12)$$

which must be a constant because t_1 and Δ are arbitrary.

Second-Order and Wide-Sense Stationarity

A process is called *stationary to order two* if its second-order density function satisfies

$$f_X(x_1, x_2; t_1, t_2) = f_X(x_1, x_2; t_1 + \Delta, t_2 + \Delta) \qquad (6.2\text{-}13)$$

for all t_1, t_2, and Δ. After some thought, the reader will conclude that (6.2-13) is a function of time differences $t_2 - t_1$ and not absolute time (let arbitrary $\Delta = -t_1$). A second-order stationary process is also first-order stationary because the second-order density function determines the lower, first-order, density.

Now the correlation $E[X_1 X_2] = E[X(t_1)X(t_2)]$ of a random process will, in general, be a function of t_1 and t_2. Let us denote this function by $R_{XX}(t_1, t_2)$ and call it the *autocorrelation function* of the random process $X(t)$:

$$R_{XX}(t_1, t_2) = E[X(t_1)X(t_2)]. \qquad (6.2\text{-}14)$$

A consequence of (6.2-13), however, is that the autocorrelation function of a second-order stationary process is a function only of time differences and not absolute time; that is, if

$$\tau = t_2 - t_1 \qquad (6.2\text{-}15)$$

then (6.2-14) becomes

$$R_{XX}(t_1, t_1 + \tau) = E[X(t_1)X(t_1 + \tau)] = R_{XX}(\tau) \qquad (6.2\text{-}16)$$

Proof of (6.2-16) uses (6.2-13); it is left as a reader exercise (see Problem 6-6).

Many practical problems require that we deal with the autocorrelation function and mean value of a random process. Problem solutions are greatly

---

† Note that the variable x_2 of integration has been replaced by the alternative variable x_1 for convenience.

simplified if these quantities are not dependent on absolute time. Of course, second-order stationarity is sufficient to guarantee these characteristics. However, it is often more restrictive than necessary, and a more relaxed form of stationarity is desirable. The most useful form is the *wide-sense stationary process*, defined as that for which two conditions are true:

$$E[X(t)] = \bar{X} = \text{constant} \tag{6.2-17a}$$

$$E[X(t)X(t + \tau)] = R_{XX}(\tau) \tag{6.2-17b}$$

A process stationary to order 2 is clearly wide-sense stationary. However, the converse is not necessarily true.

Example 6.2-1 We show that the random process

$$X(t) = A \cos(\omega_0 t + \Theta)$$

is wide-sense stationary if it is assumed that A and ω_0 are constants and Θ is a uniformly distributed random variable on the interval $(0, 2\pi)$. The mean value is

$$E[X(t)] = \int_0^{2\pi} A \cos(\omega_0 t + \theta) \frac{1}{2\pi} d\theta = 0$$

The autocorrelation function, from (6.2-14) with $t_1 = t$ and $t_2 = t + \tau$, becomes

$$R_{XX}(t, t + \tau) = E[A \cos(\omega_0 t + \Theta) A \cos(\omega_0 t + \omega_0 \tau + \Theta)]$$

$$= \frac{A^2}{2} E[\cos(\omega_0 \tau) + \cos(2\omega_0 t + \omega_0 \tau + 2\Theta)]$$

$$= \frac{A^2}{2} \cos(\omega_0 \tau) + \frac{A^2}{2} E[\cos(2\omega_0 t + \omega_0 \tau + 2\Theta)]$$

The second term easily evaluates to 0. Thus, the autocorrelation function depends only on τ and the mean value is a constant, so $X(t)$ is wide-sense stationary.

When we are concerned with two random processes $X(t)$ and $Y(t)$, we say they are *jointly* wide-sense stationary if each satisfies (6.2-17) and their *cross-correlation function*, defined in general by

$$R_{XY}(t_1, t_2) = E[X(t_1)Y(t_2)] \tag{6.2-18}$$

is a function only of time difference $\tau = t_2 - t_1$ and not absolute time; that is, if

$$R_{XY}(t, t + \tau) = E[X(t)Y(t + \tau)] = R_{XY}(\tau) \tag{6.2-19}$$

N-Order and Strict-Sense Stationarity

By extending the above reasoning to N random variables $X_i = X(t_i)$, $i = 1, 2, \ldots,$ N, we say a random process is *stationary to order N* if its Nth-order density function is invariant to a time origin shift; that is, if

$$f_X(x_1, \ldots, x_N; t_1, \ldots, t_N) = f_X(x_1, \ldots, x_N; t_1 + \Delta, \ldots, t_N + \Delta) \quad (6.2\text{-}20)$$

for all $t_1, \ldots, t_N$ and Δ. Stationarity of order N implies stationarity to all orders $k \leq N$. A process stationary to *all* orders $N = 1, 2, \ldots,$ is called *strict-sense stationary*.

Time Averages and Ergodicity

The time average of a quantity is defined as

$$A[\cdot] = \lim_{T \to \infty} \frac{1}{2T} \int_{-T}^{T} [\cdot] \, dt \quad (6.2\text{-}21)$$

Here A is used to denote time average in a manner analogous to E for the statistical average. Time average is taken over all time because, as applied to random processes, sample functions of processes are presumed to exist for all time.

Specific averages of interest are the mean value $\bar{x} = A[x(t)]$ of a sample function (a lower case letter is used to imply a sample function), and the *time autocorrelation function*, denoted $\mathcal{R}_{xx}(\tau) = A[x(t)x(t + \tau)]$. These functions are defined by

$$\bar{x} = A[x(t)] = \lim_{T \to \infty} \frac{1}{2T} \int_{-T}^{T} x(t) \, dt \quad (6.2\text{-}22)$$

$$\mathcal{R}_{xx}(\tau) = A[x(t)x(t + \tau)]$$

$$= \lim_{T \to \infty} \frac{1}{2T} \int_{-T}^{T} x(t)x(t + \tau) \, dt \quad (6.2\text{-}23)$$

For any *one* sample function of the process $X(t)$, these last two integrals simply produce two numbers (for a fixed value of τ). However, when all sample functions are considered, we see that $\bar{x}$ and $\mathcal{R}_{xx}(\tau)$ are actually *random variables*. By taking the expected value on both sides of (6.2-22) and (6.2-23), and assuming the expectation can be brought inside the integrals, we obtain†

$$E[\bar{x}] = \bar{X} \quad (6.2\text{-}24)$$

$$E[\mathcal{R}_{xx}(\tau)] = R_{XX}(\tau) \quad (6.2\text{-}25)$$

Now suppose by some theorem the random variables $\bar{x}$ and $\mathcal{R}_{xx}(\tau)$ could be made to have zero variances; that is, $\bar{x}$ and $\mathcal{R}_{xx}(\tau)$ actually become constants.

---

† We assume also that $X(t)$ is a stationary process so that the mean and the autocorrelation function are not time-dependent.

Then we could write

$$\bar{x} = \bar{X} \tag{6.2-26}$$

$$\mathcal{R}_{xx}(\tau) = R_{XX}(\tau) \tag{6.2-27}$$

In other words, the time averages $\bar{x}$ and $\mathcal{R}_{xx}(\tau)$ equal the statistical averages $\bar{X}$ and $R_{XX}(\tau)$ respectively. The *ergodic theorem* allows the validity of (6.2-26) and (6.2-27). Stated in loose terms, it more generally allows all time averages to equal the corresponding statistical averages. Processes that satisfy the ergodic theorem are called *ergodic processes*.

Ergodicity is a very restrictive form of stationarity and it may be difficult to prove that it constitutes a reasonable assumption in any physical situation. Nevertheless, we shall often assume a process is ergodic to simplify problems. In the real world, we are usually forced to work with only one sample function of a process and therefore must, like it or not, derive mean value, correlation functions, etc. from the time waveform. By assuming ergodicity, we may infer the similar statistical characteristics of the process. The reader may feel that our theory is on shaky ground based on these comments. However, it must be remembered that all our theory only serves to model real-world conditions. Therefore, what difference do our assumptions really make provided the assumed model does truly reflect real conditions?

Two random processes are called *jointly ergodic* if they are individually ergodic and also have a *time cross-correlation function* that equals the statistical cross-correlation function:†

$$\mathcal{R}_{xy}(\tau) = \lim_{T \to \infty} \frac{1}{2T} \int_{-T}^{T} x(t)y(t + \tau) \, dt = R_{XY}(\tau) \tag{6.2-28}$$

6.3 CORRELATION FUNCTIONS

The autocorrelation and cross-correlation functions were introduced in the previous section. These functions are examined further in this section, along with their properties. In addition, other correlation-type functions are introduced that are important to the study of random processes.

Autocorrelation Function and Its Properties

Recall that the autocorrelation function of a random process $X(t)$ is the correlation $E[X_1 X_2]$ of two random variables $X_1 = X(t_1)$ and $X_2 = X(t_2)$ defined by the process at times t_1 and t_2. Mathematically,

$$R_{XX}(t_1, t_2) = E[X(t_1)X(t_2)] \tag{6.3-1}$$

---

† As in ordinary stationarity, there are various *orders* of ergodic stationarity. For more detail on ergodic processes, the reader is referred to Papoulis (1965), pp. 323–332.

For time assignments, $t_1 = t$ and $t_2 = t_1 + \tau$, with τ a real number, (6.3-1) assumes the convenient form

$$R_{XX}(t, t + \tau) = E[X(t)X(t + \tau)] \qquad (6.3-2)$$

If $X(t)$ is at least wide-sense stationary, it was noted in Section 6.2 that $R_{XX}(t, t + \tau)$ must be a function only of time difference $\tau = t_2 - t_1$. Thus, for wide-sense stationary processes

$$R_{XX}(\tau) = E[X(t)X(t + \tau)] \qquad (6.3-3)$$

For such processes the autocorrelation function exhibits the following properties:

$$(1) \quad |R_{XX}(\tau)| \leq R_{XX}(0) \qquad (6.3-4)$$

$$(2) \quad R_{XX}(-\tau) = R_{XX}(\tau) \qquad (6.3-5)$$

$$(3) \quad R_{XX}(0) = E[X^2(t)] \qquad (6.3-6)$$

The first property shows that $R_{XX}(\tau)$ is bounded by its value at the origin, while the third property states that this bound is equal to the mean-squared value called the *power* in the process. The second property indicates that an autocorrelation function has even symmetry.

Other properties of stationary processes may also be stated [see Cooper and McGillem (1971), p. 113, and Melsa and Sage (1973), pp. 207–208]:

(4) If $E[X(t)] = \bar{X} \neq 0$ and $X(t)$ has no periodic components then

$$\lim_{|\tau| \to \infty} R_{XX}(\tau) = \bar{X}^2 \qquad (6.3-7)$$

(5) If $X(t)$ has a periodic component, then $R_{XX}(\tau)$ will have a periodic component with the same period. (6.3-8)

(6) If $X(t)$ is ergodic, zero-mean, and has no periodic component, then

$$\lim_{|\tau| \to \infty} R_{XX}(\tau) = 0 \qquad (6.3-9)$$

(7) $R_{XX}(\tau)$ cannot have an arbitrary shape. (6.3-10)

Properties 4 through 6 are more or less self-explanatory. Property 7 simply says that any arbitrary function cannot be an autocorrelation function. This fact will be more apparent when the *power density spectrum* is introduced in Chapter 7. It will be shown there that $R_{XX}(\tau)$ is related to the power density spectrum through the Fourier transform and the form of the spectrum is not arbitrary.

Example 6.3-1 Given the autocorrelation function for a stationary process is

$$R_{XX}(\tau) = 25 + \frac{4}{1 + 6\tau^2}$$

we shall find the mean value and variance of the process $X(t)$. From property

4, the mean value is $E[X(t)] = \bar{X} = \sqrt{25} = \pm 5$. The variance is given by (3.2-6), so

$$\sigma_X^2 = E[X^2(t)] - (E[X(t)])^2$$

But $E[X^2(t)] = R_{XX}(0) = 25 + 4 = 29$ from property 3, so

$$\sigma_X^2 = 29 - 25 = 4$$

Cross-Correlation Function and Its Properties

The cross-correlation function of two random processes $X(t)$ and $Y(t)$ was defined in (6.2-18). Setting $t_1 = t$ and $\tau = t_2 - t_1$, we may write (6.2-18) as

$$R_{XY}(t, t + \tau) = E[X(t)Y(t + \tau)] \tag{6.3-11}$$

If $X(t)$ and $Y(t)$ are at least jointly wide-sense stationary, $R_{XY}(t, t + \tau)$ is independent of absolute time and we can write

$$R_{XY}(\tau) = E[X(t)Y(t + \tau)] \tag{6.3-12}$$

If

$$R_{XY}(t, t + \tau) = 0 \tag{6.3-13}$$

then $X(t)$ and $Y(t)$ are called *orthogonal processes*. If the two processes are statistically independent, the cross-correlation function becomes

$$R_{XY}(t, t + \tau) = E[X(t)]E[Y(t + \tau)] \tag{6.3-14}$$

If, in addition to being independent, $X(t)$ and $Y(t)$ are at least wide-sense stationary, (6.3-14) becomes

$$R_{XY}(\tau) = \bar{X}\bar{Y} \tag{6.3-15}$$

which is a constant.

We may list some properties of the cross-correlation function applicable to processes that are at least wide-sense stationary:

$$(1) \quad R_{XY}(-\tau) = R_{YX}(\tau) \tag{6.3-16}$$

$$(2) \quad |R_{XY}(\tau)| \leq \sqrt{R_{XX}(0)R_{YY}(0)} \tag{6.3-17}$$

$$(3) \quad |R_{XY}(\tau)| \leq \tfrac{1}{2}[R_{XX}(0) + R_{YY}(0)] \tag{6.3-18}$$

Property 1 follows from the definition (6.3-12). It describes the symmetry of $R_{XY}(\tau)$. Property 2 can be proven by expanding the inequality

$$E[\{Y(t + \tau) + \alpha X(t)\}^2] \geq 0 \tag{6.3-19}$$

where α is a real number (see Problem 6-27). Properties 2 and 3 both constitute bounds on the magnitude of $R_{XY}(\tau)$. Equation (6.3-17) represents a tighter bound than that of (6.3-18), because the geometric mean of two positive numbers cannot exceed their arithmetic mean; that is

$$\sqrt{R_{XX}(0)R_{YY}(0)} \leq \tfrac{1}{2}[R_{XX}(0) + R_{YY}(0)] \tag{6.3-20}$$

Example 6.3-2 Let two random processes $X(t)$ and $Y(t)$ be defined by

$$X(t) = A \cos (\omega_0 t) + B \sin (\omega_0 t)$$

$$Y(t) = B \cos (\omega_0 t) - A \sin (\omega_0 t)$$

where A and B are random variables and ω_0 is a constant. It can be shown (Problem 6-12) that $X(t)$ is wide-sense stationary if A and B are uncorrelated, zero-mean random variables with the same variance (they may have different density functions, however). With these same constraints on A and B, $Y(t)$ is also wide-sense stationary. We shall now find the cross-correlation function $R_{XY}(t, t + \tau)$ and show that $X(t)$ and $Y(t)$ are *jointly* wide-sense stationary. By use of (6.3-11) we have

$$
\begin{aligned}
R_{XY}(t, t + \tau) &= E[X(t)Y(t + \tau)] \\
&= E[AB \cos (\omega_0 t) \cos (\omega_0 t + \omega_0 \tau) \\
&\quad + B^2 \sin (\omega_0 t) \cos (\omega_0 t + \omega_0 \tau) \\
&\quad - A^2 \cos (\omega_0 t) \sin (\omega_0 t + \omega_0 \tau) \\
&\quad - AB \sin (\omega_0 t) \sin (\omega_0 t + \omega_0 \tau)] \\
&= E[AB] \cos (2\omega_0 t + \omega_0 \tau) \\
&\quad + E[B^2] \sin (\omega_0 t) \cos (\omega_0 t + \omega_0 \tau) \\
&\quad - E[A^2] \cos (\omega_0 t) \sin (\omega_0 t + \omega_0 \tau)
\end{aligned}
$$

Since A and B are assumed to be zero-mean, uncorrelated random variables, $E[AB] = 0$. Also, since A and B are assumed to have equal variances, $E[A^2] = E[B^2] = \sigma^2$ and we obtain

$$R_{XY}(t, t + \tau) = -\sigma^2 \sin (\omega_0 \tau)$$

Thus, $X(t)$ and $Y(t)$ are jointly wide-sense stationary because $R_{XY}(t, t + \tau)$ depends only on τ.

Note from the above result that cross-correlation functions are not necessarily even functions of τ with the maximum at $\tau = 0$, as is the case with autocorrelation functions.

Covariance Functions

The concept of the covariance of two random variables, as defined by (5.1-13), can be extended to random processes. The *autocovariance function* is defined by

$$C_{XX}(t, t + \tau) = E[\{X(t) - E[X(t)]\}\{X(t + \tau) - E[X(t + \tau)]\}] \tag{6.3-21}$$

which can also be put in the form

$$C_{XX}(t, t + \tau) = R_{XX}(t, t + \tau) - E[X(t)]E[X(t + \tau)] \qquad (6.3-22)$$

The *cross-covariance function* for two processes $X(t)$ and $Y(t)$ is defined by

$$C_{XY}(t, t + \tau) = E[\{X(t) - E[X(t)]\}\{Y(t + \tau) - E[Y(t + \tau)]\}] \qquad (6.3-23)$$

or, alternatively,

$$C_{XY}(t, t + \tau) = R_{XY}(t, t + \tau) - E[X(t)]E[Y(t + \tau)] \qquad (6.3-24)$$

For processes that are at least jointly wide-sense stationary, (6.3-22) and (6.3-24) reduce to

$$C_{XX}(\tau) = R_{XX}(\tau) - \bar{X}^2 \qquad (6.3-25)$$

and

$$C_{XY}(\tau) = R_{XY}(\tau) - \bar{X}\bar{Y} \qquad (6.3-26)$$

The *variance* of a random process is given in general by (6.3-21) with $\tau = 0$. For a wide-sense stationary process, variance does not depend on time and is given by (6.3-25) with $\tau = 0$:

$$\sigma_X^2 = E[\{X(t) - E[X(t)]\}^2] = R_{XX}(0) - \bar{X}^2 \qquad (6.3-27)$$

For two random processes, if

$$C_{XY}(t, t + \tau) = 0 \qquad (6.3-28)$$

they are called *uncorrelated*. From (6.3-24) this means that

$$R_{XY}(t, t + \tau) = E[X(t)]E[Y(t + \tau)] \qquad (6.3-29)$$

Since this result is the same as (6.3-14), which applies to independent processes, we conclude that independent processes are uncorrelated. The converse case is not necessarily true, although it is true for *jointly gaussian processes*, which we consider in Section 6.5.

6.4 MEASUREMENT OF CORRELATION FUNCTIONS

In the real world, we can never measure the true correlation functions of two random processes $X(t)$ and $Y(t)$ because we never have *all* sample functions of the ensemble at our disposal. Indeed, we may typically have available for measurements only a portion of one sample function from each process. Thus, our only recourse is to determine time averages based on finite time portions of single sample functions, taken large enough to approximate true results for ergodic processes. Because we are able to work only with time functions, we are forced, like it or not, to presume that given processes are ergodic. This fact should not prove too disconcerting, however, if we remember that assumptions only reflect the details of our mathematical model of a real-world situation. Provided that the model gives consistent agreement with the real situation, it is of little importance whether ergodicity is assumed or not.

Figure 6.4-1 A time cross-correlation function measurement system. Autocorrelation function measurement is possible by connecting points A and B and applying either $x(t)$ or $y(t)$.

Figure 6.4-1 illustrates the block diagram of a possible system for measuring the approximate time cross-correlation function of two jointly ergodic random processes $X(t)$ and $Y(t)$. Sample functions $x(t)$ and $y(t)$ are delayed by amounts T and $T - \tau$, respectively, and the product of the delayed waveforms is formed. This product is then integrated to form the output which equals the integral at time $t_1 + 2T$, where t_1 is arbitrary and $2T$ is the integration period. The integrator can be of the integrate-and-dump variety described by Peebles (1976, p. 361).

If we assume $x(t)$ and $y(t)$ exist at least during the interval $-T < t$ and t_1 is an arbitrary time except $0 \leq t_1$, then the output is easily found to be

$$R_o(t_1 + 2T) = \frac{1}{2T} \int_{t_1 - T}^{t_1 + T} x(t)y(t + \tau)\, dt \tag{6.4-1}$$

Now if we choose $t_1 = 0$† and assume T is large, then we have

$$R_o(2T) = \frac{1}{2T} \int_{-T}^{T} x(t)y(t + \tau)\, dt \approx \mathcal{R}_{xy}(\tau) = R_{XY}(\tau) \tag{6.4-2}$$

Thus, for jointly ergodic processes, the system of Figure 6.4-1 can approximately measure their cross-correlation function (τ is varied to obtain the complete function).

Clearly, by connecting points A and B and applying either $x(t)$ or $y(t)$ to the system, we can also measure the autocorrelation functions $R_{XX}(\tau)$ and $R_{YY}(\tau)$.

Example 6.4-1 We connect points A and B together in Figure 6.4-1 and use the system to measure the autocorrelation function of the process $X(t)$ of Example 6.2-1. From (6.4-2)

$$R_o(2T) = \frac{1}{2T} \int_{-T}^{T} A^2 \cos(\omega_0 t + \theta) \cos(\omega_0 t + \theta + \omega_0 \tau)\, dt$$

$$= \frac{A^2}{4T} \int_{-T}^{T} [\cos(\omega_0 \tau) + \cos(2\omega_0 t + 2\theta + \omega_0 \tau)]\, dt$$

In writing this result θ represents a specific value of the random variable Θ;

---

† Since the processes are assumed jointly ergodic and therefore jointly stationary, the integral (6.4-1) will tend to be independent of t_1 if T is large enough.

the value that corresponds to the specific ensemble member being used in (6.4-2). On straightforward reduction of the above integral we obtain

$$R_o(2T) = R_{XX}(\tau) + \varepsilon(T)$$

where

$$R_{XX}(\tau) = (A^2/2) \cos (\omega_0 \tau)$$

is the true autocorrelation function of $X(t)$, and

$$\varepsilon(T) = (A^2/2) \cos (\omega_0 \tau + 2\theta) \frac{\sin (2\omega_0 T)}{2\omega_0 T}$$

is an error term. If we require the error term's magnitude to be at least 20 times smaller than the largest value of the true autocorrelation function then $|\varepsilon(T)| < 0.05 R_{XX}(0)$ is necessary. Thus, we must have $1/2\omega_0 T \le 0.05$ or

$$T \ge 10/\omega_0$$

In other words, if $T \ge 10/\omega_0$ the error in using Figure 6.4-1 to measure the autocorrelation function of the process $X(t) = A \cos (\omega_0 t + \Theta)$ will be 5% or less of the largest value of the true autocorrelation function.

6.5 GAUSSIAN RANDOM PROCESSES

A number of random processes are important enough to have been given names. We shall discuss only the most important of these, the *gaussian random process*.

Consider a continuous random process such as illustrated in Figure 6.1-1 and define N random variables $X_1 = X(t_1), \ldots, X_i = X(t_i), \ldots, X_N = X(t_N)$ corresponding to N time instants $t_1, \ldots, t_i, \ldots, t_N$. If, for any $N = 1, 2, \ldots$ and any times $t_1, \ldots, t_N$, these random variables are jointly gaussian, that is, they have a joint density as given by (5.3-12), the process is called gaussian. Equation (5.3-12) can be written in the form

$$f_X(x_1, \ldots, x_N; t_1, \ldots, t_N) = \frac{\exp \left\{ -(1/2)[x - \bar{X}]'[C_X]^{-1}[x - \bar{X}] \right\}}{\sqrt{(2\pi)^N |[C_X]|}} \qquad (6.5-1)$$

where matrices $[x - \bar{X}]$ and $[C_X]$ are defined in (5.3-13) and (5.3-14) and (5.3-15), respectively. The mean values $\bar{X}_i$ of $X(t_i)$ are

$$\bar{X}_i = E[X_i] = E[X(t_i)] \qquad (6.5-2)$$

The elements of the covariance matrix $[C_X]$ are

$$
\begin{aligned}
C_{ik} = C_{X_i X_k} &= E[(X_i - \bar{X}_i)(X_k - \bar{X}_k)] \\
&= E[\{X(t_i) - E[X(t_i)]\}\{X(t_k) - E[X(t_k)]\}] \\
&= C_{XX}(t_i, t_k) \qquad (6.5-3)
\end{aligned}
$$

which is the autocovariance of $X(t_i)$ and $X(t_k)$ from (6.3-21).

From (6.5-2) and (6.5-3), when used in (6.5-1), we see that the mean and auto-covariance functions are all that are needed to completely specify a gaussian random process. By expanding (6.5-3) to get

$$C_{XX}(t_i, t_k) = R_{XX}(t_i, t_k) - E[X(t_i)]E[X(t_k)] \qquad (6.5\text{-}4)$$

we see that an alternative specification using only the mean and autocorrelation function $R_{XX}(t_i, t_k)$ is possible.

If the gaussian process is not stationary the mean and autocovariance functions will, in general, depend on absolute time. However, for the important case where the process is wide-sense stationary, the mean will be constant,

$$\bar{X}_i = E[X(t_i)] = \bar{X} \qquad \text{(constant)} \qquad (6.5\text{-}5)$$

while the autocovariance and autocorrelation functions will depend only on time differences and not absolute time,

$$C_{XX}(t_i, t_k) = C_{XX}(t_k - t_i) \qquad (6.5\text{-}6)$$

$$R_{XX}(t_i, t_k) = R_{XX}(t_k - t_i) \qquad (6.5\text{-}7)$$

It follows from the preceding discussions that a wide-sense stationary gaussian process is also strictly stationary.

We illustrate some of the above remarks with an example.

Example 6.5-1 A gaussian random process is known to be wide-sense stationary with a mean of $\bar{X} = 4$ and autocorrelation function

$$R_{XX}(\tau) = 25e^{-3|\tau|}$$

We seek to specify the joint density function for three random variables $X(t_i)$, $i = 1, 2, 3$, defined at times $t_i = t_0 + [(i - 1)/2]$, with t_0 a constant.

Here $t_k - t_i = (k - i)/2$, i and $k = 1, 2, 3$, so

$$R_{XX}(t_k - t_i) = 25e^{-3|k-i|/2}$$

and

$$C_{XX}(t_k - t_i) = 25e^{-3|k-i|/2} - 16$$

from (6.5-4) through (6.5-7). Elements of the covariance matrix are found from (6.5-3). Thus,

$$[C_X] = \begin{bmatrix} (25 - 16) & (25e^{-3/2} - 16) & (25e^{-6/2} - 16) \\ (25e^{-3/2} - 16) & (25 - 16) & (25e^{-3/2} - 16) \\ (25e^{-6/2} - 16) & (25e^{-3/2} - 16) & (25 - 16) \end{bmatrix}$$

and $\bar{X}_i = 4$ completely determine (6.5-1) for this case where $N = 3$.

Two random processes $X(t)$ and $Y(t)$ are said to be *jointly gaussian* if the random variables $X(t_1), \ldots, X(t_N), Y(t_1'), \ldots, Y(t_M')$ defined at times $t_1, \ldots, t_N$ for $X(t)$ and times $t_1', \ldots, t_M'$ for $Y(t)$, are jointly gaussian for any $N, t_1, \ldots, t_N, M, t_1', \ldots, t_M'$.

★6.6 COMPLEX RANDOM PROCESSES

If the complex random variable of Section 5.6 is generalized to include time, the result is a *complex random process* $Z(t)$ given by

$$Z(t) = X(t) + jY(t) \tag{6.6-1}$$

where $X(t)$ and $Y(t)$ are real processes. $Z(t)$ is called stationary if $X(t)$ and $Y(t)$ are jointly stationary. If $X(t)$ and $Y(t)$ are jointly wide-sense stationary, then $Z(t)$ is said to be wide-sense stationary.

Two complex processes $Z_i(t)$ and $Z_j(t)$ are jointly wide-sense stationary if each is wide-sense stationary and their cross-correlation function (defined below) is a function of time differences only and not absolute time.

We may extend the operations involving process mean value, autocorrelation function, and autocovariance function to include complex processes. The *mean value* of $Z(t)$ is

$$E[Z(t)] = E[X(t)] + jE[Y(t)] \tag{6.6-2}$$

Autocorrelation function is defined by

$$R_{ZZ}(t, t + \tau) = E[Z^*(t)Z(t + \tau)] \tag{6.6-3}$$

where the asterisk * denotes the complex conjugate. *Autocovariance function* is defined by

$$C_{ZZ}(t, t + \tau) = E[\{Z(t) - E[Z(t)]\}^*\{Z(t + \tau) - E[Z(t + \tau)]\}] \tag{6.6-4}$$

If $Z(t)$ is at least wide-sense stationary, the mean value becomes a constant

$$\bar{Z} = \bar{X} + j\bar{Y} \tag{6.6-5}$$

and the correlation functions are independent of absolute time:

$$R_{ZZ}(t, t + \tau) = R_{ZZ}(\tau) \tag{6.6-6}$$

$$C_{ZZ}(t, t + \tau) = C_{ZZ}(\tau) \tag{6.6-7}$$

For two complex processes $Z_i(t)$ and $Z_j(t)$, *cross-correlation* and *cross-covariance functions* are defined by

$$R_{Z_iZ_j}(t, t + \tau) = E[Z_i^*(t)Z_j(t + \tau)] \qquad i \neq j \tag{6.6-8}$$

and

$$C_{Z_iZ_j}(t, t + \tau) = E[\{Z_i(t) - E[Z_i(t)]\}^*\{Z_j(t + \tau) - E[Z_j(t + \tau)]\}] \qquad i \neq j \tag{6.6-9}$$

respectively. If the two processes are at least jointly wide-sense stationary, we obtain

$$R_{Z_iZ_j}(t, t + \tau) = R_{Z_iZ_j}(\tau) \qquad i \neq j \qquad (6.6\text{-}10)$$

$$C_{Z_iZ_j}(t, t + \tau) = C_{Z_iZ_j}(\tau) \qquad i \neq j \qquad (6.6\text{-}11)$$

$Z_i(t)$ and $Z_j(t)$ are said to be *uncorrelated* processes if $C_{Z_iZ_j}(t, t + \tau) = 0,$ $i \neq j$. They are called *orthogonal* processes if $R_{Z_iZ_j}(t, t + \tau) = 0, i \neq j$.

Example 6.6-1 A complex random process $V(t)$ is comprised of a sum of N complex signals:

$$V(t) = \sum_{n=1}^{N} A_n e^{j\omega_0 t + j\Theta_n}$$

Here $\omega_0/2\pi$ is the (constant) frequency of each signal. A_n is a random variable representing the random amplitude of the nth signal. Similarly, Θ_n is a random variable representing a random phase angle. We assume all the variables A_n and Θ_n, for $n = 1, 2, \ldots, N$, are statistically independent and the Θ_n are uniformly distributed on $(0, 2\pi)$. We find the autocorrelation function of $V(t)$.

From (6.6-3):

$$R_{VV}(t, t + \tau) = E[V^*(t)V(t + \tau)]$$

$$= E\left[\sum_{n=1}^{N} A_n e^{-j\omega_0 t - j\Theta_n} \sum_{m=1}^{N} A_m e^{j\omega_0 t + j\omega_0 \tau + j\Theta_m} \right]$$

$$= \sum_{n=1}^{N} \sum_{m=1}^{N} e^{j\omega_0 \tau} E[A_n A_m e^{j(\Theta_m - \Theta_n)}] = R_{VV}(\tau)$$

From statistical independence:

$$R_{VV}(\tau) = e^{j\omega_0 \tau} \sum_{n=1}^{N} \sum_{m=1}^{N} E[A_n A_m]E[\exp\{j(\Theta_m - \Theta_n)\}]$$

However,

$$E[\exp\{j(\Theta_m - \Theta_n)\}] = E[\cos(\Theta_m - \Theta_n)] + jE[\sin(\Theta_m - \Theta_n)]$$

$$= \int_0^{2\pi} \int_0^{2\pi} \frac{1}{(2\pi)^2} [\cos(\theta_m - \theta_n) + j\sin(\theta_m - \theta_n)] \, d\theta_n \, d\theta_m$$

$$= \begin{cases} 0 & m \neq n \\ 1 & m = n \end{cases}$$

so

$$R_{VV}(\tau) = e^{j\omega_0 \tau} \sum_{n=1}^{N} \overline{A_n^2}$$

PROBLEMS

6-1 A random experiment consists of selecting a point on some city street that has two-way automobile traffic. Define and classify a random process for this experiment that is related to traffic flow.

6-2 A 10-meter section of a busy downtown sidewalk is actually the platform of a scale that produces a voltage proportional to the total weight of people on the scale at any time.

    (a) Sketch a typical sample function for this process.

    (b) What is the underlying random experiment for the process?

    (c) Classify the process.

****6-3** An experiment consists of measuring the weight W of some person each 10 minutes. The person is randomly male or female (which is not known though) with equal probability. A two-level discrete random process $X(t)$ is generated where

$$X(t) = \pm 10$$

The level -10 is generated in the period following a measurement if the measured weight does not exceed W_0 (some constant). Level $+10$ is generated if weight exceeds W_0. Let the weight of men in kg be a random variable having the gaussian density

$$f_W(w\,|\,\text{male}) = \frac{1}{\sqrt{2\pi}11.3} \exp\left[-(w - 77.1)^2/2(11.3)^2\right]$$

Similarly, for women

$$f_W(w\,|\,\text{female}) = \frac{1}{\sqrt{2\pi}6.8} \exp\left[-(w - 54.4)^2/2(6.8)^2\right]$$

    (a) Find W_0 so that $P\{W > W_0\,|\,\text{male}\}$ is equal to $P\{W \le W_0\,|\,\text{female}\}$.

    (b) If the levels ± 10 are interpreted as "decisions" about whether the weight measurement of a person corresponds to a male or female, give a physical significance to their generation.

    (c) Sketch a possible sample function.

6-4 The two-level *semirandom binary process* is defined by

$$X(t) = A \text{ or } -A \qquad (n - 1)T < t < nT$$

where the levels A and $-A$ occur with equal probability, T is a positive constant, and $n = 0, \pm 1, \pm 2, \ldots$.

    (a) Sketch a typical sample function.

    (b) Classify the process.

    (c) Is the process deterministic?

6-5 Sample functions in a discrete random process are constants; that is

$$X(t) = C = \text{constant}$$

where C is a discrete random variable having possible values $c_1 = 1$, $c_2 = 2$, and $c_3 = 3$ occurring with probabilities 0.6, 0.3, and 0.1 respectively.

(a) Is $X(t)$ deterministic?

(b) Find the first-order density function of $X(t)$ at any time t.

6-6 Utilize (6.2-13) to prove (6.2-16).

*$\star$6-7** A random process $X(t)$ has periodic sample functions as shown in Figure P6-7 where B, T, and $4t_0 \leq T$ are constants but ε is a random variable uniformly distributed on the interval $(0, T)$.

(a) Find the first-order distribution function of $X(t)$.

(b) Find the first-order density function.

(c) Find $E[X(t)]$, $E[X^2(t)]$, and σ_X^2.

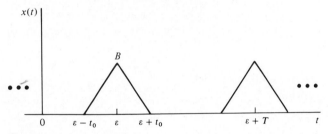

Figure P6-7

*$\star$6-8** Work Problem 6-7 for the waveform of Figure P6-8. Assume $2t_0 < T$.

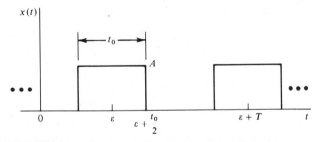

Figure P6-8

*$\star$6-9** Work Problem 6-7 for the waveform of Figure P6-9. Assume $4t_0 \leq T$.

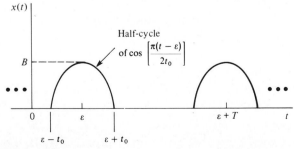

Figure P6-9

6-10 Given the random process

$$X(t) = A \sin (\omega_0 t + \Theta)$$

where A and ω_0 are constants and Θ is a random variable uniformly distributed on the interval $(-\pi, \pi)$. Define a new random process $Y(t) = X^2(t)$.
   (a) Find the autocorrelation function of $Y(t)$.
   (b) Find the cross-correlation function of $X(t)$ and $Y(t)$.
   (c) Are $X(t)$ and $Y(t)$ wide-sense stationary?
   (d) Are $X(t)$ and $Y(t)$ jointly wide-sense stationary?

6-11 A random process is defined by

$$Y(t) = X(t) \cos (\omega_0 t + \Theta)$$

where $X(t)$ is a wide-sense stationary random process that amplitude-modulates a carrier of constant angular frequency ω_0 with a random phase Θ independent of $X(t)$ and uniformly distributed on $(-\pi, \pi)$.
   (a) Find $E[Y(t)]$.
   (b) Find the autocorrelation function of $Y(t)$.
   (c) Is $Y(t)$ wide-sense stationary?

6-12 Given the random process

$$X(t) = A \cos (\omega_0 t) + B \sin (\omega_0 t)$$

where ω_0 is a constant, and A and B are uncorrelated zero-mean random variables having different density functions but the same variances σ^2. Show that $X(t)$ is wide-sense stationary but not strictly stationary.

6-13 If $X(t)$ is a stationary random process having a mean value $E[X(t)] = 3$ and autocorrelation function $R_{XX}(\tau) = 9 + 2e^{-|\tau|}$, find:
   (a) the mean value and
   (b) the variance of the random variable

$$Y = \int_0^2 X(t) \, dt$$

   (*Hint:* Assume expectation and integration operations are interchangeable.)

6-14 Define a random process by

$$X(t) = A \cos (\pi t)$$

where A is a gaussian random variable with zero mean and variance σ_A^2.
   (a) Find the density functions of $X(0)$ and $X(1)$.
   (b) Is $X(t)$ stationary in any sense?

6-15 For the random process of Problem 6-4, calculate:
   (a) the mean value $E[X(t)]$ (b) $R_{XX}(t_1 = 0.5T, \ t_2 = 0.7T)$ (c) $R_{XX}(t_1 = 0.2T, t_2 = 1.2T)$.

6-16 A random process consists of three sample functions $X(t, s_1) = 2$, $X(t, s_2) = 2 \cos (t)$, and $X(t, s_3) = 3 \sin (t)$, each occurring with equal probability. Is the process stationary in any sense?

6-17 Statistically independent, zero-mean, random processes $X(t)$ and $Y(t)$ have autocorrelation functions

$$R_{XX}(\tau) = e^{-|\tau|}$$

and

$$R_{YY}(\tau) = \cos(2\pi\tau)$$

respectively.

    (*a*) Find the autocorrelation function of the sum $W_1(t) = X(t) + Y(t)$.

    (*b*) Find the autocorrelation function of the difference $W_2(t) = X(t) - Y(t)$.

    (*c*) Find the cross-correlation function of $W_1(t)$ and $W_2(t)$.

6-18 Define a random process as $X(t) = p(t + \varepsilon)$, where $p(t)$ is any periodic waveform with period T and ε is a random variable uniformly distributed on the interval $(0, T)$. Show that

$$E[X(t)X(t + \tau)] = \frac{1}{T} \int_0^T p(\xi)p(\xi + \tau) \, d\xi = R_{XX}(\tau)$$

****6-19** Use the result of Problem 6-18 to find the autocorrelation function of random processes having periodic sample function waveforms $p(t)$ defined

    (*a*) by Figure P6-7 with $\varepsilon = 0$ and $4t_0 \le T$, and

    (*b*) by Figure P6-8 with $\varepsilon = 0$ and $2t_0 \le T$.

6-20 Define two random processes by $X(t) = p_1(t + \varepsilon)$ and $Y(t) = p_2(t + \varepsilon)$ when $p_1(t)$ and $p_2(t)$ are both periodic waveforms with period T and ε is a random variable uniformly distributed on the interval $(0, T)$. Find an expression for the cross-correlation function $E[X(t)Y(t + \tau)]$.

6-21 Prove:

    (*a*) (6.3-4)   and   (*b*) (6.3-5).

6-22 Give arguments to justify (6.3-9).

6-23 For the random process having the autocorrelation function shown in Figure P6-23, find:

    (*a*) $E[X(t)]$ (*b*) $E[X^2(t)]$ and   (*c*) σ_X^2.

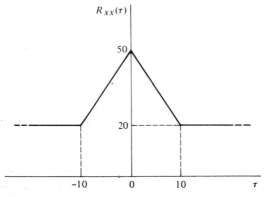

Figure P6-23

6-24 A random process $Y(t) = X(t) - X(t + \tau)$ is defined in terms of a process $X(t)$ that is at least wide-sense stationary.

(a) Show that the mean value of $Y(t)$ is 0 even if $X(t)$ has a nonzero mean value.

(b) Show that

$$\sigma_Y^2 = 2[R_{XX}(0) - R_{XX}(\tau)]$$

(c) If $Y(t) = X(t) + X(t + \tau)$, find $E[Y(t)]$ and σ_Y^2. How do these results compare to those of parts (a) and (b)?

6-25 For two zero-mean, jointly wide-sense stationary random processes $X(t)$ and $Y(t)$, it is known that $\sigma_X^2 = 5$ and $\sigma_Y^2 = 10$. Explain why each of the following functions cannot apply to the processes if they have no periodic components.

(a) $R_{XX}(\tau) = 6u(\tau) \exp(-3\tau)$ (b) $R_{XX}(\tau) = 5 \sin(5\tau)$

(c) $R_{XY}(\tau) = 9(1 + 2\tau^2)^{-1}$ (d) $R_{YY}(\tau) = -\cos(6\tau) \exp(-|\tau|)$

(e) $R_{YY}(\tau) = 5\left[\dfrac{\sin(3\tau)}{3\tau}\right]^2$ (f) $R_{YY}(\tau) = 6 + 4\left[\dfrac{\sin(10\tau)}{10\tau}\right]$

6-26 Given two random processes $X(t)$ and $Y(t)$. Find expressions for the auto-correlation function of $W(t) = X(t) + Y(t)$ if:

(a) $X(t)$ and $Y(t)$ are correlated.

(b) They are uncorrelated.

(c) They are uncorrelated with zero means.

6-27 Use (6.3-19) to prove (6.3-17).

6-28 Let $X(t)$ be a stationary continuous random process that is differentiable. Denote its time-derivative by $\dot{X}(t)$.

(a) Show that $E[\dot{X}(t)] = 0$.

(b) Find $R_{X\dot{X}}(\tau)$ in terms of $R_{XX}(\tau)$.

(c) Find $R_{\dot{X}\dot{X}}(\tau)$ in terms of $R_{XX}(\tau)$. (*Hint:* Use the definition of the derivative

$$\dot{X}(t) = \lim_{\varepsilon \to 0} \frac{X(t + \varepsilon) - X(t)}{\varepsilon}$$

and assume the order of the limit and expectation operations can be interchanged.)

6-29 A gaussian random process has an autocorrelation function

$$R_{XX}(\tau) = 6 \exp(-|\tau|/2)$$

Determine a covariance matrix for the random variables $X(t)$, $X(t + 1)$, $X(t + 2)$, and $X(t + 3)$.

6-30 Work Problem 6-29 if

$$R_{XX}(\tau) = 6 \frac{\sin(\pi\tau)}{\pi\tau}$$

6-31 An ensemble member of a stationary random process $X(t)$ is sampled at N times t_i, $i = 1, 2, \ldots, N$. By treating the samples as random variables $X_i = X(t_i)$, an *estimate* or *measurement* $\hat{\bar{X}}$ of the mean value $\bar{X} = E[X(t)]$ of the process is sometimes formed by averaging the samples:

$$\hat{\bar{X}} = \frac{1}{N} \sum_{i=1}^{N} X_i$$

(a) Show that $E[\hat{\bar{X}}] = \bar{X}$.

(b) If the samples are separated far enough in time so that the random variables X_i can be considered statistically independent, show that the variance of the estimate of the process mean is

$$(\sigma_{\hat{\bar{X}}})^2 = \sigma_X^2 / N$$

6-32 For the random process and samples defined in Problem 6-31, let an estimate of the variance of the process be defined by

$$\hat{\sigma_X^2} = \frac{1}{N} \sum_{i=1}^{N} (X_i - \hat{\bar{X}})^2$$

Show that the mean value of this estimate is

$$E[\hat{\sigma_X^2}] = \frac{N-1}{N} \sigma_X^2$$

6-33 Assume that $X(t)$ of Problem 6-31 is a zero-mean stationary gaussian process and let

$$\hat{\sigma_X^2} = \frac{1}{N} \sum_{i=1}^{N} X_i^2$$

be an estimate of the variance σ_X^2 of $X(t)$ formed from the samples. Show that the variance of the estimate is

$$\text{variance of } \hat{\sigma_X^2} = \frac{2\sigma_X^4}{N}$$

(*Hint:* Use the facts that $E[X^2] = \sigma_X^2$, $E[X^3] = 0$, and $E[X^4] = 3\sigma_X^4$ for a gaussian random variable having mean zero.)

6-34 How many samples must be taken in Problem 6-33 if the standard deviation of the estimate of the variance of $X(t)$ is to not exceed 5% of σ_X^2?

*__6-35__ A complex random process $Z(t) = X(t) + jY(t)$ is defined by jointly stationary real processes $X(t)$ and $Y(t)$. Show that

$$E[|Z(t)|^2] = R_{XX}(0) + R_{YY}(0)$$

*__6-36__ Let $X_1(t)$, $X_2(t)$, $Y_1(t)$ and $Y_2(t)$ be real random processes and define

$$Z_1(t) = X_1(t) + jY_1(t) \qquad Z_2(t) = X_2(t) - jY_2(t)$$

Find expressions for the cross-correlation function of $Z_1(t)$ and $Z_2(t)$ if:

    (a) All the real processes are correlated.

    (b) They are uncorrelated.

    (c) They are uncorrelated with zero means.

*6-37 Let $Z(t)$ be a stationary complex random process with an autocorrelation function $R_{ZZ}(\tau)$. Define the random variable

$$W = \int_a^{a+T} Z(t) \, dt$$

where $T > 0$ and a are real numbers. Show that

$$E[|W|^2] = \int_{-T}^{T} (T - |\tau|) R_{ZZ}(\tau) \, d\tau$$

ADDITIONAL PROBLEMS

6-38 For a random process $X(t)$ it is known that $f_X(x_1, x_2, x_3; t_1, t_2, t_3) = f_X(x_1, x_2, x_3; t_1 + \Delta, t_2 + \Delta, t_3 + \Delta)$ for any t_1, t_2, t_3 and Δ. Indicate which of the following statements are unequivocally true: $X(t)$ is (a) stationary to order 1, (b) stationary to order 2, (c) stationary to order 3, (d) strictly stationary, (e) wide-sense stationary, (f) not stationary in any sense, and (g) ergodic.

6-39 A random process is defined by $X(t) = X_0 + Vt$ where X_0 and V are statistically independent random variables uniformly distributed on intervals $[X_{01}, X_{02}]$ and $[V_1, V_2]$, respectively. Find (a) the mean, (b) the autocorrelation, and (c) the autocovariance functions of $X(t)$. (d) Is $X(t)$ stationary in any sense? If so, state the type.

*6-40 (a) Find the first-order density of the random process of Problem 6-39. (b) Plot the density for $t = k(X_{02} - X_{01})/(V_2 - V_1)$ with $k = 0, \frac{1}{2}, 1$, and 2. Assume $V_2 = 3V_1$ in all plots.

6-41 Assume a wide-sense stationary process $X(t)$ has a known mean $\bar{X}$ and a known autocorrelation function $R_{XX}(\tau)$. Now suppose the process is observed at time t_1 and we wish to *estimate*, that is, *predict*, what the process will be at time $t_1 + \tau$ with $\tau > 0$. We assume the estimate has the form

$$\hat{X}(t_1 + \tau) = \alpha \hat{X}(t_1) + \beta$$

where α and β are constants.

    (a) Find α and β so that the mean-squared prediction error

$$\overline{\varepsilon^2} = E[\{X(t_1 + \tau) - \hat{X}(t_1 + \tau)\}^2]$$

is minimum.

    (b) Find the minimum mean-squared error in terms of $R_{XX}(\tau)$. Develop an alternative form in terms of the autocovariance function.

6-42 Find the time average and time autocorrelation function of the random process of Example 6.2-1. Compare these results with the statistical mean and autocorrelation found in the example.

6-43 Assume that an ergodic random process $X(t)$ has an autocorrelation function

$$R_{XX}(\tau) = 18 + \frac{2}{6 + \tau^2} [1 + 4 \cos (12\tau)]$$

(a) Find $|\bar{X}|$.
(b) Does this process have a periodic component?
(c) What is the average power in $X(t)$?

6-44 Define a random process $X(t)$ as follows: (1) $X(t)$ assumes only one of two possible levels 1 or -1 at any time, (2) $X(t)$ switches back and forth between its two levels randomly with time, (3) the number of level transitions in any time interval τ is a Poisson random variable, that is, the probability of exactly k transitions, when the average rate of transitions is λ, is given by $[(\lambda\tau)^k/k!] \exp(-\lambda\tau)$, (4) transitions occurring in any time interval are statistically independent of transitions in any other interval, and (5) the levels at the start of any interval are equally probable. $X(t)$ is usually called the *random telegraph process*. It is an example of a discrete random process.

(a) Find the autocorrelation function of the process.
(b) Find probabilities $P\{X(t) = 1\}$ and $P\{X(t) = -1\}$ for any t.
(c) What is $E[X(t)]$?
(d) Discuss the stationarity of $X(t)$.

6-45 Work Problem 6-44 assuming the random telegraph signal has levels 0 and 1.

6-46 $\bar{X} = 6$ and $R_{XX}(t, t + \tau) = 36 + 25 \exp(-|\tau|)$ for a random process $X(t)$. Indicate which of the following statements are true based on what is known with certainty. $X(t)$ (a) is first-order stationary, (b) has total average power of 61 W, (c) is ergodic, (d) is wide-sense stationary, (e) has a periodic component, and (f) has an ac power of 36 W.

6-47 A zero-mean random process $X(t)$ is ergodic, has average power of 24 W, and has no periodic components. Which of the following can be a valid autocorrelation function? If one cannot, state at least one reason why. (a) $16 + 18 \cos (3\tau)$, (b) $24\text{Sa}^2(2\tau)$, (c) $[1 + 3\tau^2]^{-1} \exp(-6\tau)$, and (d) $24\delta(t - \tau)$.

6-48 Use the result of Problem 6-18 to find the autocorrelation function of a random process with periodic sample function waveform $p(t)$ defined by

$$p(t) = A \cos^2 (2\pi t/T)$$

where A and $T > 0$ are constants.

6-49 An engineer wants to measure the mean value of a noise signal that can be well-modeled as a sample function of a gaussian process. He uses the sampling estimator of Problem 6-31. After 100 samples he wishes his estimate to be within ± 0.1 V of the true mean with probability 0.9606. What is the largest variance the process can have such that his wishes will be true?

6-50 Let $X(t)$ be the sum of a deterministic signal $s(t)$ and a wide-sense stationary noise process $N(t)$. Find the mean value, and autocorrelation and autocovariance functions of $X(t)$. Discuss the stationarity of $X(t)$.

6-51 Random processes $X(t)$ and $Y(t)$ are defined by

$$X(t) = A \cos (\omega_0 t + \Theta)$$

$$Y(t) = B \cos (\omega_0 t + \Theta)$$

where A, B, and ω_0 are constants while Θ is a random variable uniform on $(0, 2\pi)$. By the procedures of Example 6.2-1 it is easy to find that $X(t)$ and $Y(t)$ are zero-mean, wide-sense stationary with autocorrelation functions

$$R_{XX}(\tau) = (A^2/2) \cos (\omega_0 \tau)$$

$$R_{YY}(\tau) = (B^2/2) \cos (\omega_0 \tau)$$

(a) Find the cross-correlation function $R_{XY}(t, t + \tau)$ and show that $X(t)$ and $Y(t)$ are jointly wide-sense stationary.

(b) Solve (6.4-2) and show that the response of the system of Figure 6.4-1 equals the true cross-correlation function plus an error term $\varepsilon(T)$ that decreases as T increases.

(c) Sketch $|\varepsilon(T)|$ versus T to show its behavior. How large must T be to make $|\varepsilon(T)|$ less than 1% of the largest value the correct cross-correlation function can have?

6-52 Consider random processes

$$X(t) = A \cos (\omega_0 t + \Theta)$$

$$Y(t) = B \cos (\omega_1 t + \Phi)$$

where A, B, ω_1, and ω_0 are constants, while Θ and Φ are statistically independent random variables uniform on $(0, 2\pi)$.

(a) Show that $X(t)$ and $Y(t)$ are jointly wide-sense stationary.

(b) If $\Theta = \Phi$ show that $X(t)$ and $Y(t)$ are not jointly wide-sense stationary unless $\omega_1 = \omega_0$.

6-53 A zero-mean gaussian random process has an autocorrelation function

$$R_{XX}(\tau) = \begin{cases} 13[1 - (|\tau|/6)] & |\tau| \leq 6 \\ 0 & \text{elsewhere} \end{cases}$$

Find the covariance function necessary to specify the joint density of random variables defined at times $t_i = 2(i - 1)$, $i = 1, 2, \ldots, 5$. Give the covariance matrix for the $X_i = X(t_i)$.

6-54 If the gaussian process of Problem 6-53 is shifted to have a constant mean $\bar{X} = -2$ but all else is unchanged, discuss how the autocorrelation function and covariance matrix change. What is the effect on the joint density of the five random variables?

*6-55 Extend Example 6.6-1 to allow the sum of complex-amplitude unequal-frequency phasors. Let Z_i, $i = 1, 2, \ldots, N$ be N complex zero-mean, uncorrelated random variables with variances $\sigma_{Z_i}^2$. Form a random process

$$Z(t) = \sum_{i=1}^{N} Z_i e^{j\omega_i t}$$

where ω_i are the frequencies of the phasors.

(a) Show that $E[Z(t)] = 0$.

(b) Derive the autocorrelation function and show that $Z(t)$ is wide-sense stationary.

*6-56 A complex random process is defined by

$$Z(t) = \exp{(j\Omega t)}$$

where Ω is a zero-mean random variable uniformly distributed on the interval from $\omega_0 - \Delta\omega$ to $\omega_0 + \Delta\omega$, where ω_0 and $\Delta\omega$ are positive constants. Find:

(a) the mean value, and (b) the autocorrelation function of $Z(t)$.

(c) Is $Z(t)$ wide-sense stationary?

*6-57 Work Problem 6-56 except assume the process

$$Z(t) = e^{j\Omega t} + e^{-j\Omega t} = 2\cos{(\Omega t)}$$

*6-58 Let $X(t)$ and $Y(t)$ be statistically independent wide-sense stationary real processes having the same autocorrelation function $R(\tau)$. Define the complex process

$$Z(t) = X(t)\cos{(\omega_0 t)} + jY(t)\sin{(\omega_0 t)}$$

where ω_0 is a positive constant. Find the autocorrelation function of $Z(t)$. Is $Z(t)$ wide-sense stationary?

SEVEN

SPECTRAL CHARACTERISTICS OF RANDOM PROCESSES

7.0 INTRODUCTION

All of the foregoing discussions concerning random processes have involved the time domain. That is, we have characterized processes by means of autocorrelation, cross-correlation, and covariance functions without any consideration of spectral properties. As is well known, both time domain *and* frequency domain analysis methods exist for analyzing linear systems and deterministic waveforms. But what about random waveforms? Is there some way to describe random processes in the frequency domain? The answer is yes, and it is the purpose of this chapter to introduce the most important concepts that apply to characterizing random processes in the frequency domain.

The spectral description of a deterministic waveform is obtained by Fourier transforming the waveform, and the reader would be correct in concluding that Fourier transforms play an important role in the spectral characterization of random waveforms. However, the direct transformation approach is not attractive for random waveforms because the transform may not exist. Thus, spectral analysis of random processes requires a bit more subtlety than do deterministic signals.

An appropriate spectrum to be associated with a random process is introduced in the following section. The concepts rely heavily on theory of Fourier transforms. Readers wishing to refresh their background on Fourier theory are referred to Appendix D where a short review is given.

7.1 POWER DENSITY SPECTRUM AND ITS PROPERTIES

The spectral properties of a *deterministic* signal $x(t)$ are contained in its *Fourier transform* $X(\omega)$ given by

$$X(\omega) = \int_{-\infty}^{\infty} x(t)e^{-j\omega t}\, dt \qquad (7.1\text{-}1)$$

The function $X(\omega)$, sometimes called simply the *spectrum* of $x(t)$, has the unit of volts per hertz and describes the way in which relative signal voltage is distributed with frequency. The Fourier transform can, therefore, be considered to be a *voltage density spectrum* applicable to $x(t)$. Both the amplitudes and phases of the frequencies present in $x(t)$ are described by $X(\omega)$. For this reason, if $X(\omega)$ is known then $x(t)$ can be recovered by means of the *inverse Fourier transform*

$$x(t) = \frac{1}{2\pi} \int_{-\infty}^{\infty} X(\omega)e^{j\omega t}\, d\omega \qquad (7.1\text{-}2)$$

In other words, $X(\omega)$ forms a complete description of $x(t)$ and vice versa.

In attempting to apply (7.1-1) to a random process, we immediately encounter problems. The principal problem is the fact that $X(\omega)$ may not exist for most sample functions of the process. Thus, we conclude that a spectral description of a random process utilizing a voltage density spectrum (Fourier transform) is not feasible because such a spectrum may not exist. Other problems arise if Laplace transforms are considered (Cooper and McGillem, 1971, p. 132).

On the other hand, if we turn our attention to the description of the *power* in the random process as a function of frequency, instead of voltage, it results that such a function does exist. We next proceed to develop this function, called the *power density spectrum*† of the random process.

The Power Density Spectrum

For a random process $X(t)$, let $x_T(t)$ be defined as that portion of a sample function $x(t)$ that exists between $-T$ and T; that is

$$x_T(t) = \begin{cases} x(t) & -T < t < T \\ 0 & \text{elsewhere} \end{cases} \qquad (7.1\text{-}3)$$

Now so long as T is finite, $x_T(t)$ will satisfy

$$\int_{-T}^{T} |x_T(t)|\, dt < \infty \qquad (7.1\text{-}4)$$

---

† Many books call this function a *power spectral density*. We shall occasionally use also the names *power density* or *power spectrum*.

and will have a Fourier transform (see Appendix D for conditions sufficient for the existence of Fourier transforms), which we denote $X_T(\omega)$, given by

$$X_T(\omega) = \int_{-T}^{T} x_T(t)e^{-j\omega t} \, dt = \int_{-T}^{T} x(t)e^{-j\omega t} \, dt \tag{7.1-5}$$

The energy contained in $x(t)$ in the interval $(-T, T)$ is[†]

$$E(T) = \int_{-T}^{T} x_T^2(t) \, dt = \int_{-T}^{T} x^2(t) \, dt \tag{7.1-6}$$

Since $x_T(t)$ is Fourier transformable, its energy must also be related to $X_T(\omega)$ by Parseval's theorem. Thus, from (7.1-6) and (D-21) of Appendix D

$$E(T) = \int_{-T}^{T} x^2(t) \, dt = \frac{1}{2\pi} \int_{-\infty}^{\infty} |X_T(\omega)|^2 \, d\omega \tag{7.1-7}$$

By dividing the expressions in (7.1-7) by $2T$, we obtain the average power $P(T)$ in $x(t)$ over the interval $(-T, T)$:

$$P(T) = \frac{1}{2T} \int_{-T}^{T} x^2(t) \, dt = \frac{1}{2\pi} \int_{-\infty}^{\infty} \frac{|X_T(\omega)|^2}{2T} \, d\omega \tag{7.1-8}$$

At this point we observe that $|X_T(\omega)|^2/2T$ is a power density spectrum because power results through its integration. However, it is not the function that we seek for two reasons. One is the fact that (7.1-8) does not represent the power in an entire sample function. There remains the step of letting T become arbitrarily large so as to include all power in the ensemble member. The second reason is that (7.1-8) is only the power in one sample function and does not represent the process. In other words, $P(T)$ is actually a random variable with respect to the random process. By taking the expected value in (7.1-8), we can obtain an average power P_{XX} for the random process.[‡]

From the above discussion it is clear that we must still form the limit as $T \to \infty$ and take the expected value of (7.1-8) to obtain a suitable power density spectrum for the random process. It is important that the limiting operation be done last (Thomas, 1969, p. 98, or Cooper and McGillem, 1971, p. 134). After these operations are performed, (7.1-8) can be written

$$P_{XX} = \lim_{T \to \infty} \frac{1}{2T} \int_{-T}^{T} E[X^2(t)] \, dt = \frac{1}{2\pi} \int_{-\infty}^{\infty} \lim_{T \to \infty} \frac{E[|X_T(\omega)|^2]}{2T} \, d\omega \tag{7.1-9}$$

[†] We assume a real process $X(t)$ and interpret $x(t)$ as either the voltage across a 1-Ω impedance or the current through 1 Ω. In other words, we shall assume a 1-Ω real impedance whenever we discuss energy or power in subsequent work, unless specifically stated otherwise.

[‡] In taking the expected value we replace $x(t)$ by $X(t)$ in (7.1-8) since the integral of $x^2(t)$ is an operation performed on all sample functions of $X(t)$.

Equation (7.1-9) establishes two important facts. First, average power P_{XX} in a random process $X(t)$ is given by the time average of its second moment:

$$P_{XX} = \lim_{T \to \infty} \frac{1}{2T} \int_{-T}^{T} E[X^2(t)] \, dt = A\{E[X^2(t)]\} \qquad (7.1\text{-}10)$$

For a process that is at least wide-sense stationary, $E[X^2(t)] = \overline{X^2}$, a constant, and $P_{XX} = \overline{X^2}$. Second, P_{XX} can be obtained by a frequency domain integration. If we define the *power density spectrum* for the random process by

$$S_{XX}(\omega) = \lim_{T \to \infty} \frac{E[|X_T(\omega)|^2]}{2T} \qquad (7.1\text{-}11)$$

the applicable integral is

$$P_{XX} = \frac{1}{2\pi} \int_{-\infty}^{\infty} S_{XX}(\omega) \, d\omega \qquad (7.1\text{-}12)$$

from (7.1-9). Two examples will illustrate the above concepts.

Example 7.1-1 Consider the random process

$$X(t) = A \cos(\omega_0 t + \Theta)$$

where A and ω_0 are real constants and Θ is a random variable uniformly distributed on the interval $(0, \pi/2)$. We shall find the average power P_{XX} in $X(t)$ by use of (7.1-10). Mean-squared value is

$$E[X^2(t)] = E[A^2 \cos^2(\omega_0 t + \Theta)] = E\left[\frac{A^2}{2} + \frac{A^2}{2} \cos(2\omega_0 t + 2\Theta)\right]$$

$$= \frac{A^2}{2} + \frac{A^2}{2} \int_{0}^{\pi/2} \frac{2}{\pi} \cos(2\omega_0 t + 2\theta) \, d\theta$$

$$= \frac{A^2}{2} - \frac{A^2}{\pi} \sin(2\omega_0 t)$$

This process is not even wide-sense stationary, since the above function is time-dependent. The time average of the above expression is

$$A\{E[X^2(t)]\} = \lim_{T \to \infty} \frac{1}{2T} \int_{-T}^{T} \left[\frac{A^2}{2} - \frac{A^2}{\pi} \sin(2\omega_0 t)\right] dt$$

which easily evaluates to

$$P_{XX} = A\{E[X^2(t)]\} = A^2/2$$

Example 7.1-2 We reconsider the process of the above example to find $S_{XX}(\omega)$ and average power P_{XX} by use of (7.1-11) and (7.1-12), respectively. First we find $X_T(\omega)$:

$$X_T(\omega) = \int_{-T}^{T} A \cos(\omega_0 t + \Theta) \exp(-j\omega t)\, dt$$

$$= \frac{A}{2} \exp(j\Theta) \int_{-T}^{T} \exp[j(\omega_0 - \omega)t]\, dt$$

$$+ \frac{A}{2} \exp(-j\Theta) \int_{-T}^{T} \exp[-j(\omega_0 + \omega)t]\, dt$$

$$= AT \exp(j\Theta) \frac{\sin[(\omega - \omega_0)T]}{(\omega - \omega_0)T}$$

$$+ AT \exp(-j\Theta) \frac{\sin[(\omega + \omega_0)T]}{(\omega + \omega_0)T}$$

Next we determine $|X_T(\omega)|^2 = X_T(\omega)X_T^*(\omega)$ and find its expected value. After some simple algebraic reduction we obtain

$$\frac{E[|X_T(\omega)|^2]}{2T} = \frac{A^2\pi}{2} \left\{ \frac{T}{\pi} \frac{\sin^2[(\omega - \omega_0)T]}{[(\omega - \omega_0)T]^2} + \frac{T}{\pi} \frac{\sin^2[(\omega + \omega_0)T]}{[(\omega + \omega_0)T]^2} \right\}$$

Now it is known that

$$\lim_{T \to \infty} \frac{T}{\pi} \left[\frac{\sin(\alpha T)}{\alpha T} \right]^2 = \delta(\alpha)$$

(Lathi, 1968, p. 24), so (7.1-11) and the above result give

$$S_{XX}(\omega) = \frac{A^2\pi}{2} [\delta(\omega - \omega_0) + \delta(\omega + \omega_0)]$$

Finally, we use this result to obtain average power from (7.1-12):

$$P_{XX} = \frac{1}{2\pi} \int_{-\infty}^{\infty} \frac{A^2\pi}{2} [\delta(\omega - \omega_0) + \delta(\omega + \omega_0)]\, d\omega = \frac{A^2}{2}$$

Thus, P_{XX} found here agrees with that of the earlier Example 7.1-1.

Properties of the Power Density Spectrum

The power density spectrum possesses a number of important properties:

(1) $\quad S_{XX}(\omega) \geq 0$ $\hspace{6cm}$ (7.1-13)

(2) $\quad S_{XX}(-\omega) = S_{XX}(\omega) \qquad X(t)$ real $\hspace{2.7cm}$ (7.1-14)

(3) $\quad S_{XX}(\omega)$ is real $\hspace{6cm}$ (7.1-15)

(4) $\quad \dfrac{1}{2\pi} \displaystyle\int_{-\infty}^{\infty} S_{XX}(\omega)\, d\omega = A\{E[X^2(t)]\}$ $\hspace{2cm}$ (7.1-16)

Property 1 follows from the definition (7.1-11) and the fact that the expected value of a nonnegative function is nonnegative. Similarly, property 3 is true from (7.1-11) since $|X_T(\omega)|^2$ is real. Some reflection on the properties of Fourier transforms of real functions will verify property 2 (see Problem 7-9). Property 4 is just another statement of (7.1-9).

Sometimes another property is included in a list of properties:

$$(5) \quad S_{\dot{X}\dot{X}}(\omega) = \omega^2 S_{XX}(\omega) \tag{7.1-17}$$

It says that the power density spectrum of the derivative $\dot{X}(t) = dX(t)/dt$ is ω^2 times the power spectrum of $X(t)$. Proof of this property is left as a reader exercise (Problem 7-10).

A final property we list is

$$(6) \quad \frac{1}{2\pi} \int_{-\infty}^{\infty} S_{XX}(\omega) e^{j\omega\tau} \, d\omega = A[R_{XX}(t, t + \tau)] \tag{7.1-18}$$

$$S_{XX}(\omega) = \int_{-\infty}^{\infty} A[R_{XX}(t, t + \tau)] e^{-j\omega\tau} \, d\tau \tag{7.1-19}$$

It states that the power density spectrum and the time average of the autocorrelation function form a Fourier transform pair. We prove this very important property in Section 7.2. Of course, if $X(t)$ is at least wide-sense stationary, $A[R_{XX}(t, t + \tau)] = R_{XX}(\tau)$, and property 6 indicates that the power spectrum and the autocorrelation function form a Fourier transform pair. Thus

$$S_{XX}(\omega) = \int_{-\infty}^{\infty} R_{XX}(\tau) e^{-j\omega\tau} \, d\tau \tag{7.1-20}$$

$$R_{XX}(\tau) = \frac{1}{2\pi} \int_{-\infty}^{\infty} S_{XX}(\omega) e^{j\omega\tau} \, d\omega \tag{7.1-21}$$

for a wide-sense stationary process.

Bandwidth of the Power Density Spectrum

Assume that $X(t)$ is a lowpass process; that is, its spectral components are clustered near $\omega = 0$ and have decreasing magnitudes at higher frequencies. Except for the fact that the area of $S_{XX}(\omega)$ is not necessarily unity, $S_{XX}(\omega)$ has characteristics similar to a probability density function (it is nonnegative and real). Indeed, by dividing $S_{XX}(\omega)$ by its area, a new function is formed with area of unity that is analogous to a density function.

Recall that standard deviation is a measure of the spread in a density function. The analogous quantity for the normalized power spectrum is a measure of its spread that we call *rms bandwidth*,† which we denote W_{rms}(rad/s). Now since $S_{XX}(\omega)$ is an even function for a real process, its "mean value" is zero and its

---

† The notation rms bandwidth stands for *root-mean-squared* bandwidth.

"standard deviation" is the square root of its second moment. Thus, upon normalization, the rms bandwidth is given by

$$W_{rms}^2 = \frac{\int_{-\infty}^{\infty} \omega^2 S_{XX}(\omega)\, d\omega}{\int_{-\infty}^{\infty} S_{XX}(\omega)\, d\omega} \tag{7.1-22}$$

Example 7.1-3 Given the power spectrum

$$S_{XX}(\omega) = \frac{10}{[1 + (\omega/10)^2]^2}$$

where the 6-dB bandwidth is 10 radians per second, we find W_{rms}. First, using (C-28) from Appendix C,

$$\int_{-\infty}^{\infty} \frac{10\, d\omega}{[1 + (\omega/10)^2]^2} = 10^5 \int_{-\infty}^{\infty} \frac{d\omega}{(100 + \omega^2)^2}$$

$$= 10^5 \left\{ \frac{\omega}{200(100 + \omega^2)} \bigg|_{-\infty}^{\infty} + \frac{1}{2000} \tan^{-1}\left(\frac{\omega}{10}\right) \bigg|_{-\infty}^{\infty} \right\}$$

$$= 50\pi$$

Next, from (C-30) of Appendix C:

$$\int_{-\infty}^{\infty} \frac{10\omega^2\, d\omega}{[1 + (\omega/10)^2]^2} = 10^5 \int_{-\infty}^{\infty} \frac{\omega^2\, d\omega}{(100 + \omega^2)^2}$$

$$= 10^5 \left\{ \frac{-\omega}{2(100 + \omega^2)} \bigg|_{-\infty}^{\infty} + \frac{1}{20} \tan^{-1}\left(\frac{\omega}{10}\right) \bigg|_{-\infty}^{\infty} \right\}$$

$$= 5000\pi$$

Thus

$$W_{rms} = \sqrt{\frac{5000\pi}{50\pi}} = 10 \text{ rad/s}$$

Although W_{rms} and the 6-dB bandwidth of $S_{XX}(\omega)$ are equal in this case, they are not equal in general.

The above concept is readily extended to a process that has a bandpass form of power spectrum; that is, its significant spectral components cluster near some frequencies $\bar{\omega}_0$ and $-\bar{\omega}_0$. If we assume that the process $X(t)$ is real, $S_{XX}(\omega)$ will be real and have even symmetry about $\omega = 0$. With this assumption we define a *mean* frequency $\bar{\omega}_0$ by

$$\bar{\omega}_0 = \frac{\int_0^{\infty} \omega S_{XX}(\omega)\, d\omega}{\int_0^{\infty} S_{XX}(\omega)\, d\omega} \tag{7.1-23}$$

and rms bandwidth by

$$W_{\text{rms}}^2 = \frac{4 \int_0^\infty (\omega - \bar{\omega}_0)^2 S_{XX}(\omega) \, d\omega}{\int_0^\infty S_{XX}(\omega) \, d\omega} \tag{7.1-24}$$

The reader is encouraged to sketch a few lowpass and bandpass power spectrums and justify for himself why the factor of 4 appears in (7.1-24).

7.2 RELATIONSHIP BETWEEN POWER SPECTRUM AND AUTOCORRELATION FUNCTION

In Section 7.1 it was stated that the inverse Fourier transform of the power density spectrum is the time average of the autocorrelation function; that is

$$\frac{1}{2\pi} \int_{-\infty}^{\infty} S_{XX}(\omega)e^{j\omega\tau} \, d\omega = A[R_{XX}(t, t + \tau)] \tag{7.2-1}$$

This expression will now be proved.

If we use (7.1-5), which is the definition of $X_T(\omega)$, in the defining equation (7.1-11) for the power spectrum we have†

$$S_{XX}(\omega) = \lim_{T \to \infty} E\left[\frac{1}{2T} \int_{-T}^{T} X(t_1)e^{j\omega t_1} \, dt_1 \int_{-T}^{T} X(t_2)e^{-j\omega t_2} \, dt_2 \right]$$

$$= \lim_{T \to \infty} \frac{1}{2T} \int_{-T}^{T} \int_{-T}^{T} E[X(t_1)X(t_2)]e^{-j\omega(t_2 - t_1)} \, dt_2 \, dt_1 \tag{7.2-2}$$

The expectation in the integrand of (7.2-2) is identified as the autocorrelation function of $X(t)$:

$$E[X(t_1)X(t_2)] = R_{XX}(t_1, t_2) \qquad -T < (t_1 \text{ and } t_2) < T \tag{7.2-3}$$

Thus, (7.2-2) becomes

$$S_{XX}(\omega) = \lim_{T \to \infty} \frac{1}{2T} \int_{-T}^{T} \int_{-T}^{T} R_{XX}(t_1, t_2)e^{-j\omega(t_2 - t_1)} \, dt_1 \, dt_2 \tag{7.2-4}$$

Suppose we next make the variable changes

$$t = t_1 \qquad\qquad\qquad dt = dt_1 \tag{7.2-5a}$$

$$\tau = t_2 - t_1 = t_2 - t \qquad d\tau = dt_2 \tag{7.2-5b}$$

in (7.2-4); we obtain

$$S_{XX}(\omega) = \lim_{T \to \infty} \frac{1}{2T} \int_{-T-t}^{T-t} \int_{-T}^{T} R_{XX}(t, t + \tau) \, dt \, e^{-j\omega\tau} \, d\tau \tag{7.2-6}$$

---

† We use $X(t)$ in (7.1-5), rather than $x(t)$, to imply that the operations performed take place on the process as opposed to one sample function.

Next, taking the limit with respect to the τ integral first will allow us to interchange the limit and τ integral operations to get

$$\mathcal{S}_{XX}(\omega) = \int_{-\infty}^{\infty} \left\{ \lim_{T \to \infty} \frac{1}{2T} \int_{-T}^{T} R_{XX}(t, t + \tau)\, dt \right\} e^{-j\omega\tau}\, d\tau \tag{7.2-7}$$

The quantity within braces is recognized as the time average of the process autocorrelation function

$$A[R_{XX}(t, t + \tau)] = \lim_{T \to \infty} \frac{1}{2T} \int_{-T}^{T} R_{XX}(t, t + \tau)\, dt \tag{7.2-8}$$

Thus, (7.2-7) becomes

$$\mathcal{S}_{XX}(\omega) = \int_{-\infty}^{\infty} A[R_{XX}(t, t + \tau)] e^{-j\omega\tau}\, d\tau \tag{7.2-9}$$

which shows that $\mathcal{S}_{XX}(\omega)$ and $A[R_{XX}(t, t + \tau)]$ form a Fourier transform pair:

$$A[R_{XX}(t, t + \tau)] \leftrightarrow \mathcal{S}_{XX}(\omega) \tag{7.2-10}$$

This expression implies (7.2-1), which we started out to prove.

For the important case where $X(t)$ is at least wide-sense stationary, $A[R_{XX}(t, t + \tau)] = R_{XX}(\tau)$ and we get

$$\mathcal{S}_{XX}(\omega) = \int_{-\infty}^{\infty} R_{XX}(\tau) e^{-j\omega\tau}\, d\tau \tag{7.2-11}$$

$$R_{XX}(\tau) = \frac{1}{2\pi} \int_{-\infty}^{\infty} \mathcal{S}_{XX}(\omega) e^{j\omega\tau}\, d\omega \tag{7.2-12}$$

or

$$R_{XX}(\tau) \leftrightarrow \mathcal{S}_{XX}(\omega) \tag{7.2-13}$$

The expressions (7.2-11) and (7.2-12) are usually called the *Wiener-Khinchin relations* after the great American mathematician Norbert Wiener (1894–1964) and the German mathematician A. I. Khinchin (1894–1959). They form the basic link between the time domain description (correlation functions) of processes and their description in the frequency domain (power spectrum).

From (7.2-13), it is clear that knowledge of the power spectrum of a process allows complete recovery of the autocorrelation function when $X(t)$ is at least wide-sense stationary; for a nonstationary process, only the time average of the autocorrelation function is recoverable from (7.2-10).

Example 7.2-1 The power spectrum will be found for the random process of Example 6.2-1 that has the autocorrelation function

$$R_{XX}(\tau) = (A^2/2) \cos(\omega_0 \tau)$$

where A and ω_0 are constants. This equation can be written in the form

$$R_{XX}(\tau) = \frac{A^2}{4} \left(e^{j\omega_0\tau} + e^{-j\omega_0\tau} \right)$$

Now we note that the inverse transform of a frequency domain impulse function is

$$\frac{1}{2\pi} \int_{-\infty}^{\infty} \delta(\omega) e^{j\omega\tau} \, d\omega = \frac{1}{2\pi}$$

from (A-2) of Appendix A. Thus

$$1 \leftrightarrow 2\pi\delta(\omega)$$

and, from the frequency-shifting property of Fourier transforms given by (D-7) of Appendix D, we get

$$e^{j\omega_0\tau} \leftrightarrow 2\pi\delta(\omega - \omega_0)$$

By using this last result, the Fourier transform of $R_{XX}(\tau)$ becomes

$$S_{XX}(\omega) = \frac{A^2\pi}{2} \left[\delta(\omega - \omega_0) + \delta(\omega + \omega_0) \right]$$

This function and $R_{XX}(\tau)$ are illustrated in Figure 7.2-1.

(a)

(b)

Figure 7.2-1 The autocorrelation function (a) and power density spectrum (b) of the wide-sense stationary random process of Example 7.2-1.

7.3 CROSS-POWER DENSITY SPECTRUM AND ITS PROPERTIES

Consider a real random process $W(t)$ given by the sum of two other real processes $X(t)$ and $Y(t)$:

$$W(t) = X(t) + Y(t) \tag{7.3-1}$$

The autocorrelation function of $W(t)$ is

$$\begin{aligned}
R_{WW}(t, t + \tau) &= E[W(t)W(t + \tau)] \\
&= E\{[X(t) + Y(t)][X(t + \tau) + Y(t + \tau)]\} \\
&= R_{XX}(t, t + \tau) + R_{YY}(t, t + \tau) \\
&\quad + R_{XY}(t, t + \tau) + R_{YX}(t, t + \tau)
\end{aligned} \tag{7.3-2}$$

Now if we take the time average of both sides of (7.3-2) and Fourier transform the resulting expression by applying (7.2-9), we have

$$\mathcal{S}_{WW}(\omega) = \mathcal{S}_{XX}(\omega) + \mathcal{S}_{YY}(\omega) + \mathcal{F}\{A[R_{XY}(t, t + \tau)]\} + \mathcal{F}\{A[R_{YX}(t, t + \tau)]\} \tag{7.3-3}$$

where $\mathcal{F}\{\cdot\}$ represents the Fourier transform. It is clear that the left side of (7.3-3) is just the power spectrum of $W(t)$. Similarly, the first two right-side terms are the power spectrums of $X(t)$ and $Y(t)$, respectively. The second two right-side terms are new quantities that are the subjects of this section. It will be shown that they are *cross-power density spectrums* defined by (7.3-12) and (7.3-14) below.

The Cross-Power Density Spectrum

For two real random processes $X(t)$ and $Y(t)$, we define $x_T(t)$ and $y_T(t)$ as truncated ensemble members; that is

$$x_T(t) = \begin{cases} x(t) & -T < t < T \\ 0 & \text{elsewhere} \end{cases} \tag{7.3-4}$$

and

$$y_T(t) = \begin{cases} y(t) & -T < t < T \\ 0 & \text{elsewhere} \end{cases} \tag{7.3-5}$$

Both $x_T(t)$ and $y_T(t)$ are assumed to be magnitude integrable over the interval $(-T, T)$ as indicated by (7.1-4). As a consequence, they will possess Fourier transforms that we denote by $X_T(\omega)$ and $Y_T(\omega)$, respectively:

$$x_T(t) \leftrightarrow X_T(\omega) \tag{7.3-6}$$

$$y_T(t) \leftrightarrow Y_T(\omega) \tag{7.3-7}$$

We next define the *cross power* $P_{XY}(T)$ in the two processes within the interval $(-T, T)$ by

$$P_{XY}(T) = \frac{1}{2T} \int_{-T}^{T} x_T(t) y_T(t) \, dt = \frac{1}{2T} \int_{-T}^{T} x(t) y(t) \, dt \qquad (7.3\text{-}8)$$

Since $x_T(t)$ and $y_T(t)$ are Fourier transformable, Parseval's theorem (D-20) applies; its left side is the same as (7.3-8). Thus, we may write

$$P_{XY}(T) = \frac{1}{2T} \int_{-T}^{T} x(t) y(t) \, dt = \frac{1}{2\pi} \int_{-\infty}^{\infty} \frac{X_T^*(\omega) Y_T(\omega)}{2T} \, d\omega \qquad (7.3\text{-}9)$$

This cross power is a random quantity since its value will vary depending on which ensemble member is considered. We form the average cross power, denoted $\bar{P}_{XY}(T)$, by taking the expected value in (7.3-9). The result is

$$\bar{P}_{XY}(T) = \frac{1}{2T} \int_{-T}^{T} R_{XY}(t, t) \, dt = \frac{1}{2\pi} \int_{-\infty}^{\infty} \frac{E[X_T^*(\omega) Y_T(\omega)]}{2T} \, d\omega \qquad (7.3\text{-}10)$$

Finally, we form the total average cross power P_{XY} by letting $T \to \infty$:

$$P_{XY} = \lim_{T \to \infty} \frac{1}{2T} \int_{-T}^{T} R_{XY}(t, t) \, dt = \frac{1}{2\pi} \int_{-\infty}^{\infty} \lim_{T \to \infty} \frac{E[X_T^*(\omega) Y_T(\omega)]}{2T} \, d\omega \qquad (7.3\text{-}11)$$

It is clear that the integrand involving ω can be defined as a *cross-power density spectrum*; it is a function of ω which we denote

$$S_{XY}(\omega) = \lim_{T \to \infty} \frac{E[X_T^*(\omega) Y_T(\omega)]}{2T} \qquad (7.3\text{-}12)$$

Thus

$$P_{XY} = \frac{1}{2\pi} \int_{-\infty}^{\infty} S_{XY}(\omega) \, d\omega \qquad (7.3\text{-}13)$$

By repeating the above procedure, we can also define another cross-power density spectrum by

$$S_{YX}(\omega) = \lim_{T \to \infty} \frac{E[Y_T^*(\omega) X_T(\omega)]}{2T} \qquad (7.3\text{-}14)$$

Cross power is given by

$$P_{YX} = \frac{1}{2\pi} \int_{-\infty}^{\infty} S_{YX}(\omega) \, d\omega = P_{XY} \qquad (7.3\text{-}15)$$

Total cross power $P_{XY} + P_{YX}$ can be interpreted as the additional power two processes are capable of generating, over and above their individual powers, due to the fact that they are correlated.

Properties of the Cross-Power Density Spectrum

Some properties of the cross-power spectrum of real random processes $X(t)$ and $Y(t)$ are listed below without formal proofs.

(1) $\mathcal{S}_{XY}(\omega) = \mathcal{S}_{YX}(-\omega) = \mathcal{S}_{YX}^*(\omega)$ (7.3-16)

(2) Re $[\mathcal{S}_{XY}(\omega)]$ and Re $[\mathcal{S}_{YX}(\omega)]$ are even functions of ω (see Problem 7-40).

(7.3-17)

(3) Im $[\mathcal{S}_{XY}(\omega)]$ and Im $[\mathcal{S}_{YX}(\omega)]$ are odd functions of ω (see Problem 7-40).

(7.3-18)

(4) $\mathcal{S}_{XY}(\omega) = 0$ and $\mathcal{S}_{YX}(\omega) = 0$ if $X(t)$ and $Y(t)$ are orthogonal. (7.3-19)

(5) If $X(t)$ and $Y(t)$ are uncorrelated and have constant means $\bar{X}$ and $\bar{Y}$

$$\mathcal{S}_{XY}(\omega) = \mathcal{S}_{YX}(\omega) = 2\pi \bar{X} \bar{Y} \delta(\omega)$$ (7.3-20)

(6) $$A[R_{XY}(t, t + \tau)] \leftrightarrow \mathcal{S}_{XY}(\omega)$$ (7.3-21)

$$A[R_{YX}(t, t + \tau)] \leftrightarrow \mathcal{S}_{YX}(\omega)$$ (7.3-22)

In the above properties, Re $[\cdot]$ and Im $[\cdot]$ represent the real and imaginary parts, respectively, and $A[\cdot]$ represents the time average, as usual, defined by (6.2-21).

Property 1 follows from (7.3-12) and (7.3-14). Properties 2 and 3 are proved by considering the symmetry that $X_T(\omega)$ and $Y_T(\omega)$ must possess for real processes. Properties 4 and 5 may be proved by substituting the integral (Fourier transform) forms for $X_T(\omega)$ and $Y_T(\omega)$ into $E[X_T^*(\omega)Y_T(\omega)]$ and showing that the function has the necessary behavior under the stated assumptions.

Property 6 states that the cross-power density spectrum and the time average of the cross-correlation function are a Fourier transform pair; its development is given in Section 7.4. For the case of jointly wide-sense stationary processes, (7.3-21) and (7.3-22) reduce to the especially useful forms

$$\mathcal{S}_{XY}(\omega) = \int_{-\infty}^{\infty} R_{XY}(\tau)e^{-j\omega\tau} \, d\tau$$ (7.3-23)

$$\mathcal{S}_{YX}(\omega) = \int_{-\infty}^{\infty} R_{YX}(\tau)e^{-j\omega\tau} \, d\tau$$ (7.3-24)

$$R_{XY}(\tau) = \frac{1}{2\pi} \int_{-\infty}^{\infty} \mathcal{S}_{XY}(\omega)e^{j\omega\tau} \, d\omega$$ (7.3-25)

$$R_{YX}(\tau) = \frac{1}{2\pi} \int_{-\infty}^{\infty} \mathcal{S}_{YX}(\omega)e^{j\omega\tau} \, d\omega$$ (7.3-26)

Example 7.3-1 Suppose we are given a cross-power spectrum defined by

$$\mathcal{S}_{XY}(\omega) = \begin{cases} a + jb\omega/W & -W < \omega < W \\ 0 & \text{elsewhere} \end{cases}$$

where $W > 0$, a and b are real constants. We use (7.3-25) to find the cross-correlation function. It is

$$R_{XY}(\tau) = \frac{1}{2\pi} \int_{-W}^{W} \left(a + j\, \frac{b\omega}{W} \right) e^{j\omega\tau}\, d\omega$$

$$= \frac{a}{2\pi} \int_{-W}^{W} e^{j\omega\tau}\, d\omega + j\, \frac{b}{2\pi W} \int_{-W}^{W} \omega e^{j\omega\tau}\, d\omega$$

On using (C-45) and (C-46) this expression will readily reduce to

$$R_{XY}(\tau) = \frac{a}{2\pi} \left[\frac{e^{j\omega\tau}}{j\tau} \Big|_{-W}^{W} \right] + j\, \frac{b}{2\pi W} \left\{ e^{j\omega\tau} \left[\frac{\omega}{j\tau} - \frac{1}{(j\tau)^2} \right] \Big|_{-W}^{W} \right\}$$

$$= \frac{1}{\pi W \tau^2} \left[(aW\tau - b) \sin(W\tau) + bW\tau \cos(W\tau) \right]$$

★7.4 RELATIONSHIP BETWEEN CROSS-POWER SPECTRUM AND CROSS-CORRELATION FUNCTION

In the following discussion we show that

$$S_{XY}(\omega) = \int_{-\infty}^{\infty} \left\{ \lim_{T \to \infty} \frac{1}{2T} \int_{-T}^{T} R_{XY}(t, t+\tau)\, dt \right\} e^{-j\omega\tau}\, d\tau \qquad (7.4\text{-}1)$$

as indicated in (7.3-21).

The development consists of using the transforms of the truncated ensemble members, given by

$$X_T(\omega) = \int_{-T}^{T} x(t) e^{-j\omega t}\, dt \qquad (7.4\text{-}2)$$

$$Y_T(\omega) = \int_{-T}^{T} y(t) e^{-j\omega t}\, dt \qquad (7.4\text{-}3)$$

in (7.3-12) and then taking the expected value and limit as indicated to obtain $S_{XY}(\omega)$. From (7.4-2) and (7.4-3):

$$X_T^*(\omega) Y_T(\omega) = \int_{-T}^{T} x(t_1) e^{j\omega t_1}\, dt_1 \int_{-T}^{T} y(t_2) e^{-j\omega t_2}\, dt_2$$

$$= \int_{-T}^{T} \int_{-T}^{T} x(t_1) y(t_2) e^{-j\omega(t_2 - t_1)}\, dt_1\, dt_2 \qquad (7.4\text{-}4)$$

Now by changing variables according to (7.2-5), dividing by $2T$, and taking the expected value, (7.4-4) becomes

$$\frac{E[X_T^*(\omega) Y_T(\omega)]}{2T} = E\left[\int_{-T-t}^{T-t} \left\{ \frac{1}{2T} \int_{-T}^{T} x(t) y(t+\tau)\, dt \right\} e^{-j\omega\tau}\, d\tau \right]$$

$$= \int_{-T-t}^{T-t} \left\{ \frac{1}{2T} \int_{-T}^{T} R_{XY}(t, t+\tau)\, dt \right\} e^{-j\omega\tau}\, d\tau \qquad (7.4\text{-}5)$$

After the limit is taken:

$$S_{XY}(\omega) = \lim_{T \to \infty} \frac{E[X_T^*(\omega)Y_T(\omega)]}{2T}$$

$$= \lim_{T \to \infty} \int_{-T-t}^{T-t} \left\{ \frac{1}{2T} \int_{-T}^{T} R_{XY}(t, t+\tau)\, dt \right\} e^{-j\omega\tau}\, d\tau$$

$$= \int_{-\infty}^{\infty} \left\{ \lim_{T \to \infty} \frac{1}{2T} \int_{-T}^{T} R_{XY}(t, t+\tau)\, dt \right\} e^{-j\omega\tau}\, d\tau \qquad (7.4\text{-}6)$$

which is the same as (7.4-1). Since (7.4-6) is a Fourier transform, and such transforms are unique, the inverse transform applies:

$$\lim_{T \to \infty} \frac{1}{2T} \int_{-T}^{T} R_{XY}(t, t+\tau)\, dt = \frac{1}{2\pi} \int_{-\infty}^{\infty} S_{XY}(\omega)e^{j\omega\tau}\, d\omega \qquad (7.4\text{-}7)$$

It should be noted from (7.4-7) that, given the cross-power spectrum, the cross-correlation function cannot in general be recovered, only its time average can. For jointly wide-sense stationary processes, however, the cross-correlation function $R_{XY}(\tau)$ *can* be found from $S_{XY}(\omega)$ since its time average is just $R_{XY}(\tau)$.

Although we shall not give the proof, a development similar to the above shows that (7.3-22) is true.

Example 7.4-1 Let the cross-correlation function of two processes $X(t)$ and $Y(t)$ be

$$R_{XY}(t, t+\tau) = \frac{AB}{2} \{ \sin(\omega_0\tau) + \cos[\omega_0(2t+\tau)] \}$$

where A, B, and ω_0 are constants. We find the cross-power spectrum by use of (7.4-1). First, the time average is formed

$$\lim_{T \to \infty} \frac{1}{2T} \int_{-T}^{T} R_{XY}(t, t+\tau)\, dt$$

$$= \frac{AB}{2} \sin(\omega_0\tau) + \frac{AB}{2} \lim_{T \to \infty} \frac{1}{2T} \int_{-T}^{T} \cos[\omega_0(2t+\tau)]\, dt$$

The integral is readily evaluated and is found to be zero. Finally we Fourier transform the time-averaged cross-correlation function with the aid of pair 12 of Appendix E:

$$S_{XY}(\omega) = \mathcal{F}\left\{ \frac{AB}{2} \sin(\omega_0\tau) \right\}$$

$$= \frac{-j\pi AB}{2} [\delta(\omega - \omega_0) - \delta(\omega + \omega_0)]$$

7.5 SOME NOISE DEFINITIONS AND OTHER TOPICS

In many practical problems it is helpful to sometimes characterize noise through its power density spectrum. Indeed, in the following discussions we *define* two forms of noise on the basis of their power spectrums. We also consider the response of a product device when one of its input waveforms is a random signal or noise.

White and Colored Noise

A sample function $n(t)$ of a wide-sense stationary noise random process $N(t)$ is called *white noise* if the power density spectrum of $N(t)$ is a constant at all frequencies. Thus, we define

$$S_{NN}(\omega) = \mathcal{N}_0/2 \qquad (7.5\text{-}1)$$

for white noise, where $\mathcal{N}_0$ is a real positive constant. By inverse Fourier transformation of (7.5-1), the autocorrelation function of $N(t)$ is found to be

$$R_{NN}(\tau) = (\mathcal{N}_0/2)\,\delta(\tau) \qquad (7.5\text{-}2)$$

The above two functions are illustrated in Figure 7.5-1. White noise derives its name by analogy with "white" light, which contains all visible light frequencies in its spectrum.

White noise is unrealizable as can be seen by the fact that it possesses infinite average power:

$$\frac{1}{2\pi} \int_{-\infty}^{\infty} S_{NN}(\omega)\, d\omega = \infty \qquad (7.5\text{-}3)$$

However, one type of real-world noise closely approximates white noise. *Thermal noise* generated by thermal agitation of electrons in any electrical conductor has a power spectrum that is constant up to very high frequencies and then decreases. For example, a resistor at temperature T in kelvin produces a noise voltage

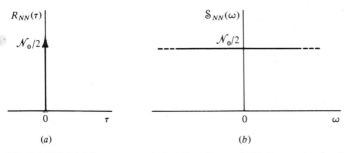

Figure 7.5-1 (*a*) The autocorrelation function and (*b*) the power density spectrum of white noise. [*Adapted from Peebles (1976) with permission of publishers Addison-Wesley, Advanced Book Program.*]

across its open-circuited terminals having the power spectrum† (Carlson, 1975, p. 118)

$$S_{NN}(\omega) = \frac{(\mathcal{N}_0/2)(\alpha|\omega|/T)}{e^{\alpha|\omega|/T} - 1} \tag{7.5-4}$$

where $\alpha = 7.64(10^{-12})$ kelvin-seconds is a constant. At a temperature of $T = 290$ K (usually called *room temperature* although it corresponds to a rather cool room at 63°F), this function remains above 0.9 $(\mathcal{N}_0/2)$ for frequencies up to 10^{12} Hz or 1000 GHz. Thus, thermal noise has a nearly flat spectrum at all frequencies that are likely to ever be used in radio, microwave, or millimeter-wave systems.‡

Noise having a nonzero and constant power spectrum over a *finite* frequency band and zero everywhere else is called *band-limited white noise*. Figure 7.5-2a

† The unit of $S_{NN}(\omega)$ is actually volts squared per hertz. According to our convention, we obtain watts per hertz by presuming the voltage exists across a 1-Ω resistor.

‡ This statement must be reexamined for $T < 290$ K, such as in some superconducting systems or other low-temperature devices (masers).

(a)

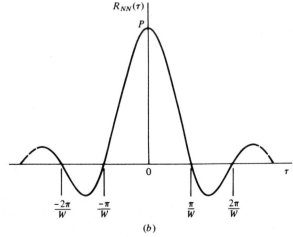

(b)

Figure 7.5-2 Power density spectrum (a) and autocorrelation function (b) of lowpass band-limited white noise.

depicts such a power spectrum that is lowpass. Here

$$S_{NN}(\omega) = \begin{cases} \dfrac{P\pi}{W} & -W < \omega < W \\ 0 & \text{elsewhere} \end{cases} \tag{7.5-5}$$

Inverse transformation of (7.5-5) gives the autocorrelation function shown in Figure 7.5-2b:

$$R_{NN}(\tau) = P \frac{\sin (W\tau)}{W\tau} \tag{7.5-6}$$

The constant P equals the power in the noise.

Band-limited white noise can also be bandpass as illustrated in Figure 7.5-3.

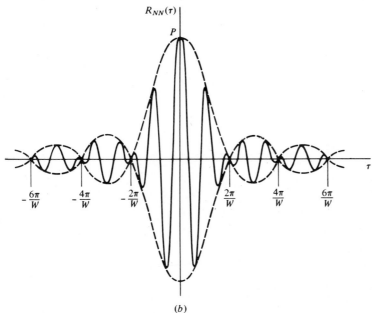

Figure 7.5-3 Power density spectrum (a) and autocorrelation function (b) for bandpass band-limited white noise. [*Adapted from Peebles (1976) with permission of publishers Addison-Wesley, Advanced Book Program.*]

The applicable power spectrum and autocorrelation function are:

$$\mathcal{S}_{NN}(\omega) = \begin{cases} P\pi/W & \omega_0 - (W/2) < |\omega| < \omega_0 + (W/2) \\ 0 & \text{elsewhere} \end{cases} \tag{7.5-7}$$

and

$$R_{NN}(\tau) = P \frac{\sin (W\tau/2)}{(W\tau/2)} \cos (\omega_0 \tau) \tag{7.5-8}$$

where ω_0 and W are constants and P is the power in the noise.

Again, by analogy with colored light that has only a portion of the visible light frequencies in its spectrum, we define *colored noise* as any noise that is not white. An example serves to illustrate colored noise.

Example 7.5-1 A wide-sense stationary noise process $N(t)$ has an autocorrelation function

$$R_{NN}(\tau) = Pe^{-3|\tau|}$$

where P is a constant. We find its power spectrum. It is

$$\mathcal{S}_{NN}(\omega) = \int_{-\infty}^{\infty} Pe^{-3|\tau|}e^{-j\omega\tau}\, d\tau$$

$$= P \int_{0}^{\infty} e^{-(3 + j\omega)\tau}\, d\tau + P \int_{-\infty}^{0} e^{(3 - j\omega)\tau}\, d\tau$$

(a)

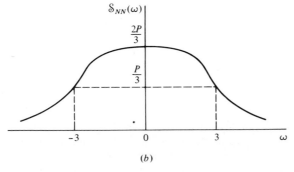

(b)

Figure 7.5-4 The autocorrelation function (a) and power spectrum (b) of the colored noise of Example 7.5-1. [*Adapted from Peebles (1976) with permission of publishers Addison-Wesley, Advanced Book Program.*]

These integrals easily evaluate using (C-45) to give

$$S_{NN}(\omega) = \frac{P}{3 + j\omega} + \frac{P}{3 - j\omega} = \frac{6P}{9 + \omega^2}$$

This power spectrum is sketched in Figure 7.5-4 along with the preceding autocorrelation function.

Product Device Response to a Random Signal

Product devices are frequently encountered in electrical systems. Often they involve the product of a random waveform $X(t)$ (either signal or noise or the sum of signal and noise) with a cosine (or sine) "carrier" wave as illustrated in Figure 7.5-5. The response is the new process

$$Y(t) = X(t)A_0 \cos(\omega_0 t) \qquad (7.5\text{-}9)$$

where A_0 and ω_0 are constants. We seek to find the power spectrum $S_{YY}(\omega)$ of $Y(t)$ in terms of the power spectrum $S_{XX}(\omega)$ of $X(t)$.

The autocorrelation function of $Y(t)$ is

$$
\begin{aligned}
R_{YY}(t, t + \tau) &= E[Y(t)Y(t + \tau)] \\
&= E[A_0^2 X(t)X(t + \tau) \cos(\omega_0 t) \cos(\omega_0 t + \omega_0 \tau)] \\
&= \frac{A_0^2}{2} R_{XX}(t, t + \tau)[\cos(\omega_0 \tau) + \cos(2\omega_0 t + \omega_0 \tau)] \quad (7.5\text{-}10)
\end{aligned}
$$

Even if $X(t)$ is wide-sense stationary $Y(t)$ is not since $R_{YY}(t, t + \tau)$ depends on t. Thus, we apply (7.1-19) to obtain $S_{YY}(\omega)$ after we take the time average of $R_{YY}(t, t + \tau)$. Let $X(t)$ be assumed wide-sense stationary. Then (7.5-10) becomes

$$A[R_{YY}(t, t + \tau)] = \frac{A_0^2}{2} R_{XX}(\tau) \cos(\omega_0 \tau) \qquad (7.5\text{-}11)$$

On Fourier transforming (7.5-11) we have

$$S_{YY}(\omega) = \frac{A_0^2}{4} [S_{XX}(\omega - \omega_0) + S_{XX}(\omega + \omega_0)] \qquad (7.5\text{-}12)$$

A possible power density spectrum of $X(t)$ and that given by (7.5-12) are illustrated in Figure 7.5-6. It presumes that $X(t)$ is a lowpass process, although this is not a constraint in applying (7.5-12).

Figure 7.5-5 A product of interest in electrical systems. [*Adapted from Peebles (1976) with permission of publishers Addison-Wesley, Advanced Book Program.*]

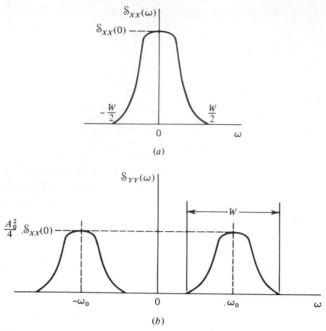

Figure 7.5-6 Power density spectrums applicable to Figure 7.5-5; (*a*) at the input and (*b*) at the output. [*Adapted from Peebles (1976) with permission of publishers Addison-Wesley, Advanced Book Program.*]

Example 7.5-2 One important use of the product device is in recovery (demodulation) of the information signal (music, speech, etc.) conveyed in the wave transmitted from a conventional broadcast radio station that uses AM (*amplitude modulation*). The wave received by a receiver tuned to a station with frequency $\omega_0/2\pi$ is one input to the product device. The other is a "local oscillator" signal $A_0 \cos(\omega_0 t)$ generated within the receiver. The product device output passes through a lowpass filter which has as its output the desired information signal. Unfortunately, this signal also contains noise because noise is also present at the input to the product device; the input noise is added to the received radio wave. We shall calculate the power in the output noise of the product demodulator.

Let the power spectrum of the input noise, denoted $X(t)$, be approximated by an idealized (rectangular) function with bandwidth W_{RF} centered at $\pm \omega_0$. Thus,

$$S_{XX}(\omega) = \begin{cases} \mathcal{N}_0/2 & -\omega_0 - (W_{RF}/2) < \omega < -\omega_0 + (W_{RF}/2) \\ \mathcal{N}_0/2 & \omega_0 - (W_{RF}/2) < \omega < \omega_0 + (W_{RF}/2) \\ 0 & \text{elsewhere} \end{cases}$$

where $\mathcal{N}_0/2$ is the power density within the noise band. By applying (7.5-12) the power density spectrum of the output noise $Y(t)$ of the product device is readily found (by sketch) to be

$$S_{YY}(\omega) = \begin{cases} \mathcal{N}_0 A_0^2/8 & -2\omega_0 - (W_{RF}/2) < \omega < -2\omega_0 + (W_{RF}/2) \\ \mathcal{N}_0 A_0^2/4 & -W_{RF}/2 < \omega < W_{RF}/2 \\ \mathcal{N}_0 A_0^2/8 & 2\omega_0 - (W_{RF}/2) < \omega < 2\omega_0 + (W_{RF}/2) \\ 0 & \text{elsewhere} \end{cases}$$

Now only the noise in the band $-W_{RF}/2 < \omega < W_{RF}/2$ cannot be removed by a lowpass filter (which usually follows the product device to remove unwanted noise and other undesired outputs) because the desired signal is in the same band. This remaining component of $S_{YY}(\omega)$ gives rise to the final output noise power, denoted N_o,

$$N_o = \frac{1}{2\pi} \int_{-W_{RF}/2}^{W_{RF}/2} \frac{\mathcal{N}_0 A_0^2}{4} \, d\omega = \frac{\mathcal{N}_0 A_0^2 W_{RF}}{8\pi}$$

*7.6 POWER SPECTRUMS OF COMPLEX PROCESSES

Power spectrums may readily be defined for complex processes. We consider only those processes that are at least wide-sense stationary. In terms of the autocorrelation function $R_{ZZ}(\tau)$ of a complex random process $Z(t)$, the power density spectrum is defined as its Fourier transform

$$S_{ZZ}(\omega) = \int_{-\infty}^{\infty} R_{ZZ}(\tau)e^{-j\omega\tau} \, d\tau \tag{7.6-1}$$

The inverse transform applies, so

$$R_{ZZ}(\tau) = \frac{1}{2\pi} \int_{-\infty}^{\infty} S_{ZZ}(\omega)e^{j\omega\tau} \, d\omega \tag{7.6-2}$$

For two jointly wide-sense stationary complex processes $Z_m(t)$ and $Z_n(t)$, their cross-power density spectrum and cross-correlation function are a Fourier transform pair:

$$S_{Z_m Z_n}(\omega) = \int_{-\infty}^{\infty} R_{Z_m Z_n}(\tau)e^{-j\omega\tau} \, d\tau \tag{7.6-3}$$

$$R_{Z_m Z_n}(\tau) = \frac{1}{2\pi} \int_{-\infty}^{\infty} S_{Z_m Z_n}(\omega)e^{j\omega\tau} \, d\omega \tag{7.6-4}$$

An equivalent statement is:

$$R_{Z_m Z_n}(\tau) \leftrightarrow S_{Z_m Z_n}(\omega) \tag{7.6-5}$$

Example 7.6-1 We reconsider the complex process $V(t)$ of Example 6.6-1 and find its power spectrum. From the previous example

$$R_{VV}(\tau) = e^{j\omega_0\tau} \sum_{n=1}^{N} \overline{A_n^2}$$

On Fourier transforming this autocorrelation function we obtain

$$S_{VV}(\omega) = \mathcal{F}\left\{ e^{j\omega_0\tau} \sum_{n=1}^{N} \overline{A_n^2} \right\}$$

$$= \sum_{n=1}^{N} \overline{A_n^2} \mathcal{F}\{e^{j\omega_0\tau}\}$$

$$= 2\pi\delta(\omega - \omega_0) \sum_{n=1}^{N} \overline{A_n^2}$$

after using pair 9 of Appendix E.

PROBLEMS

7-1 We are given the random process

$$X(t) = A \cos(\omega_0 t + \Theta)$$

where A and ω_0 are constants and Θ is a random variable uniformly distributed on the interval $(0, \pi)$.
   (a) Is $X(t)$ wide-sense stationary?
   (b) Find the power in $X(t)$ by using (7.1-10).
   (c) Find the power spectrum of $X(t)$ by using (7.1-11) and calculate power from (7.1-12). Do your two powers agree?

7-2 Work Problem 7-1 if the process is defined by

$$X(t) = u(t)A \cos(\omega_0 t + \Theta)$$

where $u(t)$ is the unit-step function.

★7-3 Work Problem 7-2 assuming Θ is uniform on the interval $(0, \pi/2)$.

7-4 Work Problem 7-1 if the random process is given by $X(t) = A \sin(\omega_0 t + \Theta)$.

★7-5 Work Problem 7-1 if the random process is

$$X(t) = A^2 \cos^2(\omega_0 t + \Theta)$$

7-6 Let A and B be random variables. We form the random process

$$X(t) = A \cos(\omega_0 t) + B \sin(\omega_0 t)$$

where ω_0 is a real constant.

(a) Show that if A and B are uncorrelated with zero means and equal variances, then $X(t)$ is wide-sense stationary.

(b) Find the autocorrelation function of $X(t)$.

(c) Find the power density spectrum.

7-7 A limiting form for the impulse function was given in Example 7.1-2. Give arguments to show that the following are also true:

(a) $\lim_{T \to \infty} T \exp[-\pi \alpha^2 T^2] = \delta(\alpha)$

(b) $\lim_{T \to \infty} \dfrac{T}{2} \exp[-|\alpha| T] = \delta(\alpha)$

7-8 Work Problem 7-7 for the following cases:

(a) $\lim_{T \to \infty} \dfrac{T}{\pi} \dfrac{\sin(\alpha T)}{\alpha T} = \delta(\alpha)$

(b) $\lim_{\substack{T \to \infty \\ |\alpha| < 1/T}} T[1 - |\alpha| T] = \delta(\alpha)$

7-9 Show that (7.1-14) is true.

7-10 Prove (7.1-17). [*Hint:* Use (D-6) of Appendix D and the definition of the derivative.]

7-11 A random process is defined by

$$Y(t) = X(t) \cos(\omega_0 t + \Theta)$$

where $X(t)$ is a lowpass wide-sense stationary process, ω_0 is a real constant, and Θ is a random variable uniformly distributed on the interval $(0, 2\pi)$. Find and sketch the power density spectrum of $Y(t)$ in terms of that of $X(t)$. Assume Θ is independent of $X(t)$.

7-12 Determine which of the following functions can and cannot be valid power density spectrums. For those that are not, explain why.

(a) $\dfrac{\omega^2}{\omega^6 + 3\omega^2 + 3}$ (b) $\exp[-(\omega - 1)^2]$

(c) $\dfrac{\omega^2}{\omega^4 + 1} - \delta(\omega)$ (d) $\dfrac{\omega^4}{1 + \omega^2 + j\omega^6}$

7-13 Work Problem 7-12 for the following functions.

(a) $\dfrac{\cos(3\omega)}{1 + \omega^2}$ (b) $\dfrac{1}{(1 + \omega^2)^2}$

(c) $\dfrac{|\omega|}{1 + 2\omega + \omega^2}$ (d) $\dfrac{1}{\sqrt{1 - 3\omega^2}}$

7-14 Given that $X(t) = \sum_{i=1}^{N} \alpha_i X_i(t)$ where $\{\alpha_i\}$ is a set of real constants and the processes $X_i(t)$ are stationary and orthogonal, show that

$$S_{XX}(\omega) = \sum_{i=1}^{N} \alpha_i^2 S_{X_i X_i}(\omega)$$

7-15 A random process is given by

$$X(t) = A \cos(\Omega t + \Theta)$$

where A is a real constant, Ω is a random variable with density function $f_\Omega(\cdot)$, and Θ is a random variable uniformly distributed on the interval $(0, 2\pi)$ independent of Ω. Show that the power spectrum of $X(t)$ is

$$S_{XX}(\omega) = \frac{\pi A^2}{2} [f_\Omega(\omega) + f_\Omega(-\omega)]$$

7-16 If $X(t)$ is a stationary process, find the power spectrum of

$$Y(t) = A + BX(t)$$

in terms of the power spectrum of $X(t)$ if A and B are real constants.

7-17 Find the power density spectrum of the random process for which

$$R_{XX}(\tau) = P \cos^4(\omega_0 \tau)$$

if P and ω_0 are constants. Determine the power in the process by use of (7.1-12).

7-18 A random process has the power density spectrum

$$S_{XX}(\omega) = \frac{6\omega^2}{1 + \omega^4}$$

Find the average power in the process.

7-19 Work Problem 7-18 for the power spectrum

$$S_{XX}(\omega) = \frac{6\omega^2}{[1 + \omega^2]^3}$$

7-20 Work Problem 7-18 for the power spectrum

$$S_{XX}(\omega) = \frac{6\omega^2}{(1 + \omega^2)^4}$$

7-21 Assume $X(t)$ is a wide-sense stationary process with nonzero mean value $\bar{X} \neq 0$. Show that

$$S_{XX}(\omega) = 2\pi \bar{X}^2 \delta(\omega) + \int_{-\infty}^{\infty} C_{XX}(\tau) e^{-j\omega\tau} \, d\tau$$

where $C_{XX}(\tau)$ is the autocovariance function of $X(t)$.

7-22 For a random process $X(t)$, assume that

$$R_{XX}(\tau) = P e^{-\tau^2/2a^2}$$

where $P > 0$ and $a > 0$ are constants. Find the power density spectrum of $X(t)$. [*Hint:* Use Appendix E to evaluate the Fourier transform of $R_{XX}(\tau)$.]

7-23 A random process has an autocorrelation function

$$R_{XX}(\tau) = \begin{cases} P[1 - (2\tau/T)] & 0 < \tau \leq T/2 \\ P[1 + (2\tau/T)] & -T/2 \leq \tau \leq 0 \\ 0 & \tau < -T/2 \quad \text{and} \quad \tau > T/2 \end{cases}$$

Find and sketch its power density spectrum. (*Hint:* Use Appendix E.)

****7-24** A random process $X(t)$ has a periodic autocorrelation function where the function of Problem 7-23 forms the central period of duration T. Find and sketch the power spectrum.

7-25 If the random processes of Problem 7-14 are stationary, zero-mean, statistically independent processes, show that the power spectrum of the sum is the same as for orthogonal processes. For stationary independent processes with nonzero means, what is $S_{XX}(\omega)$?

7-26 Given that a process $X(t)$ has the autocorrelation function

$$R_{XX}(\tau) = Ae^{-\alpha|\tau|} \cos(\omega_0 \tau)$$

where $A > 0$, $\alpha > 0$, and ω_0 are real constants, find the power spectrum of $X(t)$.

7-27 A random process $X(t)$ having the power spectrum of Problem 7-19 is applied to an ideal differentiator.

(*a*) Find the power spectrum of the differentiator's output.

(*b*) What is the power in the derivative?

7-28 Work Problem 7-27 for the power spectrum of Problem 7-20.

7-29 A wide-sense stationary random process $X(t)$ is used to define another process by

$$Y(t) = \int_{-\infty}^{\infty} h(\xi)X(t - \xi) \, d\xi$$

where $h(t)$ is some real function having a Fourier transform $H(\omega)$. Show that the power spectrum of $Y(t)$ is given by

$$S_{YY}(\omega) = S_{XX}(\omega)|H(\omega)|^2$$

7-30 A deterministic signal $A \cos(\omega_0 t)$, where A and ω_0 are real constants, is added to a noise process $N(t)$ for which

$$S_{NN}(\omega) = \frac{W^2}{W^2 + \omega^2}$$

and $W > 0$ is a constant.

(*a*) Find the ratio of average signal power to average noise power.

(*b*) What value of W maximizes the signal-to-noise ratio? What is the consequence of choosing this value of W?

7-31 Find the rms bandwidth of the power spectrum

$$
\mathcal{S}_{XX}(\omega) = \begin{cases} \dfrac{P}{1 + (\omega/W)^2} & |\omega| < KW \\ 0 & |\omega| > KW \end{cases}
$$

where P, W, and K are real positive constants. If $K \to \infty$, what happens?

7-32 Find the rms bandwidth of the power spectrum

$$
\mathcal{S}_{XX}(\omega) = \begin{cases} P \cos{(\pi\omega/2W)} & |\omega| \le W \\ 0 & |\omega| > W \end{cases}
$$

where $W > 0$ and $P > 0$ are constants.

7-33 Determine the rms bandwidth of the power spectrums given by:

(a) $\mathcal{S}_{XX}(\omega) = \begin{cases} P & |\omega| < W \\ 0 & |\omega| > W \end{cases}$

(b) $\mathcal{S}_{XX}(\omega) = \begin{cases} P[1 - |\omega/W|] & |\omega| \le W \\ 0 & |\omega| > W \end{cases}$

where P and W are real positive constants.

*7-34 Given the power spectrum

$$
\mathcal{S}_{XX}(\omega) = \frac{P}{\left[1 + \left(\dfrac{\omega - \alpha}{W}\right)^2\right]^2} + \frac{P}{\left[1 + \left(\dfrac{\omega + \alpha}{W}\right)^2\right]^2}
$$

where P, α, and W are real positive constants, find the mean frequency and rms bandwidth.

7-35 Show that the rms bandwidth of the power spectrum of a real bandpass process $X(t)$ is given by

$$
W_{\text{rms}}^2 = 4[\overline{W^2} - \bar{\omega}_0^2]
$$

where $\bar{\omega}_0$ is given by (7.1-23) and $\overline{W^2}$ is given by the right side of (7.1-22).

*7-36 Jointly wide-sense stationary random processes $X(t)$ and $Y(t)$ define a process $W(t)$ by

$$
W(t) = X(t) \cos{(\omega_0 t)} + Y(t) \sin{(\omega_0 t)}
$$

where ω_0 is a real positive constant.

(a) Develop some conditions on the mean values and correlation functions of $X(t)$ and $Y(t)$ such that $W(t)$ is wide-sense stationary.

(b) With the conditions of part (a) applied to $W(t)$, find its power spectrum in terms of power spectrums of $X(t)$ and $Y(t)$.

(c) If $X(t)$ and $Y(t)$ are also uncorrelated, what is the power spectrum of $W(t)$?

7-37 A random process is given by

$$W(t) = AX(t) + BY(t)$$

where A and B are real constants and $X(t)$ and $Y(t)$ are jointly wide-sense station-ary processes.

    (a) Find the power spectrum $S_{WW}(\omega)$ of $W(t)$.
    (b) Find $S_{WW}(\omega)$ if $X(t)$ and $Y(t)$ are uncorrelated.
    (c) Find the cross-power spectrums $S_{XW}(\omega)$ and $S_{YW}(\omega)$.

*7-38 Define two random processes by

$$X(t) = A \cos (\omega_0 t + \Theta)$$

$$Y(t) = W(t) \cos (\omega_0 t + \Theta)$$

where A and ω_0 are real positive constants, Θ is a random variable independent of $W(t)$, and $W(t)$ is a random process with a constant mean value $\bar{W}$. By using (7.3-12), show that

$$S_{XY}(\omega) = \frac{A\bar{W}\pi}{2} [\delta(\omega - \omega_0) + \delta(\omega + \omega_0)]$$

regardless of the form of the probability density function of Θ.

*7-39 Again consider the random processes of Problem 7-38.

    (a) Use (6.3-11) to show that the cross-correlation function is given by

$$R_{XY}(t, t + \tau) = \frac{A\bar{W}}{2} \{\cos (\omega_0 \tau) + E[\cos (2\Theta)] \cos (2\omega_0 t + \omega_0 \tau)$$

$$- E[\sin (2\Theta)] \sin (2\omega_0 t + \omega_0 \tau)\}$$

where the expectation is with respect to Θ only.

    (b) Find the time average of $R_{XY}(t, t + \tau)$ and determine the cross-power density spectrum $S_{XY}(\omega)$.

7-40 Decompose the cross-power spectrums into real and imaginary parts according to

$$S_{XY}(\omega) = R_{XY}(\omega) + jI_{XY}(\omega)$$

$$S_{YX}(\omega) = R_{YX}(\omega) + jI_{YX}(\omega)$$

and prove that

$$R_{XY}(\omega) = R_{YX}(-\omega) = R_{YX}(\omega)$$

$$I_{XY}(\omega) = I_{YX}(-\omega) = -I_{YX}(\omega)$$

7-41 From the results of Problem 7-40, prove (7.3-16).

7-42 Show that (7.3-19) and (7.3-20) are true.

7-43 (a) Sketch the power spectrum of (7.5-4) as a function of $\alpha\omega/T$.

(b) For what values of ω will $S_{NN}(\omega)$ remain above $0.5(\mathcal{N}_0/2)$ when $T = 4.2$ K (the value of liquid helium at one atmosphere of pressure)? These values form the region where thermal noise is approximately white in some amplifiers operated at very low temperatures, such as a *maser*.

7-44 For the power spectrum given in Figure 7.5-2a, show that (7.5-6) defines the corresponding band-limited noise autocorrelation function.

7-45 Show that (7.5-8) gives the autocorrelation function of the bandpass band-limited noise defined by Figure 7.5-3a.

7-46 A lowpass random process $X(t)$ has a continuous power spectrum $S_{XX}(\omega)$ and $S_{XX}(0) \neq 0$. Find the bandwidth W of a lowpass band-limited white-noise power spectrum having a density $S_{XX}(0)$ and the same total power as in $X(t)$.

7-47 Work Problem 7-46 for a bandpass process assuming $S_{XX}(\omega_0) \neq 0$, where ω_0 is some convenient frequency about which the spectral components of $X(t)$ cluster.

★7-48 A complex random process is given by

$$Z(t) = Ae^{j\Omega t}$$

where Ω is a random variable with probability density function $f_\Omega(\cdot)$ and A is a complex constant. Show that the power spectrum of $Z(t)$ is

$$S_{ZZ}(\omega) = 2\pi |A|^2 f_\Omega(\omega)$$

ADDITIONAL PROBLEMS

7-49 The autocorrelation function of a random process $X(t)$ is

$$R_{XX}(\tau) = 3 + 2 \exp(-4\tau^2)$$

(a) Find the power spectrum of $X(t)$.
(b) What is the average power in $X(t)$?
(c) What fraction of the power lies in the frequency band $-1/\sqrt{2} \leq \omega \leq 1/\sqrt{2}$?

7-50 State whether or not each of the following functions can be a valid power density spectrum. For those that cannot, explain why.

(a) $\dfrac{|\omega| \exp(-4\omega^2)}{1 + j\omega}$

(b) $\cos(3\omega) \exp(-\omega^2 + j2\omega)$

(c) $\dfrac{\omega^6}{(12 + \omega^2)^6}$

(d) $6 \tan[12\omega/(1 + \omega^2)]$

(e) $\cos^2(\omega) \exp(-8\omega^2)$

(f) $(-j\omega)(j\omega)/(3 - j\omega)^2(3 + j\omega)^2$

7-51 If $S_{XX}(\omega)$ is a valid power spectrum of a random process $X(t)$, discuss whether the functions $dS_{XX}(\omega)/d\omega$ and $d^2S_{XX}(\omega)/d\omega^2$ can be valid power spectrums.

7-52 (a) Rework Problem 7-15 and show that even if Θ is a constant (not random) the power spectrum is still given by

$$S_{XX}(\omega) = (\pi A^2/2)[f_\Omega(\omega) + f_\Omega(-\omega)]$$

[*Hint:* Time-average the autocorrelation function before Fourier transforming to obtain $S_{XX}(\omega)$.]

(b) Find the total power in $X(t)$ and show that it is independent of the form of the density function $f_\Omega(\omega)$.

7-53 Find the rms bandwidth of the power spectrum

$$S_{XX}(\omega) = 1/[1 + (\omega/W)^2]^3$$

where $W > 0$ is a constant.

7-54 Work Problem 7-53 for the power spectrum

$$S_{XX}(\omega) = \omega^2/[1 + (\omega/W)^2]^3$$

7-55 Work Problem 7-53 for the power spectrum

$$S_{XX}(\omega) = 1/[1 + (\omega/W)^2]^4$$

7-56 Work Problem 7-53 for the power spectrum

$$S_{XX}(\omega) = \omega^2/[1 + (\omega/W)^2]^4$$

7-57 Generalize Problems 7-53 and 7-55 by finding the rms bandwidth of the power spectrum

$$S_{XX}(\omega) = 1/[1 + (\omega/W)^2]^N$$

where $N \geq 2$ is an integer.

7-58 Generalize Problems 7-54 and 7-56 by finding the rms bandwidth of the power spectrum

$$S_{XX}(\omega) = \omega^2/[1 + (\omega/W)^2]^N$$

where $N \geq 3$ is an integer.

7-59 Assume a random process has a power spectrum

$$S_{XX}(\omega) = \begin{cases} 4 - (\omega^2/9) & |\omega| \leq 6 \\ 0 & \text{elsewhere} \end{cases}$$

Find (a) the average power, (b) the rms bandwidth, and (c) the autocorrelation function of the process.

7-60 Show that rms bandwidth of a lowpass random process $X(t)$, as given by (7.1-22), can also be obtained from

$$W_{\text{rms}}^2 = \frac{-1}{R_{XX}(0)} \frac{d^2 R_{XX}(\tau)}{d\tau^2}\bigg|_{\tau=0}$$

where $R_{XX}(\tau)$ is the autocorrelation function of $X(t)$.

7-61 A random process has the autocorrelation function

$$R_{XX}(\tau) = B \cos^2 (\omega_0 \tau) \exp (-W|\tau|)$$

where B, ω_0, and W are positive constants.

(a) Find and sketch the power spectrum of $X(t)$ when ω_0 is at least several times larger than W.

(b) Compute the average power in the lowpass part of the power spectrum. Repeat for the bandpass part. In each case assume $\omega_0 \gg W$.

*7-62 Generalize Problem 7-61 by replacing $\cos^2 (\omega_0 \tau)$ with $\cos^N (\omega_0 \tau)$ where $N \geq 0$ is an integer. What is the resulting power spectrum when N is (a) odd, and (b) even?

*7-63 The product of a wide-sense stationary gaussian random process $X(t)$ with itself delayed by T seconds forms a new process $Y(t) = X(t)X(t - T)$. Determine (a) the autocorrelation function, and (b) the power spectrum of $Y(t)$. {Hint: Use the fact that $E[X_1 X_2 X_3 X_4] = E[X_1 X_2]E[X_3 X_4] + E[X_1 X_3]E[X_2 X_4] + E[X_1 X_4]E[X_2 X_3] - 2E[X_1]E[X_2]E[X_3]E[X_4]$ for gaussian random variables X_1, X_2, X_3, and X_4. (Thomas, 1969, p. 64.)}

7-64 Find the cross-correlation function $R_{XY}(t, t + \tau)$ and cross-power spectrum $\mathcal{S}_{XY}(\omega)$ for the delay-and-multiply device of Problem 7-63. {Hint: Use the fact that $E[X_1 X_2 X_3] = E[X_1]E[X_2 X_3] + E[X_2]E[X_3 X_1] + E[X_3]E[X_1 X_2] - 2E[X_1]E[X_2]E[X_3]$ for three gaussian random variables X_1, X_2, and X_3. (Thomas, 1969, p. 64.)}

7-65 If $X(t)$ and $Y(t)$ are real random processes determine which of the following functions can be valid. For those that are not, state at least one reason why.

(a) $R_{XX}(\tau) = \exp (-|\tau|)$

(b) $|R_{XY}(\tau)| \leq j\sqrt{R_{XX}(0)R_{YY}(0)}$

(c) $R_{XX}(\tau) = 2 \sin (3\tau)$

(d) $\mathcal{S}_{XX}(\omega) = 6/(6 + 7\omega^3)$

(e) $\mathcal{S}_{XX}(\omega) = \dfrac{4 \exp (-3|\tau|)}{1 + \omega^2}$

(f) $\mathcal{S}_{XY}(\omega) = 3 + j\omega^2$

(g) $\mathcal{S}_{XY}(\omega) = 18\delta(\omega)$

7-66 Form the product of two statistically independent jointly wide-sense stationary random processes $X(t)$ and $Y(t)$ as

$$W(t) = X(t)Y(t)$$

Find general expressions for the following correlation functions and power spectrums in terms of those of $X(t)$ and $Y(t)$: (a) $R_{WW}(t, t + \tau)$ and $\mathcal{S}_{WW}(\omega)$, (b) $R_{XW}(t, t + \tau)$ and $\mathcal{S}_{XW}(\omega)$, and (c) $R_{WX}(t, t + \tau)$ and $\mathcal{S}_{WX}(\omega)$. (d) If

$$R_{XX}(\tau) = (W_1/\pi)\text{Sa}(W_1 \tau)$$

and

$$R_{YY}(\tau) = (W_2/\pi)\text{Sa}(W_2 \tau)$$

with constants $W_2 > W_1$, find explicit functions for $R_{WW}(t, t + \tau)$ and $\mathcal{S}_{WW}(\omega)$.

7-67 An engineer is working with the function

$$R_{XY}(\tau) = P(1 + \tau) \exp(-W^2\tau^2)$$

where $P > 0$ and $W > 0$ are constants. He suspects that the function may not be a valid cross-correlation for two jointly stationary processes $X(t)$ and $Y(t)$, as he has been told. Determine if his suspicions are true. [*Hint:* Find the cross-power spectrum and see if it satisfies properties (7.3-16) through (7.3-18).]

7-68 A wide-sense stationary process $X(t)$ is applied to an ideal differentiator having the response $Y(t) = dX(t)/dt$. The cross-correlation of the input-output processes is known to be

$$R_{XY}(\tau) = dR_{XX}(\tau)/d\tau$$

(a) Determine $S_{XY}(\omega)$ and $S_{YX}(\omega)$ in terms of the power spectrum $S_{XX}(\omega)$ of $X(t)$.

(b) Since $S_{XX}(\omega)$ must be real, nonnegative, and have even symmetry, what are the properties of $S_{XY}(\omega)$?

7-69 The cross-correlation of jointly wide-sense stationary processes $X(t)$ and $Y(t)$ is assumed to be

$$R_{XY}(\tau) = Bu(\tau) \exp(-W\tau)$$

where $B > 0$ and $W > 0$ are constants.

(a) Find $R_{YX}(\tau)$.

(b) Find $S_{XY}(\omega)$ and $S_{YX}(\omega)$.

7-70 Work Problem 7-69 for the function

$$R_{XY}(\tau) = Bu(\tau)\tau \exp(-W\tau)$$

7-71 The cross-power spectrum for random processes $X(t)$ and $Y(t)$ can be written as

$$S_{XY}(\omega) = S_{XX}(\omega)H(\omega)$$

where $S_{XX}(\omega)$ is the power spectrum of $X(t)$ and $H(\omega)$ is a function with an inverse Fourier transform $h(\tau)$. Derive expressions for $R_{XY}(\tau)$ and $R_{YX}(\tau)$ in terms of $R_{XX}(\tau)$ and $h(\tau)$.

7-72 The power spectrum of a bandpass process $X(t)$ is shown in Figure P7-72. $X(t)$ is applied to a product device where the second multiplying input is $3 \cos(\omega_0 t)$. Plot the power spectrum of the device's output $3X(t) \cos(\omega_0 t)$.

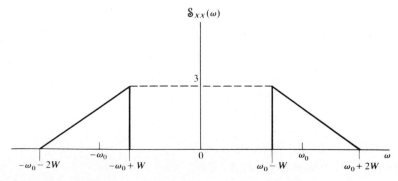

Figure P7-72

7-73 Let the "carrier" $A_0 \cos(\omega_0 t)$ in Figure 7.5-5 be modified to add a phase random variable Θ so that $Y(t) = A_0 X(t) \cos(\omega_0 t + \Theta)$. If Θ is uniformly distributed on $(0, 2\pi)$ and is independent of $X(t)$, find $R_{YY}(t, t + \tau)$ and $\mathcal{S}_{YY}(\omega)$ when $X(t)$ is wide-sense stationary.

7-74 Assume a stationary bandpass process $X(t)$ is adequately approximated by the power spectrum

$$\mathcal{S}_{XX}(\omega) = Pu(\omega - \omega_0)(\omega - \omega_0) \exp\left[-(\omega - \omega_0)^2/b\right]$$

$$+ Pu(-\omega - \omega_0)(-\omega - \omega_0) \exp\left[-(\omega + \omega_0)^2/b\right]$$

where ω_0, $P > 0$, and $b > 0$ are constants. The product $Y(t) = X(t) \cos(\omega_0 t)$ is formed.

    (a) Find and sketch the power spectrum of $Y(t)$.

    (b) Determine the average power in $X(t)$ and $Y(t)$.

***7-75** Compute the power spectrum of the complex process of Problem 6-55.

***7-76** Let $X(t)$ and $Y(t)$ be statistically independent processes with power spectrums

$$\mathcal{S}_{XX}(\omega) = 2\delta(\omega) + 1/[1 + (\omega/10)^2]$$

and

$$\mathcal{S}_{YY}(\omega) = 4/[1 + (\omega/2)^2]$$

A complex process

$$Z(t) = [X(t) + jY(t)] \exp(j\omega_0 t)$$

is formed where ω_0 is a constant much larger than 10.

    (a) Determine the autocorrelation function of $Z(t)$.

    (b) Find and sketch the power spectrum of $Z(t)$.

EIGHT

LINEAR SYSTEMS WITH RANDOM INPUTS

8.0 INTRODUCTION

A large part of our preceding work has been aimed at describing a random signal by modeling it as a sample function of a random process. We have found that time domain methods based on correlation functions, and frequency domain techniques based on power spectrums, constitute powerful ways of defining the behavior of random signals. Our work must not stop here, however, because one of the most important aspects of random signals is how they interact with linear systems. The knowledge of how to describe a random waveform would be of little value to a communication or control system engineer, for example, unless he was also able to determine how such a waveform will alter the desired output of his system.

In this chapter, we explore methods of describing the response of a linear system when the applied waveform is random. We begin by discussing some basic aspects of linear systems in the following section. Those readers well-versed in linear system theory can proceed directly to Section 8.2 without loss. For others, the topics of Section 8.1 should serve as a brief review and summary.

8.1 LINEAR SYSTEM FUNDAMENTALS

In this section, a brief summary of the basic aspects of linear systems is given. Attention will be limited to a system having only one input and one output, or response, as illustrated in Figure 8.1-1. It is assumed that the input signal $x(t)$ and

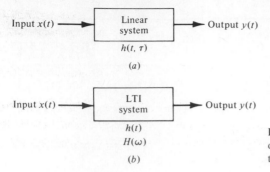

Figure 8.1-1 (*a*) A general single-input single-output linear system, and (*b*) a similar linear, time-invariant (LTI) system.

the response $y(t)$ are deterministic signals, even though some of the topics discussed apply to random waveforms. Which topics are applicable to random signals will be made clear when they are used in later sections.

The General Linear System

Clearly, the linear system (Figure 8.1-1*a*) will, in general, cause the response $y(t)$ to be different from the input signal $x(t)$. We think of the system as *operating* on $x(t)$ to cause $y(t)$ and write

$$y(t) = L[x(t)] \tag{8.1-1}$$

Here L is an *operator* representing the action of the system on $x(t)$.

A system is said to be linear if its response to a sum of inputs $x_n(t)$, $n = 1$, $2, \ldots, N$, is equal to the sum of responses taken separately. Thus, if $x_n(t)$ causes a response $y_n(t)$, $n = 1, 2, \ldots, N$, then for a linear system

$$y(t) = L\left[\sum_{n=1}^{N} \alpha_n x_n(t)\right] = \sum_{n=1}^{N} \alpha_n L[x_n(t)] = \sum_{n=1}^{N} \alpha_n y_n(t) \tag{8.1-2}$$

must hold, where the α_n are arbitrary constants and N may be infinite.

From the definition (2.3-2) and properties of the impulse function we may write

$$x(t) = \int_{-\infty}^{\infty} x(\xi)\delta(t - \xi) \, d\xi \tag{8.1-3}$$

By substituting (8.1-3) into (8.1-1) and observing that the operator operates on the time function, we obtain

$$y(t) = L[x(t)] = L\left[\int_{-\infty}^{\infty} x(\xi)\delta(t - \xi) \, d\xi\right] = \int_{-\infty}^{\infty} x(\xi)L[\delta(t - \xi)] \, d\xi \tag{8.1-4}$$

We now *define* a new function $h(t, \xi)$ as the *impulse response* of the linear system; that is,

$$L[\delta(t - \xi)] = h(t, \xi) \tag{8.1-5}$$

Equation (8.1-4) becomes

$$y(t) = \int_{-\infty}^{\infty} x(\xi)h(t, \xi) \, d\xi \tag{8.1-6}$$

which shows that the response of a general linear system is completely determined by its impulse response through (8.1-6).

Linear Time-Invariant Systems

A general linear system is said to be also time-invariant if the *form* of its impulse response $h(t, \xi)$ does not depend on the time that the impulse is applied. Thus, if an impulse $\delta(t)$, occurring at $t = 0$, causes the response $h(t)$, then an impulse $\delta(t - \xi)$, occurring at $t = \xi$, must cause the response $h(t - \xi)$ if the system is time-invariant. This fact means that

$$h(t, \xi) = h(t - \xi) \tag{8.1-7}$$

for a linear, time-invariant system, so (8.1-6) becomes

$$y(t) = \int_{-\infty}^{\infty} x(\xi)h(t - \xi) \, d\xi \tag{8.1-8}$$

Equation (8.1-8) is known as the *convolution integral* of $x(t)$ and $h(t)$; it is sometimes written in the short form

$$y(t) = x(t) * h(t) \tag{8.1-9}$$

By a suitable change of variables, (8.1-8) can be put in the alternative form

$$y(t) = \int_{-\infty}^{\infty} h(\xi)x(t - \xi) \, d\xi \tag{8.1-10}$$

Time-Invariant System Transfer Function

Either (8.1-8) or (8.1-10) shows that a linear time-invariant system is completely characterized by its impulse response, which is a temporal characterization. By Fourier transformation of $y(t)$, we may derive an equivalent characterization in the frequency domain. Hence, if $X(\omega)$, $Y(\omega)$ and $H(\omega)$ are the respective Fourier transforms of $x(t)$, $y(t)$, and $h(t)$, then

$$Y(\omega) = \int_{-\infty}^{\infty} y(t)e^{-j\omega t} \, dt = \int_{-\infty}^{\infty} \left[\int_{-\infty}^{\infty} x(\xi)h(t - \xi) \, d\xi \right] e^{-j\omega t} \, dt$$

$$= \int_{-\infty}^{\infty} x(\xi) \left[\int_{-\infty}^{\infty} h(t - \xi)e^{-j\omega(t - \xi)} \, dt \right] e^{-j\omega \xi} \, d\xi$$

$$= \int_{-\infty}^{\infty} x(\xi)H(\omega)e^{-j\omega \xi} \, d\xi = X(\omega)H(\omega) \tag{8.1-11}$$

The function $H(\omega)$ is called the *transfer function* of the system. Equation (8.1-11) shows that the Fourier transform of the response of any linear time-invariant system is equal to the product of the transform of the input signal and the transform of the network impulse response.

In the actual calculation of a transfer function for a given network, an alternative definition based on the response of the system to an exponential signal

$$x(t) = e^{j\omega t} \tag{8.1-12}$$

may be more convenient. It can be shown (Thomas, 1969, p. 142, or Papoulis, 1962, p. 83) that†

$$H(\omega) = \frac{L[e^{j\omega t}]}{e^{j\omega t}} = \frac{y(t)}{x(t)} \tag{8.1-13}$$

where

$$y(t) = L[e^{j\omega t}] \tag{8.1-14}$$

An example serves to illustrate the determination of $H(\omega)$ by means of (8.1-13).

Example 8.1-1 We find $H(\omega)$ for the network shown in Figure 8.1-2. By assuming a clockwise current i (and no loading in the output circuit), we have‡

$$x(t) = L\frac{di}{dt} + y(t)$$

But $y(t) = iR$ so

$$\frac{di}{dt} = \frac{1}{R}\frac{dy(t)}{dt}$$

and

$$x(t) = \frac{L}{R}\frac{dy(t)}{dt} + y(t)$$

† It should be carefully observed that (8.1-13) holds *only* for $x(t)$ given by (8.1-12); that is, for an exponential waveform.

‡ L in the network is an inductance and should not be confused with L above, which stands for a linear system operator.

Input $x(t)$ R Output $y(t)$

Figure 8.1-2 A linear time-invariant network. [*Reproduced from Peebles (1976) with permission of publishers Addison-Wesley, Advanced Book Program.*]

With $x(t) = \exp(j\omega t)$ as the input we must have an output $y(t) = H(\omega)x(t)$ from (8.1-13). Hence, $dy(t)/dt = H(\omega)j\omega x(t)$ and

$$x(t) = \frac{L}{R} H(\omega)j\omega x(t) + H(\omega)x(t)$$

Finally, we solve for $H(\omega)$:

$$H(\omega) = \frac{1}{1 + (j\omega L/R)}$$

Idealized Systems

To simplify the analysis of many complex systems, it is often convenient to *approximate* the system's transfer function $H(\omega)$ by an idealized one. Idealized transfer functions are illustrated in Figure 8.1-3a for a lowpass system; (b) applies to a highpass system and (c) applies to a bandpass system. In every case the *idealized system* has a transfer function magnitude that is flat within its passband and

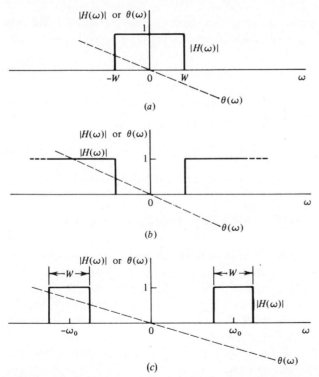

(a)

(b)

(c)

Figure 8.1-3 Ideal system transfer functions. (a) Lowpass, (b) highpass, and (c) bandpass systems. [*Reproduced from Peebles (1976) with permission of publishers Addison-Wesley, Advanced Book Program.*]

zero outside this band; its midband gain is unity and its phase $\theta(\omega)$ is defined to be a linear function of frequency.

In replacing an actual system with an idealized one, the latter would be assigned a midband gain and phase slope that approximate the actual values. The bandwidth W (in lowpass and bandpass cases) is chosen according to some convenient basis. For example, W could be made equal to the 3-dB bandwidth of the actual system, or alternatively, it could be chosen to satisfy a specific requirement. An example of the latter case is considered in Section 8.5 where W, called *noise bandwidth*, is selected to cause the actual and ideal systems to produce the same output noise power when each is excited by the same noise source.

Causal and Stable Systems

To complete our summary of basic topics in linear system theory, we consider two final items.

A linear time-invariant system is said to be *causal* if it does not respond prior to the application of an input signal. Mathematically, this implies $y(t) = 0$ for $t < t_0$ if $x(t) = 0$ for $t < t_0$, where t_0 is any real constant. From (8.1-10), this condition requires that

$$h(t) = 0 \qquad \text{for} \qquad t < 0 \tag{8.1-15}$$

All passive, linear time-invariant networks that can be constructed will satisfy (8.1-15). As a consequence, a system satisfying (8.1-15) is often called *physically realizable*.

A linear time-invariant system is said to be *stable* if its response to any bounded input is bounded; that is, if $|x(t)| < M$, where M is some constant, then $|y(t)| < MI$ for a stable system where I is another constant independent of the input. By considering (8.1-10), it is readily shown that

$$I = \int_{-\infty}^{\infty} |h(t)| \, dt < \infty \tag{8.1-16}$$

will ensure that a system having the impulse response $h(t)$ will be stable.

8.2 RANDOM SIGNAL RESPONSE OF LINEAR SYSTEMS

With the preceding summary of linear system theory in mind, we proceed now to determine characteristics of the response of a stable, linear, time-invariant system as illustrated in Figure 8.1-1b when the applied waveform is an ensemble member $x(t)$ of a random process $X(t)$. We assume in all work that the system's impulse response $h(t)$ is a real function.† In this section we restrict our attention to temporal characteristics such as mean value and mean-squared value of the response, its autocorrelation function, and applicable cross-correlation functions. Spectral characteristics are developed in Section 8.4.

---

† All real-world networks have real impulse responses.

System Response—Convolution

Even when $x(t)$ is a random signal, the network's response $y(t)$ is given by the convolution integral:

$$y(t) = \int_{-\infty}^{\infty} x(\xi)h(t - \xi) \, d\xi \qquad (8.2\text{-}1)$$

or

$$y(t) = \int_{-\infty}^{\infty} h(\xi)x(t - \xi) \, d\xi \qquad (8.2\text{-}2)$$

where $h(t)$ is the network's impulse response.

We may view (8.2-2) as an operation on an ensemble member $x(t)$ of the random process $X(t)$ that produces an ensemble member of a new process $Y(t)$. With this viewpoint, we may think of (8.2-2) as defining the process $Y(t)$ in terms of the process $X(t)$:

$$Y(t) = \int_{-\infty}^{\infty} h(\xi)X(t - \xi) \, d\xi \qquad (8.2\text{-}3)$$

Thus, we may envision the system as accepting the random process $X(t)$ as its input and responding with the new process $Y(t)$ according to (8.2-3).

Mean and Mean-Squared Value of System Response

We may readily apply (8.2-3) to find the mean value of the system's response. By assuming $X(t)$ is wide-sense stationary, we have†

$$E[Y(t)] = E\left[\int_{-\infty}^{\infty} h(\xi)X(t - \xi) \, d\xi \right]$$

$$= \int_{-\infty}^{\infty} h(\xi)E[X(t - \xi)] \, d\xi$$

$$= \bar{X} \int_{-\infty}^{\infty} h(\xi) \, d\xi = \bar{Y} \quad \text{(constant)} \qquad (8.2\text{-}4)$$

---

† It is known (Cooper and McGillem, 1971, p. 169) that the operation

$$E\left[\int_{t_1}^{t_2} W(t)h(t) \, dt \right] = \int_{t_1}^{t_2} E[W(t)]h(t) \, dt$$

is valid, where $W(t)$ is some bounded function of a random process [on the interval (t_1, t_2)] and $h(t)$ is a nonrandom time function, if

$$\int_{t_1}^{t_2} E[|W(t)|] |h(t)| \, dt < \infty$$

where t_1 and t_2 are real constants that may be infinite. This condition is satisfied in all *physical* cases if $W(t)$ is wide-sense stationary because $W(t)$ will be bounded and the systems are stable [see (8.1-16)].

This expression indicates that the mean value of $Y(t)$ equals the mean value of $X(t)$ times the area under the impulse response if $X(t)$ is wide-sense stationary.

For the mean-squared value of $Y(t)$, we calculate

$$E[Y^2(t)] = E\left[\int_{-\infty}^{\infty} h(\xi_1)X(t-\xi_1)\,d\xi_1 \int_{-\infty}^{\infty} h(\xi_2)X(t-\xi_2)\,d\xi_2\right]$$

$$= \int_{-\infty}^{\infty}\int_{-\infty}^{\infty} E[X(t-\xi_1)X(t-\xi_2)]h(\xi_1)h(\xi_2)\,d\xi_1\,d\xi_2 \qquad (8.2\text{-}5)$$

If we assume the input is wide-sense stationary then

$$E[X(t-\xi_1)X(t-\xi_2)] = R_{XX}(\xi_1 - \xi_2) \qquad (8.2\text{-}6)$$

and (8.2-5) becomes independent of t:

$$\overline{Y^2} = E[Y^2(t)] = \int_{-\infty}^{\infty}\int_{-\infty}^{\infty} R_{XX}(\xi_1 - \xi_2)h(\xi_1)h(\xi_2)\,d\xi_1\,d\xi_2 \qquad (8.2\text{-}7)$$

Although this expression gives the power in $Y(t)$, it may be tedious to calculate in most cases. We develop an example of its solution for a simple case.

Example 8.2-1 We find $\overline{Y^2}$ for a system having white noise at its input. Here

$$R_{XX}(\xi_1 - \xi_2) = (\mathcal{N}_0/2)\delta(\xi_1 - \xi_2)$$

where $\mathcal{N}_0$ is a positive real constant. From (8.2-7):

$$\overline{Y^2} = \int_{-\infty}^{\infty}\int_{-\infty}^{\infty} (\mathcal{N}_0/2)\delta(\xi_1 - \xi_2)h(\xi_1)\,d\xi_1\,h(\xi_2)\,d\xi_2$$

$$= (\mathcal{N}_0/2)\int_{-\infty}^{\infty} h^2(\xi_2)\,d\xi_2$$

Output power becomes proportional to the area under the square of $h(t)$ in this case.

Autocorrelation Function of Response

Let $X(t)$ be wide-sense stationary. The autocorrelation function of $Y(t)$ is

$$R_{YY}(t, t+\tau) = E[Y(t)Y(t+\tau)]$$

$$= E\left[\int_{-\infty}^{\infty} h(\xi_1)X(t-\xi_1)\,d\xi_1 \int_{-\infty}^{\infty} h(\xi_2)X(t+\tau-\xi_2)\,d\xi_2\right]$$

$$= \int_{-\infty}^{\infty}\int_{-\infty}^{\infty} E[X(t-\xi_1)X(t+\tau-\xi_2)]h(\xi_1)h(\xi_2)\,d\xi_1\,d\xi_2 \qquad (8.2\text{-}8)$$

which reduces to

$$R_{YY}(\tau) = \int_{-\infty}^{\infty} \int_{-\infty}^{\infty} R_{XX}(\tau + \xi_1 - \xi_2)h(\xi_1)h(\xi_2) \, d\xi_1 \, d\xi_2 \qquad (8.2\text{-}9)$$

because $X(t)$ is assumed wide-sense stationary.

Two facts result from (8.2-9). First, $Y(t)$ is wide-sense stationary if $X(t)$ is wide-sense stationary because $R_{YY}(\tau)$ does not depend on t and $E[Y(t)]$ is a constant from (8.2-4). Second, the form of (8.2-9) shows that $R_{YY}(\tau)$ is the two-fold convolution of the input autocorrelation function with the network's impulse response; that is

$$R_{YY}(\tau) = R_{XX}(\tau) * h(-\tau) * h(\tau) \qquad (8.2\text{-}10)$$

Cross-Correlation Functions of Input and Output

The cross-correlation function of $X(t)$ and $Y(t)$ is

$$R_{XY}(t, t + \tau) = E[X(t)Y(t + \tau)] = E\left[X(t) \int_{-\infty}^{\infty} h(\xi)X(t + \tau - \xi) \, d\xi \right]$$

$$= \int_{-\infty}^{\infty} E[X(t)X(t + \tau - \xi)]h(\xi) \, d\xi \qquad (8.2\text{-}11)$$

If $X(t)$ is wide-sense stationary, (8.2-11) reduces to

$$R_{XY}(\tau) = \int_{-\infty}^{\infty} R_{XX}(\tau - \xi)h(\xi) \, d\xi \qquad (8.2\text{-}12)$$

which is the convolution of $R_{XX}(\tau)$ with $h(\tau)$:

$$R_{XY}(\tau) = R_{XX}(\tau) * h(\tau) \qquad (8.2\text{-}13)$$

A similar development shows that

$$R_{YX}(\tau) = \int_{-\infty}^{\infty} R_{XX}(\tau - \xi)h(-\xi) \, d\xi \qquad (8.2\text{-}14)$$

or

$$R_{YX}(\tau) = R_{XX}(\tau) * h(-\tau) \qquad (8.2\text{-}15)$$

From (8.2-12) and (8.2-14), it is clear that the cross-correlation functions depend on τ and not on absolute time t. As a consequence of this fact $X(t)$ and $Y(t)$ are *jointly* wide-sense stationary if $X(t)$ is wide-sense stationary, because we have already shown $Y(t)$ to be wide-sense stationary.

By substituting (8.2-12) into (8.2-9), autocorrelation function and cross-correlation functions are seen to be related by

$$R_{YY}(\tau) = \int_{-\infty}^{\infty} R_{XY}(\tau + \xi_1)h(\xi_1) \, d\xi_1 \qquad (8.2\text{-}16)$$

or

$$R_{YY}(\tau) = R_{XY}(\tau) * h(-\tau) \qquad (8.2\text{-}17)$$

A similar substitution of (8.2-14) into (8.2-9) gives

$$R_{YY}(\tau) = \int_{-\infty}^{\infty} R_{YX}(\tau - \xi_2)h(\xi_2)\,d\xi_2 \qquad (8.2\text{-}18)$$

or

$$R_{YY}(\tau) = R_{YX}(\tau) * h(\tau) \qquad (8.2\text{-}19)$$

Example 8.2-2 We shall continue Example 8.2-1 by finding the cross-correlation functions $R_{XY}(\tau)$ and $R_{YX}(\tau)$. From (8.2-12)

$$R_{XY}(\tau) = \int_{-\infty}^{\infty} (\mathcal{N}_0/2)\delta(\tau - \xi)h(\xi)\,d\xi$$

$$= (\mathcal{N}_0/2)h(\tau)$$

From (8.2-14)

$$R_{YX}(\tau) = \int_{-\infty}^{\infty} (\mathcal{N}_0/2)\delta(\tau - \xi)h(-\xi)\,d\xi$$

$$= (\mathcal{N}_0/2)h(-\tau) = R_{XY}(-\tau)$$

These two results are seen to satisfy (6.3-16), as they should.

8.3 SYSTEM EVALUATION USING RANDOM NOISE

A practical application of the foregoing theory can be immediately developed; it is based on the cross-correlation function of (8.2-12). Suppose we desire to find the impulse response of some linear time-invariant system. If we have available a broadband (relative to the system) noise source having a flat power spectrum, and a cross-correlation measurement device, such as shown in Figure 6.4-1, $h(t)$ can easily be determined.

For the approximately white noise source

$$R_{XX}(\tau) \approx \left(\frac{\mathcal{N}_0}{2}\right)\delta(\tau) \qquad (8.3\text{-}1)$$

With this noise applied to the system, the cross-correlation function from (8.2-12) or Example 8.2-2 becomes

$$R_{XY}(\tau) \approx \int_{-\infty}^{\infty} \left(\frac{\mathcal{N}_0}{2}\right)\delta(\tau - \xi)h(\xi)\,d\xi$$

$$= \left(\frac{\mathcal{N}_0}{2}\right)h(\tau) \qquad (8.3\text{-}2)$$

Figure 8.3-1 A method for finding a system's impulse response. [*Reproduced from Peebles (1976) with permission of publishers Addison-Wesley, Advanced Book Program.*]

or

$$h(\tau) \approx \left(\frac{2}{\mathcal{N}_0}\right) R_{XY}(\tau) \qquad (8.3\text{-}3)$$

Since a measurement $\hat{R}_{XY}(\tau)$ of $R_{XY}(\tau)$ can be obtained from the cross-correlation measurement device, (8.3-3) gives us a measurement $\hat{h}(\tau)$ of $h(\tau)$

$$\hat{h}(\tau) = \left(\frac{2}{\mathcal{N}_0}\right) \hat{R}_{XY}(\tau) \approx h(\tau) \qquad (8.3\text{-}4)$$

Figure 8.3-1 illustrates the concepts described here.

8.4 SPECTRAL CHARACTERISTICS OF SYSTEM RESPONSE

Because the Fourier transform of a correlation function (autocorrelation or cross-correlation) is a power spectrum for wide-sense stationary processes, it would seem that if $R_{XX}(\tau)$ is known for the input process one can find $R_{YY}(\tau)$, $R_{XY}(\tau)$, and $R_{YX}(\tau)$ as described in Section 8.2 and therefore obtain power spectrums by transformation. Indeed, this approach is conceptually valid. However, from a practical standpoint the integrals involved may be difficult to evaluate.

In this section an alternative approach is taken where the desired power spectrum involving the system's response is related to the power spectrum of the input. In every case, the input process $X(t)$ is assumed to be wide-sense stationary, which, as previously proved, means that $Y(t)$ and $X(t)$ are jointly wide-sense stationary.

Power Density Spectrum of Response

We show now that the power density spectrum $S_{YY}(\omega)$ of the response of a linear time-invariant system having a transfer function $H(\omega)$ is given by

$$S_{YY}(\omega) = S_{XX}(\omega)|H(\omega)|^2 \qquad (8.4\text{-}1)$$

where $S_{XX}(\omega)$ is the power spectrum of the input process $X(t)$. We call $|H(\omega)|^2$ the *power transfer function* of the system.

The proof of (8.4-1) begins by writing $S_{YY}(\omega)$ as the Fourier transform of the output autocorrelation function

$$S_{YY}(\omega) = \int_{-\infty}^{\infty} R_{YY}(\tau)e^{-j\omega\tau}\, d\tau \qquad (8.4\text{-}2)$$

On substitution of (8.2-9), (8.4-2) becomes

$$S_{YY}(\omega) = \int_{-\infty}^{\infty} h(\xi_1) \int_{-\infty}^{\infty} h(\xi_2) \int_{-\infty}^{\infty} R_{XX}(\tau + \xi_1 - \xi_2) e^{-j\omega\tau} \, d\tau \, d\xi_2 \, d\xi_1 \quad (8.4\text{-}3)$$

The change of variable $\xi = \tau + \xi_1 - \xi_2, d\xi = d\tau$, produces

$$S_{YY}(\omega) = \int_{-\infty}^{\infty} h(\xi_1) e^{j\omega\xi_1} \, d\xi_1 \int_{-\infty}^{\infty} h(\xi_2) e^{-j\omega\xi_2} \, d\xi_2 \int_{-\infty}^{\infty} R_{XX}(\xi) e^{-j\omega\xi} \, d\xi \quad (8.4\text{-}4)$$

These three integrals are recognized as $H^*(\omega)$, $H(\omega)$, and $S_{XX}(\omega)$ respectively. Hence

$$S_{YY}(\omega) = H^*(\omega)H(\omega)S_{XX}(\omega) = S_{XX}(\omega) |H(\omega)|^2 \quad (8.4\text{-}5)$$

and (8.4-1) is proved.

The average power, denoted P_{YY}, in the system's response is readily found by using (8.4-5):

$$P_{YY} = \frac{1}{2\pi} \int_{-\infty}^{\infty} S_{XX}(\omega) |H(\omega)|^2 \, d\omega \quad (8.4\text{-}6)$$

Example 8.4-1 The power spectrum and average power of the response of the network of Example 8.1-1 will be found when $X(t)$ is white noise for which

$$S_{XX}(\omega) = \frac{\mathcal{N}_0}{2}$$

Here $H(\omega) = [1 + (j\omega L/R)]^{-1}$ so

$$|H(\omega)|^2 = \frac{1}{1 + (\omega L/R)^2}$$

and

$$S_{YY}(\omega) = S_{XX}(\omega) |H(\omega)|^2 = \frac{\mathcal{N}_0/2}{1 + (\omega L/R)^2}$$

Average power in $Y(t)$, from (8.4-6), is

$$P_{YY} = \frac{1}{2\pi} \int_{-\infty}^{\infty} S_{YY}(\omega) \, d\omega = \frac{\mathcal{N}_0}{4\pi} \int_{-\infty}^{\infty} \frac{d\omega}{1 + (\omega L/R)^2} = \frac{\mathcal{N}_0 R}{4L}$$

after an integral from Appendix C is used.

As a check on the calculation of P_{YY}, we note that (pair 15, Appendix E)

$$h(t) = (R/L)u(t)e^{-Rt/L} \leftrightarrow H(\omega) = \frac{1}{1 + (j\omega L/R)}$$

for this network, and, using the result of Example 8.2-1, we get

$$P_{YY} = \overline{Y^2} = \left(\frac{\mathcal{N}_0}{2}\right)\int_0^\infty \left(\frac{R}{L}\right)^2 e^{-2Rt/L}\,dt = \frac{\mathcal{N}_0 R}{4L}$$

The two powers are in agreement.

Cross-Power Density Spectrums of Input and Output

It is easily shown (see Problem 8-42) that the Fourier transforms of the cross-correlation functions of (8.2-12) and (8.2-14) may be written as

$$\mathcal{S}_{XY}(\omega) = \mathcal{S}_{XX}(\omega)H(\omega) \tag{8.4-7}$$

$$\mathcal{S}_{YX}(\omega) = \mathcal{S}_{XX}(\omega)H(-\omega) \tag{8.4-8}$$

respectively.

8.5 NOISE BANDWIDTH

Consider a system having a lowpass transfer function $H(\omega)$. Assume white noise is applied at the input. The power density of this white noise is $\mathcal{N}_0/2$ where $\mathcal{N}_0$ is a real positive constant. The total average power emerging from the network is [from (8.4-6)]

$$P_{YY} = \frac{1}{2\pi}\int_{-\infty}^\infty \left(\frac{\mathcal{N}_0}{2}\right)|H(\omega)|^2\,d\omega \tag{8.5-1}$$

By assuming the system impulse response is real,† $|H(\omega)|^2$ will be an even function of ω and (8.5-1) can be written

$$P_{YY} = \frac{\mathcal{N}_0}{2\pi}\int_0^\infty |H(\omega)|^2\,d\omega \tag{8.5-2}$$

Now consider an idealized system that is equivalent to the actual system in the sense that both produce the same output average power when they both are excited by the same white noise source, and both have the same value of power transfer function at midband; that is, $|H(0)|^2$ is the same in both systems. The principal difference between the two systems is that the idealized one has a rectangularly shaped power transfer function $|H_I(\omega)|^2$ defined by

$$|H_I(\omega)|^2 = \begin{cases} |H(0)|^2 & |\omega| < W_N \\ 0 & |\omega| > W_N \end{cases} \tag{8.5-3}$$

---

† The impulse response of any physical system is always real.

where W_N is a positive constant selected to make output powers in the two systems equal. The output power in the idealized system is

$$\frac{1}{2\pi} \int_{-\infty}^{\infty} \left(\frac{\mathcal{N}_0}{2}\right) |H_I(\omega)|^2 \, d\omega = \frac{\mathcal{N}_0}{2\pi} \int_0^{W_N} |H(0)|^2 \, d\omega = \frac{\mathcal{N}_0 |H(0)|^2 W_N}{2\pi} \qquad (8.5\text{-}4)$$

By equating (8.5-2) and (8.5-4), we require that W_N be given by

$$W_N = \frac{\int_0^\infty |H(\omega)|^2 \, d\omega}{|H(0)|^2} \qquad (8.5\text{-}5)$$

W_N is called the *noise bandwidth* of the system.

Example 8.5-1 The noise bandwidth is found for a system having the power transfer function

$$|H(\omega)|^2 = \frac{1}{1 + (\omega/W)^2}$$

where W is the 3-dB bandwidth in radians per second. Here $|H(0)|^2 = 1$, so

$$W_N = \int_0^\infty \frac{W^2 \, d\omega}{W^2 + \omega^2} = W \tan^{-1}\left(\frac{\omega}{W}\right)\Big|_0^\infty = \frac{W\pi}{2}$$

This expression shows that W_N is larger than the system 3-dB bandwidth by a factor of about 1.57.

If we repeat the above development for a bandpass transfer function with a centerband frequency ω_0 it will be found that

$$W_N = \frac{\int_0^\infty |H(\omega)|^2 \, d\omega}{|H(\omega_0)|^2} \qquad (8.5\text{-}6)$$

Proof of this result is left as a reader exercise (see Problem 8-45). The development also provides a simple expression for output noise power in terms of noise bandwidth:

$$P_{YY} = \frac{\mathcal{N}_0}{2\pi} |H(\omega_0)|^2 W_N \qquad (8.5\text{-}7)$$

For a lowpass filter, (8.5-7) applies by letting $\omega_0 = 0$.

*8.6 BANDPASS, BAND-LIMITED, AND NARROWBAND PROCESSES

A random process $N(t)$ will be called *bandpass* if its power density spectrum $S_{NN}(\omega)$ has its significant components clustered in a band of width W (rad/s) that does not include $\omega = 0$. Such a power spectrum is illustrated in Figure 8.6-1a.† Our definition does not prevent the power spectrum from being nonzero at $\omega = 0$; it only requires that $S_{NN}(0)$ be small in relation to more significant values, so as to distinguish the bandpass case from a lowpass power spectrum with significant peaking at higher frequencies.

All subsequent discussions in this section will relate to special forms of bandpass processes.

*Band-Limited Processes

If the power spectrum of a bandpass random process is *zero* outside some frequency band of width W (rad/s) that does not include $\omega = 0$, the process is called *band-limited*. The concept of a band-limited process forms a convenient approximation for physical processes that often allows analytical problem solutions that otherwise might not be possible. A band-limited bandpass process power spectrum is illustrated in Figure 8.6-1b.

† Power spectrums arising in physical systems will always decrease as frequency becomes sufficiently large, so a suitable value of W can always be found. For example, W could be chosen to include all frequencies for which $S_{NN}(\omega) \geq 0.1 S_{NN}(\omega_0)$ where ω_0 is some convenient frequency near where $S_{NN}(\omega)$ has its largest magnitude (see Figure 8.6-1).

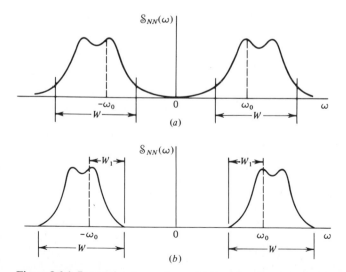

Figure 8.6-1 Power density spectrums (a) for a bandpass random process and (b) for a band-limited bandpass process.

*Narrowband Processes

A band-limited random process is said to be *narrowband* if $W \ll \omega_0$, where ω_0 is some conveniently chosen frequency near band-center or near where the power spectrum is at its maximum. A power spectrum of a narrowband process is sketched in Figure 8.6-2a. A typical sample function, if viewed on an oscilloscope, might look as shown in (b). The appearance of $n(t)$ suggests that the process might be represented by a cosine function with angular frequency ω_0 and slowly varying amplitude and phase; that is, by

$$N(t) = A(t) \cos \left[\omega_0 t + \Theta(t) \right] \tag{8.6-1}$$

where $A(t)$ is a random process representing the slowly varying amplitude and $\Theta(t)$ is a process representing the slowly varying phase. Indeed this is the case, and, for the important practical case where $N(t)$ is gaussian noise, it is known that $A(t)$ and $\Theta(t)$ have Rayleigh and uniform (over 2π) first-order probability density functions respectively. The processes $A(t)$ and $\Theta(t)$ are not statistically independent when $N(t)$ is gaussian (Davenport, 1970, p. 522, or Davenport and Root, 1958, pp. 161–165), but for any one instant in time the process *random variables* are independent.

In some problems, (8.6-1) is a preferred representation for $N(t)$. For others, it is convenient to use the equivalent form

$$N(t) = X(t) \cos (\omega_0 t) - Y(t) \sin (\omega_0 t) \tag{8.6-2}$$

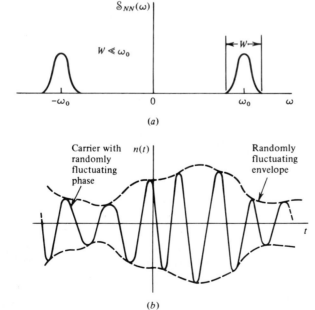

(a)

(b)

Figure 8.6-2 (a) A power spectrum of a narrowband random process $N(t)$ and (b) a typical ensemble member $n(t)$. [*Reproduced from Peebles (1976) with permission of publishers Addison-Wesley, Advanced Book Program.*]

where the processes $X(t)$ and $Y(t)$ are given by

$$X(t) = A(t) \cos [\Theta(t)] \qquad (8.6\text{-}3)$$

$$Y(t) = A(t) \sin [\Theta(t)] \qquad (8.6\text{-}4)$$

Expressions relating $A(t)$ and $\Theta(t)$ to $X(t)$ and $Y(t)$ are

$$A(t) = \sqrt{X^2(t) + Y^2(t)} \qquad (8.6\text{-}5)$$

$$\Theta(t) = \tan^{-1} [Y(t)/X(t)] \qquad (8.6\text{-}6)$$

★Properties of Band-Limited Processes

The representations (8.6-1) and (8.6-2) are actually more general than implied above; they can also be applied to any band-limited random process. For the remainder of this section we concern ourselves only with (8.6-2).

Let $N(t)$ be any band-limited wide-sense stationary real random process with a mean value of zero and a power density spectrum that satisfies

$$S_{NN}(\omega) \neq 0 \qquad 0 < \omega_0 - W_1 < |\omega| < \omega_0 - W_1 + W$$

$$S_{NN}(\omega) = 0 \qquad \text{elsewhere} \qquad (8.6\text{-}7)$$

where W_1 and W are real positive constants. Then $N(t)$ can be represented by the right side of (8.6-2),† where the random processes $X(t)$ and $Y(t)$ have the following properties:

(1) $X(t)$ and $Y(t)$ are jointly wide-sense stationary $\qquad$ (8.6-8)

(2) $E[X(t)] = 0 \qquad E[Y(t)] = 0 \qquad$ (8.6-9)

(3) $E[X^2(t)] = E[Y^2(t)] = E[N^2(t)] \qquad$ (8.6-10)

(4) $R_{XX}(\tau) = \dfrac{1}{\pi} \displaystyle\int_0^\infty S_{NN}(\omega) \cos [(\omega - \omega_0)\tau] \, d\omega \qquad$ (8.6-11)

(5) $R_{YY}(\tau) = R_{XX}(\tau) \qquad$ (8.6-12)

(6) $R_{XY}(\tau) = \dfrac{1}{\pi} \displaystyle\int_0^\infty S_{NN}(\omega) \sin [(\omega - \omega_0)\tau] \, d\omega \qquad$ (8.6-13)

(7) $R_{YX}(\tau) = -R_{XY}(\tau) \qquad R_{XY}(\tau) = -R_{XY}(-\tau) \qquad$ (8.6-14)

(8) $R_{XY}(0) = E[X(t)Y(t)] = 0 \qquad R_{YX}(0) = 0 \qquad$ (8.6-15)

---

† If we denote the right side of (8.6-2) by $\hat{N}(t)$ the equality in (8.6-2) must be interpreted in the sense of zero mean-squared error; that is $N(t)$ equals $\hat{N}(t)$ in the sense that

$$E[\{N(t) - \hat{N}(t)\}^2] = 0$$

(Ziemer and Tranter, 1976, p. 241).

$$(9) \quad S_{XX}(\omega) = L_p[S_{NN}(\omega - \omega_0) + S_{NN}(\omega + \omega_0)] \tag{8.6-16}$$

$$(10) \quad S_{YY}(\omega) = S_{XX}(\omega) \tag{8.6-17}$$

$$(11) \quad S_{XY}(\omega) = jL_p[S_{NN}(\omega - \omega_0) - S_{NN}(\omega + \omega_0)] \tag{8.6-18}$$

$$(12) \quad S_{YX}(\omega) = -S_{XY}(\omega) \tag{8.6-19}$$

In the preceding 12 results, ω_0 is any convenient frequency within the band of $S_{NN}(\omega)$; $R_{XX}(\tau)$, $R_{YY}(\tau)$, $R_{XY}(\tau)$, and $R_{YX}(\tau)$ are autocorrelation and cross-correlation functions of $X(t)$ and $Y(t)$ while $S_{XX}(\omega)$, $S_{YY}(\omega)$, $S_{XY}(\omega)$, and $S_{YX}(\omega)$ are the corresponding power spectrums; and $L_p[\cdot]$ denotes taking the lowpass part of the quantity within the brackets.

We outline the proofs of the above properties in the next subsection. Here we discuss their meaning and develop an example. We see that in addition to being zero-mean (property 2) wide-sense stationary (property 1) processes, $X(t)$ and $Y(t)$ also have equal powers (property 3), the same autocorrelation function (property 5), and therefore the same power spectrum (property 10). Random variables defined for the processes $X(t)$ and $Y(t)$ at any one time are orthogonal (property 8). If $N(t)$ has a power spectrum with components having even symmetry about $\omega = \pm\omega_0$, then $X(t)$ and $Y(t)$ will be orthogonal processes (property 6). A consequence of this last point is that the cross-power spectrums of $X(t)$ and $Y(t)$ are zero (properties 11 and 12).

Example 8.6-1 Consider the bandpass process having the power density spectrum shown in Figure 8.6-3a. We shall find $S_{XX}(\omega)$, $S_{XY}(\omega)$, and $R_{XY}(\tau)$. By shifting $S_{NN}(\omega)$ by $+\omega_0$ and $-\omega_0$ as shown in (b), we may construct $S_{XX}(\omega)$ according to (8.6-16) as the lowpass portion of $S_{NN}(\omega - \omega_0) + S_{NN}(\omega + \omega_0)$, as illustrated in (c). This function also equals $S_{YY}(\omega)$ by (8.6-17). Similarly, we form the difference of the spectrums in (b) to obtain $S_{XY}(\omega)$ according to (8.6-18) as shown in (d). This function also gives $S_{YX}(\omega)$ from (8.6-19) as shown.

To find $R_{XY}(\tau)$ we apply (8.6-13):

$$R_{XY}(\tau) = \frac{1}{\pi}\int_{\omega_0 - W_1}^{\omega_0 + W_2} P\sin[(\omega - \omega_0)\tau]\,d\omega = \frac{P}{\pi\tau}\int_{-W_1\tau}^{W_2\tau}\sin(x)\,dx$$

$$= \frac{P}{\pi\tau}[\cos(W_1\tau) - \cos(W_2\tau)]$$

$$= \frac{P}{\pi\tau}\left\{\cos\left[\frac{(W_2 + W_1)\tau}{2} - \frac{(W_2 - W_1)\tau}{2}\right]\right.$$

$$\left. - \cos\left[\frac{(W_2 + W_1)\tau}{2} + \frac{(W_2 - W_1)\tau}{2}\right]\right\}$$

$$= \frac{2P}{\pi\tau}\sin\left[\frac{(W_2 + W_1)\tau}{2}\right]\sin\left[\frac{(W_2 - W_1)\tau}{2}\right]$$

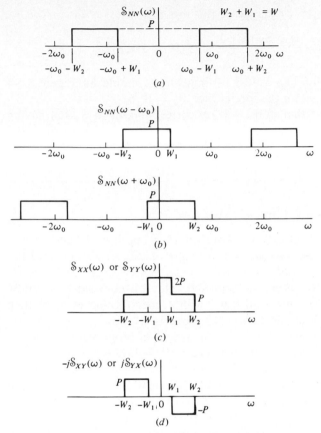

(a)

(b)

(c)

(d)

Figure 8.6-3 Power spectrums applicable to Example 8.6-1.

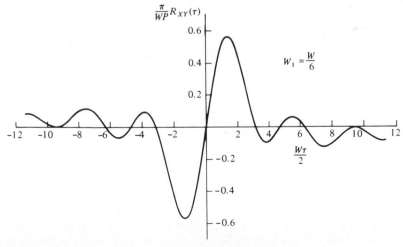

Figure 8.6-4 Cross-correlation function of Example 8.6-1.

Now since $W_1 + W_2 = W$, we may write this result as

$$R_{XY}(\tau) = \frac{WP}{\pi} \frac{\sin (W\tau/2)}{(W\tau/2)} \sin [(W - 2W_1)\tau/2]$$

which is an odd function of τ as (8.6-14) indicates it should be. Figure 8.6-4 illustrates a plot of $R_{XY}(\tau)$ for the special case $W_1 = W/6$.

It should be noted that if $W_1 = W/2$, corresponding to $S_{NN}(\omega)$ having even components about $\omega = \pm\omega_0$, we get $R_{XY}(\tau) = 0$ for all τ. In this case, $X(t)$ and $Y(t)$ are orthogonal processes; they are also independent if $N(t)$ is gaussian.

★Proof of Properties of Band-Limited Processes

It is a quite long and involved task to prove all 12 properties of band-limited processes in detail. Therefore, we shall outline most of the proofs and give the details on only a few.

Property 2 is proved by taking the expected value on both sides of (8.6-2). Since $N(t)$ is assumed wide-sense stationary with a mean value of zero, then $E[X(t)] = 0$ and $E[Y(t)] = 0$ are necessary and property 2 follows.

The sequence of developments leading to the proofs of properties 9 and 4 will now be given. We begin by assuming the usual case $W_1 = W/2$ (see Figure 8.6-1b) and observing that the network of Figure 8.6-5a gives $X(t)$ at its output if the ideal lowpass filter has a bandwidth $W/2$ and if $\omega_0 > W/2$.† We shall assume these conditions true. Thus

$$V_1(t) = 2N(t) \cos (\omega_0 t)$$

$$= 2[X(t) \cos^2 (\omega_0 t) - Y(t) \sin (\omega_0 t) \cos (\omega_0 t)]$$

$$= X(t) + [X(t) \cos (2\omega_0 t) - Y(t) \sin (2\omega_0 t)] \qquad (8.6\text{-}20)$$

The filter will remove the bandpass process contained within the brackets so that only $X(t)$ appears in the output. Next, we develop an expression for $R_{XX}(t, t + \tau)$:

$$R_{XX}(t, t + \tau) = E[X(t)X(t + \tau)]$$

$$= E\left[\int_{-\infty}^{\infty} h(u)V_1(t - u) \, du \int_{-\infty}^{\infty} h(v)V_1(t + \tau - v) \, dv \right]$$

$$= \int_{-\infty}^{\infty} \int_{-\infty}^{\infty} h(u)h(v)R_{NN}(\tau + u - v)4 \cos [\omega_0(t - u)]$$

$$\cdot \cos [\omega_0(t + \tau - v)] \, du \, dv \qquad (8.6\text{-}21)$$

---

† These are idealized values based on an ideal product device. Practical values of bandwidth and ω_0 may be considerably different. The assumption $W_1 = W/2$ is for simple definition of filter bandwidth and is not a constraint in properties 9 or 4.

Figure 8.6-5 Block diagrams of networks that realize (a) $X(t)$ and (b) $Y(t)$ from a random process $N(t) = X(t) \cos(\omega_0 t) - Y(t) \sin(\omega_0 t)$. [*Reproduced from Peebles (1976) with permission of publishers Addison-Wesley, Advanced Book Program.*]

In developing (8.6-21), we have written $X(t)$ and $X(t + \tau)$ in terms of the convolution integral involving $h(t)$, the impulse response of the lowpass filter, substituted $V_1(t)$ from (8.6-20), and used the fact that $N(t)$ is assumed wide-sense stationary. The further reduction of (8.6-21) is lengthy (Peebles, 1976, p. 157) and will only be outlined. If the cosine factors are replaced by their exponential forms and if $R_{NN}(\tau + u - v)$ is replaced by its equivalent, the inverse transform of the power spectrum $S_{NN}(\omega)$, (8.6-21) becomes the sum of four integrals. It can be shown that two of these integrals, the only two involving t, are zero. Thus, $R_{XX}(t, t + \tau)$ becomes a function of τ only and $X(t)$ is therefore wide-sense stationary, proving part of property 1. The two remaining integrals are used to prove properties 9 and 4.

A procedure exactly the same as discussed in the last paragraph can be used to prove first that $Y(t)$ is wide-sense stationary, thereby providing the proof of another part of property 1. The development also proves properties 10 and 5; it is based on the fact that $Y(t)$ is produced by the operations shown in Figure 8.6-5b.

Property 3 next results from use of property 5 with $\tau = 0$ and the integration of $S_{XX}(\omega)$ using property 9.

Properties 11, 6, 8, and the balance of property 1 are proved by considering the cross-correlation function

$$R_{XY}(t, t + \tau) = E[X(t)Y(t + \tau)]$$

$$= E\left[\int_{-\infty}^{\infty} h(u)V_1(t - u)\, du \int_{-\infty}^{\infty} h(v)V_2(t + \tau - v)\, dv\right]$$

$$= -\int_{-\infty}^{\infty}\int_{-\infty}^{\infty} h(u)h(v)R_{NN}(\tau + u - v)4\cos[\omega_0(t - u)]$$

$$\cdot \sin[\omega_0(t + \tau - v)]\, dv\, du \qquad (8.6\text{-}22)$$

which is developed in a manner analogous to (8.6-21). Reduction of (8.6-22) as discussed earlier shows that $R_{XY}(t, t + \tau)$ depends only on τ, so that $X(t)$ and $Y(t)$ are jointly wide-sense stationary (proving property 1); it also proves properties 11 and 6. Property 8 results from property 6 with $\tau = 0$.

Proofs of the remaining properties, 7 and 12, follow from consideration of the autocorrelation function of $N(t)$. It is readily found by using (8.6-2) that

$$R_{NN}(t, t + \tau) = E[N(t)N(t + \tau)]$$

$$= [R_{XX}(\tau) + R_{YY}(\tau)]\tfrac{1}{2} \cos (\omega_0 \tau)$$

$$+ [R_{XX}(\tau) - R_{YY}(\tau)]\tfrac{1}{2} \cos (2\omega_0 t + \omega_0 \tau)$$

$$- [R_{XY}(\tau) - R_{YX}(\tau)]\tfrac{1}{2} \sin (\omega_0 \tau)$$

$$- [R_{XY}(\tau) + R_{YX}(\tau)]\tfrac{1}{2} \sin (2\omega_0 t + \omega_0 \tau) \qquad (8.6\text{-}23)$$

Since $N(t)$ is wide-sense stationary by original assumption, its autocorrelation function cannot be a function of t. Thus, we require

$$R_{XX}(\tau) = R_{YY}(\tau) \qquad (8.6\text{-}24)$$

and

$$R_{XY}(\tau) = -R_{YX}(\tau) \qquad (8.6\text{-}25)$$

in (8.6-23); these results prove property 12 and the first part of property 7. Finally, recognizing that $R_{XY}(\tau) = R_{YX}(-\tau)$ for a cross-correlation function, we obtain the second part of property 7, which says that $R_{XY}(\tau)$ is an odd function of τ.

8.7 MODELING OF NOISE SOURCES

All our work in this chapter so far has related to finding the response of a linear system when a random waveform (desired signal or undesired noise) was applied at its input. In every case, the system was assumed to not contain any internal sources. In particular, the system was assumed to be free of any internally generated *noise*. In the real world, such an assumption is never justified because all networks (systems) generate one or more types of noise internally. For example, all conductors or semiconductors in a circuit are known to generate *thermal noise* (see Section 7.5) because of thermal agitation of free electrons.† The question naturally arises: How can we handle practical networks that produce internally generated noise? The remainder of this chapter is concerned with answering this question.

We shall find that, by suitable modeling techniques for both the network and for the external source that drives the network, all the internally generated network noise can be thought of as having been caused *by the external source*. In

---

† There are many other types of internally generated noise such as *shot noise, partition noise, induced grid noise, flicker noise, secondary emission noise*, etc. The reader is referred to the literature for more detail (Mumford and Scheibe, 1968; van der Ziel, 1970).

effect, we shall replace the noisy practical network with a noise-free identical network that is driven by a "more noisy" source.

Our work begins by developing models for noise sources.

Resistive (Thermal) Noise Source

Suppose we have an ideal (noise-free, infinite input impedance) voltmeter that responds to voltages that fall in a small ideal (rectangular) frequency band $d\omega/2\pi$ centered at angular frequency ω. If such a voltmeter is used to measure the voltage across a resistor of resistance R (ohms), it is found, both in practice and theoretically, that a noise voltage $e_n(t)$ would exist having a mean-squared value given by

$$\overline{e_n^2(t)} = \frac{2kTR \, d\omega}{\pi} \tag{8.7-1}$$

Here $k = 1.38(10^{-23})$ joule per Kelvin is *Boltzmann's constant*,† and T is temperature in Kelvin. This result is independent of the value of ω up to extremely high frequencies. (See Section 7.5 where $\mathcal{N}_0/2$ equals $2kTR$ here. The reader should justify this fact as an exercise.)

Now because the voltmeter does not load the resistor, $\overline{e_n^2(t)}$ is the mean-squared open-circuit voltage of the resistor which can be treated as a voltage source with internal impedance R. In other words, the noisy resistor can be modeled as a Thevenin‡ voltage source as shown in Figure 8.7-1a. An equivalent current source is shown in (b) where

$$\overline{i_n^2(t)} = \overline{e_n^2(t)}/R^2 = \frac{2kT \, d\omega}{\pi R} \tag{8.7-2}$$

is the short-circuit mean-squared current.

From Figure 8.7-1a it is found that the *incremental noise power* dN_L delivered to the load in the incremental band $d\omega$ by the noisy resistor as a source is

$$dN_L = \frac{\overline{e_n^2(t)}R_L}{(R + R_L)^2} = \frac{2kTRR_L \, d\omega}{\pi(R + R_L)^2} \tag{8.7-3}$$

The maximum delivered power occurs when $R_L = R$. We call this maximum power the *incremental available power* of the source and denote it by dN_{as}; it is given by

$$dN_{as} = \overline{e_n^2(t)}/4R = \frac{kT \, d\omega}{2\pi} \tag{8.7-4}$$

We see from (8.7-4) that the incremental power *available* from a resistor source is *independent of the resistance of the source* and depends only on its physical temperature T. These facts may be used as a basis for modeling arbitrary sources.

---

† Ludwig Boltzmann (1844–1906) was an Austrian physicist.
‡ Named for the French physicist Léon Thevenin (1857–1926).

Figure 8.7-1 Equivalent circuit models of a noisy resistor: (a) voltage model and (b) current model. [*Adapted from Peebles (1976) with permission of publishers Addison-Wesley, Advanced Book Program.*]

Arbitrary Noise Sources, Effective Noise Temperature

Suppose an actual noise source has an incremental available noise power dN_{as}, open-circuit output mean-squared voltage $\overline{e_n^2(t)}$, and impedance as measured between its output terminals of $Z_o(\omega) = R_o(\omega) + jX_o(\omega)$. The available noise power is easily found to be

$$dN_{as} = \frac{\overline{e_n^2(t)}}{4R_o(\omega)} \tag{8.7-5}$$

If we now ascribe all the source's noise to the resistive part $R_o(\omega)$ of its output impedance by *defining* an *effective noise temperature* T_s such that (8.7-1) applies, then

$$\overline{e_n^2(t)} = 2kT_s R_o(\omega) \frac{d\omega}{\pi} \tag{8.7-6}$$

As with a purely resistive source, available power is still independent of the source impedance but depends on the source's temperature

$$dN_{as} = kT_s \frac{d\omega}{2\pi} \tag{8.7-7}$$

We consider two examples that illustrate effective noise temperature.

Example 8.7-1 Two different resistors at different physical temperatures are placed in series. The effective noise temperature of the series combination as a noise source is to be found.

Figure 8.7-2 Equivalent circuits for two resistors at different temperatures in series.

Figure 8.7-2 illustrates Thevenin equivalent circuits for the combination. Since the individual resistors as sources may be considered independent, their mean-squared voltages add. Hence,

$$\overline{e_1^2(t)} + \overline{e_2^2(t)} = \overline{e_n^2(t)}$$

By applying (8.7-1) to both sides of the preceding expression, we obtain

$$2k[T_1 R_1 + T_2 R_2] \frac{d\omega}{\pi} = 2k[T_s(R_1 + R_2)] \frac{d\omega}{\pi}$$

or

$$T_s = \frac{T_1 R_1 + T_2 R_2}{R_1 + R_2}$$

Example 8.7-1 clearly shows that effective noise temperature of a source is not necessarily equal to its physical temperature. In the special case where $T_1 = T_2 = T$, then $T_s = T$. More generally, it is true that any passive, two-terminal source that contains only resistors, capacitors, and inductors, all at the same physical temperature T, will have an effective noise temperature $T_s = T$. (Ziemer and Tranter, 1976, p. 471). The next example can be used to illustrate this last point.

Example 8.7-2 We reconsider Example 8.7-1, except we now allow a capacitor to be placed across one resistor as shown in Figure 8.7-3.

By superposition, $\overline{e_n^2(t)}$ is the sum of contributions from each resistor as a noise source. The mean-squared voltage, denoted $\overline{e_{n1}^2(t)}$, due to the first resistor is readily seen to be

$$\overline{e_{n1}^2(t)} = \overline{e_1^2(t)} \left| \frac{1}{1 + j\omega R_1 C_1} \right|^2 = \frac{\overline{e_1^2(t)}}{1 + \omega^2 R_1^2 C_1^2}$$

Figure 8.7-3 Equivalent circuits for a linear, passive, two-terminal network of two resistors and one capacitor.

That due to the second resistor is

$$\overline{e_{n2}^2(t)} = \overline{e_2^2(t)}$$

Thus, by applying (8.7-1) to the two individual resistor mean-squared voltages, we have

$$\overline{e_n^2(t)} = \overline{e_{n1}^2(t)} + \overline{e_{n2}^2(t)} = 2k\left[\frac{T_1 R_1}{1 + \omega^2 R_1^2 C_1^2} + T_2 R_2\right]\frac{d\omega}{\pi}$$

Next, we find the output impedance of the network as an overall source by imagining the noise sources set to 0. We get

$$Z_o(\omega) = R_2 + \frac{R_1(1/j\omega C_1)}{R_1 + (1/j\omega C_1)} = R_2 + \frac{R_1}{1 + j\omega R_1 C_1}$$

$$= R_2 + \frac{R_1(1 - j\omega R_1 C_1)}{1 + \omega^2 R_1^2 C_1^2}$$

which has a resistive part

$$R_o(\omega) = R_2 + \frac{R_1}{1 + \omega^2 R_1^2 C_1^2}$$

By applying (8.7-6) to the equivalent source, we have

$$\overline{e_n^2(t)} = 2kT_s\left[R_2 + \frac{R_1}{1 + \omega^2 R_1^2 C_1^2}\right]\frac{d\omega}{\pi}$$

Finally, we equate $\overline{e_n^2(t)}$ for the actual and equivalent networks to find T_s:

$$T_s = \frac{T_1 R_1 + T_2 R_2(1 + \omega^2 R_1^2 C_1^2)}{R_1 + R_2(1 + \omega^2 R_1^2 C_1^2)}$$

The preceding example shows that effective noise temperature may be a function of frequency. In this case, the available noise power is also frequency dependent.

Again we see that $T_s = T$ in the above example if $T_1 = T_2 = T$, as it must because it is a linear, passive, two-terminal network with only resistors and a capacitor, as noted previously.

An Antenna as a Noise Source

In practice, all antennas produce noise at their output because of reception of electromagnetic radiation from noise sources external to the antenna.† The amount of available noise power dN_{as} in an incremental band $d\omega$ depends in a rather complicated manner on all the space surrounding the antenna. However, it is possible to model the antenna in a simple way by assigning to it an *antenna temperature* T_a chosen so that dN_{as} and T_a are related by (8.7-4). Thus,

$$dN_{as} = kT_a \frac{d\omega}{2\pi} \tag{8.7-8}$$

In general, antenna temperature may vary with frequency. However, in many applications T_a can be considered constant (with respect to ω) because its variation with frequency over a frequency band comparable to that of the desired signal being received is often small.

Example 8.7-3 A very sensitive meter that is capable of measuring noise power in a (small) frequency band 1 kHz wide at any frequency $\omega/2\pi$ is attached to a microwave antenna used in a radio relay link. It registers 2.0 (10^{-18}) W when the meter's input impedance is matched to the antenna so that its reading is maximum. We find the antenna temperature T_a.

Since maximum power is extracted from the antenna, the power is its available power and (8.7-8) gives

$$T_a = \frac{2\pi \, dN_{as}}{k \, d\omega} = \frac{2\pi(2)10^{-18}}{1.38(10^{-23})2\pi(10^3)} = \frac{200}{1.38} \approx 144.9 \text{ K}$$

8.8 INCREMENTAL MODELING OF NOISY NETWORKS

In this section we shall show how a noisy network can be modeled as a noise-free network excited by a suitably chosen external noise source. We also develop some measures of the "noisiness" of a network. All our work is applicable to an incremental band $d\omega$.

---

† There are many sources of external noise; several of these are described by Peebles (1976, pp. 463–464).

Figure 8.8-1 A linear two-port network driven by a source of impedance Z_s.

Available Power Gain

Consider first a linear, noise-free, two-port (4-terminal) network having an input impedance Z_i when the output port is open-circuited. Its output impedance, found by looking back into its output port, is Z_o when being driven by a source with source impedance Z_s. The source open-circuit voltage is $e_s(t)$ and the network's open-circuit output voltage is $e_o(t)$. The applicable network is illustrated in Figure 8.8-1.

The available power, denoted dN_{as}, of the source is

$$dN_{as} = \frac{\overline{e_s^2(t)}}{4R_s} \tag{8.8-1}$$

where R_s is the real part of Z_s. This power is independent of Z_i. The available power, denoted dN_{aos}, in the output due to the source is

$$dN_{aos} = \frac{\overline{e_o^2(t)}}{4R_o} \tag{8.8-2}$$

where R_o is the real part of Z_o. This power does depend on Z_i through its influence on the generation of $e_o(t)$ but does not depend on the load impedance Z_L. We define the *available power gain* denoted G_a of the two-port network as the ratio of the available powers

$$G_a = \frac{dN_{aos}}{dN_{as}} = \frac{R_s \overline{e_o^2(t)}}{R_o \overline{e_s^2(t)}} \tag{8.8-3}$$

When a cascade of M noise-free networks is involved where $M = 1, 2, \ldots$, it is easy to see that the overall available power gain G_a is the product of available power gains G_m, $m = 1, 2, \ldots, M$, if G_m is the gain of stage m when all preceding stages are connected and treated as its source (see Problem 8-65). Thus,

$$G_a = \prod_{m=1}^{M} G_m \tag{8.8-4}$$

Equivalent Networks, Effective Input Noise Temperature

Consider next the case of a linear two-port network *with* internally generated noise. The network is assumed to be driven from a source with effective noise temperature T_s as shown in Figure 8.8-2a. If G_a is the network's available power gain, the available output noise power due to the source alone is

$$dN_{aos} = G_a \, dN_{as} = G_a k T_s \frac{d\omega}{2\pi} \tag{8.8-5}$$

from (8.8-3) and (8.7-7).

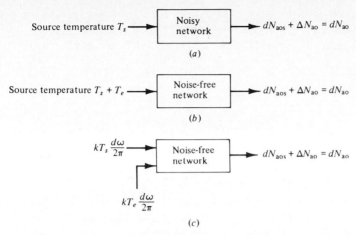

Figure 8.8-2 A network with internally generated noise driven from a noise source (*a*), and equivalent noise-free networks (*b*) and (*c*). [*Reproduced from Peebles (1976) with permission of publishers Addison-Wesley, Advanced Book Program.*]

Total available output noise power dN_{ao} is larger than dN_{aos} because of internally generated noise. Let ΔN_{ao} represent the *excess available noise power* at the output. We shall imagine that ΔN_{ao} is generated by the source by defining *effective input noise temperature* T_e as the temperature *increase* that the source would require to account for all output available noise power. It therefore follows that

$$\Delta N_{\mathrm{ao}} = G_a k T_e \frac{d\omega}{2\pi} \qquad (8.8\text{-}6)$$

With this definition, the noisy network is replaced by a noise-free network driven by a source of temperature $T_s + T_e$ as shown in Figure 8.8-2*b*.

It is somewhat helpful to model the available source noise power by use of two inputs, as shown in Figure 8.8-2*c*. The second input represents the internally generated noise due to the network. The representation is convenient in visualizing noise effects when networks are cascaded as illustrated in Figure 8.8-3. By equating expressions for output available noise powers in the cascade and equivalent network, the effective input noise temperature T_e of the cascade is determined to be

$$T_e = T_{e1} + \frac{T_{e2}}{G_1} + \frac{T_{e3}}{G_1 G_2} + \cdots + \frac{T_{eM}}{G_1 G_2 \cdots G_{M-1}} \qquad (8.8\text{-}7)$$

where T_{em} and G_m, $m = 1, 2, \ldots, M$, are the effective input noise temperature and available power gain, respectively, for the mth stage when all $m - 1$ previous stages are connected and form its source.

An especially useful application of (8.8-7) is to the cascade of stages in an amplifier. We develop an example.

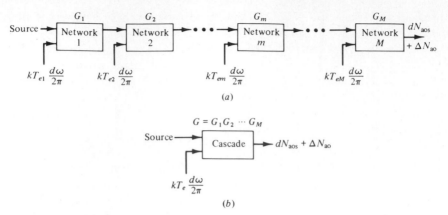

Figure 8.8-3 (a) M networks in cascade and (b) the equivalent network. [*Reproduced from Peebles (1976) with permission of publishers Addison-Wesley, Advanced Book Program.*]

Example 8.8-1 The stages in a three-stage amplifier have effective input noise temperatures $T_{e1} = 1350$ K, $T_{e2} = 1700$ K and $T_{e3} = 2600$ K. The respective available power gains are $G_1 = 16$, $G_2 = 10$, and $G_3 = 6$. We find the effective input noise temperature of the overall amplifier by use of (8.8-7):

$$T_e = 1350 + \frac{1700}{16} + \frac{2600}{16(10)} = 1350 + 106.25 + 16.25$$

$$= 1472.5 \text{ K}$$

We see that, even though T_{e2} and T_{e3} are larger than T_{e1}, the contributions to T_e by the second and third stages are much smaller than that of the first stage because of the gain of previous stages. In general, it is clear from (8.8-7) that an amplifier should have its lowest noise, highest gain stage first, followed by its next best stage, etc., for best noise performance.

Spot Noise Figures

Effective input noise temperature T_e of a network is a measure of its noise performance. Better performance corresponds to lower values of T_e. Another measure of performance is *incremental* or *spot noise figure* denoted by F and defined as the total incremental available output noise power dN_{ao} divided by the incremental available output noise power due to the source alone:

$$F = \frac{dN_{ao}}{dN_{aos}} = \frac{dN_{aos} + \Delta N_{ao}}{dN_{aos}} = 1 + \frac{\Delta N_{ao}}{dN_{aos}} \tag{8.8-8}$$

An alternative form derives from the substitution of (8.8-5) and (8.8-6):

$$F = 1 + \frac{T_e}{T_s} \tag{8.8-9}$$

In an ideal network, $T_e = 0$ so $F = 1$. For any real network, F is larger than unity.

In practice, a given network might be driven by a variety of sources. For example, an amplifier might be driven by an antenna, mixer, attenuator, other amplifier, etc. Its spot noise figure is therefore a function of the effective noise temperature of the source. However, by defining a *standard source* as having a *standard noise temperature* $T_0 = 290$ K and *standard spot noise figure* F_0, given by

$$F_0 = 1 + \frac{T_e}{T_0} \tag{8.8-10}$$

a network can be specified independent of its application.

When a network is used with the source for which it is intended to operate F will be called the *operating spot noise figure* and given the symbol F_{op}. From (8.8-9)

$$F_{op} = 1 + \frac{T_e}{T_s} \tag{8.8-11}$$

Operating and standard spot noise figures can also be developed for a cascade of networks (see Problems 8-66 and 8-68).

Example 8.8-2 An engineer purchases an amplifier that has a narrow bandwidth of 1 kHz and standard spot noise figure of 3.8 at its frequency of operation. The amplifier's available output noise power is 0.1 mW when its input is connected to a radio receiving antenna having an antenna temperature of 80 K. We find the amplifier's input effective noise temperature T_e, its operating spot noise figure F_{op}, and its available power gain G_a.

T_e derives from (8.8-10):

$$T_e = T_0(F_0 - 1) = 290(3.8\text{-}1) = 812 \text{ K}$$

We can now use (8.8-11) to obtain F_{op}:

$$F_{op} = 1 + \frac{812}{80} = 11.15$$

From (8.8-5) and (8.8-6) we add to get total available output noise power:

$$dN_{ao} = dN_{aos} + \Delta N_{ao} = \frac{k(T_s + T_e)G_a \, d\omega}{2\pi}$$

so

$$G_a = \frac{2\pi \, dN_{ao}}{k(T_s + T_e) \, d\omega} = \frac{2\pi(0.1)10^{-3}}{1.38(10^{-23})(812 + 80)2\pi(10^3)} \approx 8.12(10^{12})$$

8.9 MODELING OF PRACTICAL NOISY NETWORKS

In a realistic network, the frequency band of interest is not incremental. Therefore such quantities as available power gain, noise temperature, and noise figure
are not necessarily constant but become frequency dependent, in general. In this
section we extend the earlier concepts based on an incremental frequency band to
include practical networks, by defining *average* noise temperatures and *average*
noise figures.

Average Noise Figures

We define *average operating noise figure* $\bar{F}_{op}$ as the *total* output available noise
power N_{ao} from a network divided by the *total* output available noise power N_{aos}
due to the source alone. Thus,

$$\bar{F}_{op} = \frac{N_{ao}}{N_{aos}} \tag{8.9-1}$$

N_{aos} is found by integration of (8.8-5):

$$N_{aos} = \frac{k}{2\pi} \int_0^\infty T_s G_a \, d\omega \tag{8.9-2}$$

We may similarly use (8.8-8) with (8.8-5) to determine N_{ao}:

$$N_{ao} = \int_0^\infty dN_{ao} = \int_0^\infty F_{op} \, dN_{aos} = \frac{k}{2\pi} \int_0^\infty F_{op} T_s G_a \, d\omega \tag{8.9-3}$$

Thus, from (8.9-1)

$$\bar{F}_{op} = \frac{\int_0^\infty F_{op} T_s G_a \, d\omega}{\int_0^\infty T_s G_a \, d\omega} \tag{8.9-4}$$

In many cases the source's temperature is approximately constant. Operating
average noise figure then becomes

$$\bar{F}_{op} = \frac{\int_0^\infty F_{op} G_a \, d\omega}{\int_0^\infty G_a \, d\omega} \qquad T_s \text{ constant} \tag{8.9-5}$$

An antenna is an example of a source having an approximately constant noise
temperature (so long as the surroundings viewed by the antenna are fixed).
Another example is a standard source for which $T_s = T_0 = 290$ K is constant. We
define *average standard noise figure* $\bar{F}_0$ as that for which the source is standard. In
this case

$$\bar{F}_0 = \frac{\int_0^\infty F_0 G_a \, d\omega}{\int_0^\infty G_a \, d\omega} \tag{8.9-6}$$

as can be shown by repeating the steps leading to (8.9-4).

Average Noise Temperatures

From the definition of effective input noise temperature T_e, it follows that the incremental available output noise power from a network with available power gain G_a that is driven by a source of temperature T_s is

$$dN_{ao} = G_a k(T_s + T_e) \frac{d\omega}{2\pi} \tag{8.9-7}$$

Total available power is therefore

$$N_{ao} = \int_0^\infty dN_{ao} = \frac{k}{2\pi} \int_0^\infty G_a(T_s + T_e)\, d\omega \tag{8.9-8}$$

Next, we define *average effective source temperature* $\bar{T}_s$ and *average effective input noise temperature* $\bar{T}_e$ as *constant* temperatures that produce the same total available power as given by (8.9-8). Hence

$$N_{ao} = \frac{k}{2\pi} (\bar{T}_s + \bar{T}_e) \int_0^\infty G_a\, d\omega \tag{8.9-9}$$

By equating (8.9-9) and (8.9-8) on a term-by-term basis, we get

$$\bar{T}_s = \frac{\int_0^\infty T_s G_a\, d\omega}{\int_0^\infty G_a\, d\omega} \tag{8.9-10}$$

and

$$\bar{T}_e = \frac{\int_0^\infty T_e G_a\, d\omega}{\int_0^\infty G_a\, d\omega} \tag{8.9-11}$$

If (8.8-10) and (8.8-11) are substituted into (8.9-6) and (8.9-4), respectively, we obtain the interrelationships

$$\bar{F}_0 = 1 + \frac{\bar{T}_e}{T_0} \tag{8.9-12}$$

$$\bar{F}_{op} = 1 + \frac{\bar{T}_e}{\bar{T}_s} \tag{8.9-13}$$

By equating $\bar{T}_e$ from these last two expressions, we obtain alternative interrelationships

$$\bar{F}_0 = 1 + \frac{\bar{T}_s}{T_0} (\bar{F}_{op} - 1) \tag{8.9-14}$$

$$\bar{F}_{op} = 1 + \frac{T_0}{\bar{T}_s} (\bar{F}_0 - 1) \tag{8.9-15}$$

Average effective noise temperature is a very useful concept for modeling network noise in a simple way. To demonstrate this fact, note that (8.9-9) can be written as

$$N_{ao} = \frac{k}{2\pi} (\bar{T}_s + \bar{T}_e) G_a(\omega_0) \frac{\int_0^\infty G_a(\omega) \, d\omega}{G_a(\omega_0)} \tag{8.9-16}$$

where ω_0 is the centerband angular frequency of the function $G_a(\omega)$. Since $G_a(\omega)$ is the available power gain (or power transfer function) of the network, we identify

$$W_N = \frac{\int_0^\infty G_a(\omega) \, d\omega}{G_a(\omega_0)} \tag{8.9-17}$$

as the *noise bandwidth* of the network, by analogy with (8.5-6). Equation (8.9-16) becomes

$$N_{ao} = G_a(\omega_0) k (\bar{T}_s + \bar{T}_e) \frac{W_N}{2\pi} \tag{8.9-18}$$

which says that actual available output noise power is that due to a source with *constant* temperature $\bar{T}_s + \bar{T}_e$ driving an equivalent noise-free network with an ideal rectangular transfer function of bandwidth W_N(rad/s) and midband available power gain $G_a(\omega_0)$. This result represents a very simple network model.

Modeling of Attenuators

Consider a source of average effective temperature $\bar{T}_s$ driving an impedance-matched lossy attenuator with power loss L (a number not less than one) at all frequencies. The attenuator has a physical temperature T_L. It can be shown (Peebles, 1976, p. 463; Mumford and Scheibe, 1968, p. 23) that the average effective input noise temperature of the attenuator is

$$\bar{T}_e = T_L(L - 1) \tag{8.9-19}$$

From (8.9-12) and (8.9-13) the applicable average noise figures are

$$\bar{F}_0 = 1 + \frac{T_L}{T_0} (L - 1) \tag{8.9-20}$$

$$\bar{F}_{op} = 1 + \frac{T_L}{\bar{T}_s} (L - 1) \tag{8.9-21}$$

Note that if $T_L = T_0$ or if $T_L = \bar{T}_s$, the average noise figure of the attenuator is just equal to its loss.

Model of Example System

One of the most important applications of the theory of this and the preceding two sections is in modeling receiving systems. As illustrated in Figure 8.9-1a, consider a receiving antenna that drives a receiver amplifier through various broad-

Figure 8.9-1 A model of a receiving system (*a*) and its equivalent (*b*). [*Reproduced from Peebles (1976), with permission of publishers Addison-Wesley, Advanced Book Program.*]

band components having an overall loss L. These components (which may include microwave transmission lines, isolators, or other devices) are all assumed to have physical temperature T_L. The antenna temperature is T_a and the receiver average effective input noise temperature is $\bar{T}_R$. The receiver's noise bandwidth is W_N and it has a centerband available power gain $G_R(\omega_0)$. We demonstrate that the system is equivalent to that shown in Figure 8.9-1*b*.

The equivalent system has the same noise bandwidth as the actual system and has a centerband available power gain $G_R(\omega_0)/L$. It is driven by a simple source with *system noise temperature* $\bar{T}_{\text{sys}}$. The available output noise power in the actual system is the sum of the antenna's contribution plus those due to excess noises in the attenuator and receiver. By using earlier models, this noise power is

$$N_{\text{ao}} = k[T_a + T_L(L - 1) + \bar{T}_R L] \frac{G_R(\omega_0)W_N}{L(2\pi)} \tag{8.9-22}$$

For the equivalent system

$$N_{\text{ao}} = k\bar{T}_{\text{sys}} \frac{G_R(\omega_0)W_N}{L(2\pi)} \tag{8.9-23}$$

By equating the above two expressions, we obtain

$$\bar{T}_{\text{sys}} = T_a + T_L(L - 1) + \bar{T}_R L \tag{8.9-24}$$

From (8.9-24), the average effective input noise temperature of the system taken at point A in Figure 8.9-1*a* is

$$\bar{T}_e = T_L(L - 1) + \bar{T}_R L \tag{8.9-25}$$

From (8.9-13) the average system operating noise figure is

$$\bar{F}_{op} = 1 + \frac{T_L}{T_a}(L - 1) + \frac{\bar{T}_R}{T_a} L \tag{8.9-26}$$

Example 8.9-1 An antenna with temperature $T_a = 150$ K is connected to a receiver by means of a waveguide that is at a physical temperature of 280 K and has a loss of 1.5 (1.76 dB).† The receiver has a noise bandwidth of $W_N/2\pi = 10^6$ Hz and an average effective input noise temperature $\bar{T}_R = 700$ K. We determine the system's noise temperature $\bar{T}_{sys}$, its operating average noise figure $\bar{F}_{op}$, and its available output noise power when $G_R(\omega_0) = 10^{12}$ (120 dB).
From (8.9-24)

$$\bar{T}_{sys} = 150 + 280(1.5 - 1) + 700(1.5) = 1340 \text{ K}$$

From (8.9-26)

$$\bar{F}_{op} = 1 + \frac{280}{150}(1.5 - 1) + \frac{700}{150}(1.5) \approx 8.93 \quad \text{or} \quad 9.51 \text{ dB}$$

Finally, we use (8.9-23) to find N_{ao}:

$$N_{ao} = 1.38(10^{-23})1340.0(10^{12})\frac{10^6}{1.5} \approx 12.3 \text{ mW}$$

PROBLEMS

8-1 A signal $x(t) = u(t) \exp(-\alpha t)$ is applied to a network having an impulse response $h(t) = Wu(t) \exp(-Wt)$. Here α and W are real positive constants and $u(\cdot)$ is the unit-step function. Find the system's response by use of (8.1-10).

8-2 Work Problem 8-1 by using (8.1-11) to find the spectrum $Y(\omega)$ of the response.

8-3 A rectangular pulse of amplitude A and duration T, defined by

$$x(t) = \begin{cases} A & 0 < t < T \\ 0 & \text{elsewhere} \end{cases}$$

is applied to the system of Problem 8-1.
    (a) Find the time response $y(t)$.
    (b) Sketch your response for $W = \pi/T$ and $W = 2\pi/T$.

---

† A number L expressed in *decibels* (dB), denoted L_{dB}, is related to L as a numeric (power ratio) by $L_{dB} = 10 \log_{10}(L)$.

8-4 A filter is called *gaussian* if it has a transfer function

$$H(\omega) = \frac{1}{\sqrt{2\pi}W_{rms}} e^{-\omega^2/2W_{rms}^2}$$

where W_{rms} is the root-mean-squared (rms) bandwidth.

(a) Sketch $H(\omega)$.

(b) How is W_{rms} related to the 3-dB bandwidth?

8-5 Two systems have transfer functions $H_1(\omega)$ and $H_2(\omega)$.

(a) Show that the transfer function $H(\omega)$ of the *cascade* of the two, which means that the output of the first feeds the input of the second system, is $H(\omega) = H_1(\omega)H_2(\omega)$.

(b) For a cascade of N systems with transfer functions $H_n(\omega)$, $n = 1, 2, \ldots, N$, show that

$$H(\omega) = \prod_{n=1}^{N} H_n(\omega)$$

*8-6 Work Problem 8-1 if the output of the given network is applied to a second identical network and the response is taken from the second network.

8-7 The impulse response of a system is

$$h(t) = \begin{cases} t^3 e^{-t^2} & 0 < t \\ 0 & t < 0 \end{cases}$$

By use of (8.1-8) or (8.1-10), find the response of the network to the pulse

$$x(t) = \begin{cases} A & 0 < t < T \\ 0 & \text{elsewhere} \end{cases}$$

where A and T are real positive constants.

8-8 Work Problem 8-7 if the network's impulse response is

$$h(t) = \begin{cases} t^3 e^{-t} & 0 < t \\ 0 & t < 0 \end{cases}$$

8-9 Given the network shown in Figure P8-9.

(a) Find the impulse response $h(t)$.

(b) By Fourier transforming $h(t)$, find $H(\omega)$.

(c) Sketch $h(t)$ and $H(\omega)$.

Figure P8-9 [*Reproduced from Peebles (1976), with permission of publishers Addison-Wesley, Advanced Book Program.*]

8-10 Find the transfer function of the network of Figure P8-9 by use of (8.1-13).

8-11 By using (8.1-13), find the transfer function of the network illustrated in Figure P8-11. Assume that no loading is present due to any output circuitry.

Figure P8-11

8-12 Work Problem 8-11 for the network of Figure P8-12.

Figure P8-12

*8-13 (a) Work Problem 8-11 for the network of Figure P8-13.

(b) Under what conditions will the network behave approximately as a lowpass filter?

(c) Find a relationship between R_1, C_1, R_2, and C_2 such that the network behaves at all frequencies as a pure resistive attenuator.

Figure P8-13 [*Reproduced from Peebles (1976), with permission of publishers Addison-Wesley, Advanced Book Program.*]

8-14 Given the network shown in Figure P8-14.

(a) If the output causes no loading on the network, find the transfer function $H(\omega)$.

(b) Define $\omega_0 = 1/\sqrt{LC}$ and $Q_0 = R/\omega_0 L$. Plot $|H(\omega)|^2$ as a function of $x = (\omega - \omega_0)Q_0/\omega_0$ for Q_0 large and ω near ω_0. (*Hint:* Use the approximation $\omega \approx \omega_0$ for the most significant values of ω when Q_0 is large.)

Figure P8-14

*8-15 (a) Find the transfer function $H(\omega)$ for the network shown in Figure P8-15.

(b) Define $\omega_0 = 1/\sqrt{LC}$ and $Q_0 = 1/\omega_0(R + R_L)C$ and assume $Q_0 \gg 1$, so that the values of ω for which $H(\omega)$ is significant correspond to $\omega \approx \omega_0$. Use these facts to obtain an approximation for $H(\omega)$.

(c) If an impulse is applied to the network, find an expression for the approximate energy absorbed by R_L. (*Hint:* Use Parseval's theorem).

Input $x(t)$ R_L Output $y(t)$

                                              Figure P8-15

8-16 A class of filters called *Butterworth filters* has a power transfer function defined by

$$|H(\omega)|^2 = \frac{1}{1 + (\omega/W)^{2n}}$$

where $n = 1, 2, \ldots$, is a number related to the number of circuit elements and W is the 3-dB bandwidth in radians per second. Sketch $|H(\omega)|^2$ for $n = 1, 2, 4,$ and 8 and note the behavior. As $n \to \infty$, what does $|H(\omega)|^2$ become?

8-17 Determine which of the following impulse responses do not correspond to a system that is stable, or realizable, or both, and state why.

(a) $h(t) = u(t + 3)$
(b) $h(t) = u(t)e^{-t^2}$
(c) $h(t) = e^t \sin (\omega_0 t)$ ω_0 a real constant
(d) $h(t) = u(t)e^{-3t} \sin (\omega_0 t)$ ω_0 a real constant.

8-18 Use (8.1-10) and prove (8.1-15).

8-19 Show that (8.1-16) must be true if a linear time-invariant system is to be stable.

8-20 A system is defined by

$$y(t) = \int_{-\infty}^{t} x(\xi) \, d\xi$$

for all $x(t)$ for which the integral exists. Show that the system is linear, time-invariant, and causal.

8-21 A random process

$$X(t) = A \sin (\omega_0 t + \Theta)$$

where A and ω_0 are real positive constants and Θ is a random variable uniformly distributed on the interval $(-\pi, \pi)$, is applied to the network of Problem 8-1. Find an expression for the network's response process using (8.2-3).

8-22 Work Problem 8-21 for a network with impulse response

$$h(t) = u(t)te^{-t}$$

8-23 A random process $X(t)$ is applied to a linear time-invariant system. A response $Y(t) = X(t) - X(t - \tau)$ occurs where τ is a real constant.
    (a) Sketch a block diagram of the system.
    (b) Find the system's transfer function.
8-24 Work Problem 8-23 if the response is

$$Y(t) = X(t - \tau) + \int_{t_1}^{t_2} X(t - \xi)\, d\xi$$

where t_1 and t_2 are real constants.
8-25 A random process $X(t)$ has an autocorrelation function

$$R_{XX}(\tau) = A^2 + Be^{-|\tau|}$$

where A and B are positive constants. Find the mean value of the response of a system having an impulse response

$$h(t) = \begin{cases} e^{-Wt} & 0 < t \\ 0 & t < 0 \end{cases}$$

where W is a real positive constant, for which $X(t)$ is its input.
8-26 Work Problem 8-25 for the system for which

$$h(t) = \begin{cases} te^{-Wt} & 0 < t \\ 0 & t < 0 \end{cases}$$

8-27 Work Problem 8-25 for the system for which

$$h(t) = \begin{cases} e^{-Wt} \sin(\omega_0 t) & 0 < t \\ 0 & t < 0 \end{cases}$$

where W and ω_0 are real positive constants.
8-28 White noise with power density 5 W/Hz is applied to the system of Problem 8-25. Find the mean-squared value of the response using (8.2-7).
8-29 Work Problem 8-28 for the system of Problem 8-26.
8-30 Work Problem 8-28 for the system of Problem 8-27.
8-31 Let jointly wide-sense stationary processes $X_1(t)$ and $X_2(t)$ cause responses $Y_1(t)$ and $Y_2(t)$, respectively, from a linear time-invariant system with impulse response $h(t)$. If the sum $X(t) = X_1(t) + X_2(t)$ is applied, the response is $Y(t)$. Find expressions, in terms of $h(t)$ and characteristics of $X_1(t)$ and $X_2(t)$, for
    (a) $E[Y]$ (b) $R_{YY}(t, t + \tau)$
8-32 Show that the cross-correlation function for the output components $Y_1(t)$ and $Y_2(t)$ in Problem 8-31 is given by

$$R_{Y_1 Y_2}(t, t + \tau) = \int_{-\infty}^{\infty} \int_{-\infty}^{\infty} R_{X_1 X_2}(\tau + u - v)h(u)h(v)\, du\, dv$$

$$= R_{Y_1 Y_2}(\tau)$$

8-33 Two separate systems have impulse responses $h_1(t)$ and $h_2(t)$. A process $X_1(t)$ is applied to the first system and its response is $Y_1(t)$. Similarly, a process $X_2(t)$ invokes a response $Y_2(t)$ from the second system. Find the cross-correlation function of $Y_1(t)$ and $Y_2(t)$ in terms of $h_1(t)$, $h_2(t)$, and the cross-correlation function of $X_1(t)$ and $X_2(t)$. Assume $X_1(t)$ and $X_2(t)$ are jointly wide-sense stationary.

8-34 Two systems are cascaded. A random process $X(t)$ is applied to the input of the first system that has impulse response $h_1(t)$; its response $W(t)$ is the input to the second system having impulse response $h_2(t)$. The second system's output is $Y(t)$. Find the cross-correlation function of $W(t)$ and $\dot{Y}(t)$ in terms of $h_1(t)$ and $h_2(t)$, and the autocorrelation function of $X(t)$ if $X(t)$ is wide-sense stationary.

8-35 Let the two systems of Problem 8-34 be identical, each with the impulse response given in Problem 8-26. If $E[X(t)] = 2$ and $W = 3$ rad/s, find $E[Y(t)]$.

8-36 The random process $X(t)$ of Problem 8-21 (the signal) is added to white noise with power density $\mathcal{N}_0/2$, where $\mathcal{N}_0$ is a positive constant, and the sum is applied to the network of Example 8.1-1.

    (*a*) Find the power spectrums of the output signal and output noise.

    (*b*) Find the ratio of output signal average power to output noise average power.

    (*c*) What value of $W = R/L$ will maximize the ratio of part (*b*)?

8-37 A random process $X(t)$ having autocorrelation function

$$R_{XX}(\tau) = Pe^{-\alpha|\tau|}$$

where P and α are real positive constants, is applied to the input of a system with impulse response

$$h(t) = \begin{cases} We^{-Wt} & 0 < t \\ 0 & t < 0 \end{cases}$$

where W is a real positive constant. Find the autocorrelation function of the network's response $Y(t)$.

8-38 Find the cross-correlation function $R_{XY}(\tau)$ for Problem 8-37.

8-39 For the processes and system of Problem 8-31, show that the power spectrum of $Y(t)$ is

$$S_{YY}(\omega) = |H(\omega)|^2[S_{X_1X_2}(\omega) + S_{X_2X_2}(\omega) + S_{X_1X_1}(\omega) + S_{X_2X_1}(\omega)]$$

8-40 If $X_1(t)$ and $X_2(t)$ are statistically independent random processes in Problem 8-31, use the results of Problem 8-39 to show that the output power spectrum becomes

$$S_{YY}(\omega) = |H(\omega)|^2[S_{X_1X_1}(\omega) + S_{X_2X_2}(\omega) + 4\pi\bar{X}_1\bar{X}_2\,\delta(\omega)]$$

8-41 Rework Example 8.4-1 when the network is replaced by two identical networks in cascade, that is, when $H(\omega) = [1 + (j\omega L/R)]^{-2}$.

8-42 Show that (8.4-7) and (8.4-8) are true.

8-43 A network with transfer function $H(\omega) = j\omega$ is a *differentiator*; its input is the wide-sense stationary random process $X(t)$ and its output is $\dot{X}(t) = dX(t)/dt$.
   (a) By using (8.4-7), show that

$$R_{X\dot{X}}(\tau) = \frac{dR_{XX}(\tau)}{d\tau}$$

   (b) By using (8.4-1), show that

$$R_{\dot{X}\dot{X}}(\tau) = -\frac{d^2R_{XX}(\tau)}{d\tau^2}$$

8-44 Given the random process

$$Y(t) = \frac{1}{2T}\int_{t-T}^{t+T} X(\xi)\,d\xi$$

where $X(t)$ is a wide-sense stationary process. Use (8.2-1) to show that the power spectrum of $Y(t)$ is

$$S_{YY}(\omega) = S_{XX}(\omega)\left[\frac{\sin(\omega T)}{\omega T}\right]^2$$

8-45 Prove (8.5-6).

8-46 A random process $X(t)$ has a power spectrum $S_{XX}(\omega)$ that is nonzero only for $-W_X < \omega < W_X$, where W_X is a real positive constant. $X(t)$ is applied to a system with transfer function

$$H(\omega) = 1 + j(\omega/W_H) \qquad -W_X < \omega < W_X$$

Find the average power P_{YY} in the network's response $Y(t)$ in terms of the rms bandwidth of $S_{XX}(\omega)$, the constant W_H, and the average power P_{XX} in $X(t)$. Discuss the effect of letting $W_X \to \infty$.

8-47 Find the noise bandwidth of the system having the power transfer function

$$|H(\omega)|^2 = \frac{1}{1 + (\omega/W)^4}$$

where W is a real positive constant.

8-48 Work Problem 8-47 for the function

$$|H(\omega)|^2 = \frac{1}{[1 + (\omega/W)^2]^2}$$

8-49 Work Problem 8-47 for the function

$$|H(\omega)|^2 = \frac{1}{[1 + (\omega/W)^2]^3}$$

8-50 White noise with power density $\mathcal{N}_0/2$ is applied to a lowpass network for which $|H(0)| = 2$; it has a noise bandwidth of 2 MHz. If the average output noise power is 0.1 W in a 1-Ω resistor, what is $\mathcal{N}_0$?

8-51 White noise with power density $\mathcal{N}_0/2$ is applied to an ideal lowpass filter with bandwidth W.

(a) Find and sketch the autocorrelation function of the response.

(b) If samples of the output noise taken at times $t_n = n\pi/W$, $n = 0, \pm 1, \pm 2, \ldots$, are considered as values of random variables, what can you say about these random variables?

8-52 Work Problem 8-51 for an ideal bandpass filter centered on a frequency $\omega_0/2\pi$ that has a bandwidth W. Assume sample times are now $t_n = n2\pi/W$, $n = 0, \pm 1, \pm 2, \ldots$.

*****8-53** A band-limited random process $N(t)$ has the power density spectrum

$$S_{NN}(\omega) = \begin{cases} P \cos\left[\pi(\omega - \omega_0)/W\right] & -W/2 \le \omega - \omega_0 \le W/2 \\ P \cos\left[\pi(\omega + \omega_0)/W\right] & -W/2 \le \omega + \omega_0 \le W/2 \\ 0 & \text{elsewhere} \end{cases}$$

where P, W, and $\omega_0 > W$ are real positive constants.

(a) Find the power in $N(t)$.

(b) Find the power spectrum $S_{XX}(\omega)$ of $X(t)$ when $N(t)$ is represented by (8.6-2).

(c) Find the cross-correlation function $R_{XY}(\tau)$.

(d) Are $X(t)$ and $Y(t)$ orthogonal processes?

*****8-54** A band-limited random process is given by (8.6-2) and has the power density spectrum shown in Figure P8-54.

(a) Sketch $S_{XX}(\omega)$.

(b) Sketch $S_{XY}(\omega)$, if a sketch is possible.

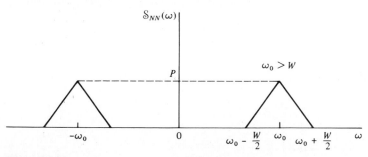

Figure P8-54

*8-55 Work Problem 8-54 for the power spectrum of Figure P8-55.

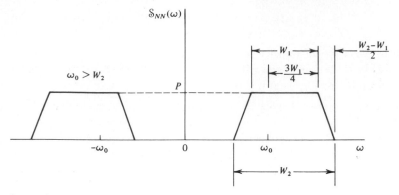

Figure P8-55

*8-56 Use (8.6-2) and derive (8.6-23).

8-57 A sonar echo system on a submarine transmits a random noise $n(t)$ to determine the distance to another "target" submarine. Distance R is given by $v\tau_R/2$ where v is the speed of the sound waves in water and τ_R is the time it takes the reflected version of $n(t)$ to return. Its block diagram is shown in Figure P8-57. Assume that $n(t)$ is a sample function of an ergodic random process $N(t)$ and T is very large.

(a) Find V in terms of a correlation function of $N(t)$.
(b) What value of the delay τ_T will cause V to be maximum?
(c) State in words how the submarine can determine the distance to the target.

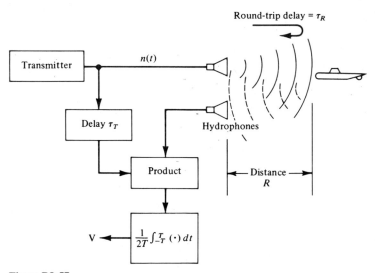

Figure P8-57

8-58 Two resistors with resistances R_1 and R_2 are connected in parallel and have physical temperatures T_1 and T_2, respectively.

(a) Find the effective noise temperature T_s of an equivalent resistor with resistance equal to the parallel combination of R_1 and R_2.

(b) If $T_1 = T_2 = T$, what is T_s?

8-59 Work Problem 8-58 for three resistances R_1, R_2, and R_3 in parallel when they have physical temperatures T_1, T_2, and T_3, respectively.

8-60 Work Example 8.7-2 if a second capacitor is placed across the resistance R_2. Is it possible to choose C_2 so that T_s is independent of frequency?

*★**8-61** Find the effective noise temperature of the network of Figure P8-61 if R_1 and R_2 are at physical temperatures T_1 and T_2, respectively.

Figure P8-61

8-62 A two-port network is illustrated in Figure P8-62. Find its available power gain.

Figure P8-62

8-63 If the two-port network of Problem 8-62 has a physical temperature T_L and is driven by a source of resistance R_s and effective noise temperature T_s, what is the effective input noise temperature of the network?

8-64 If the output of the network of Problem 8-62 is connected to the input of a second indentical network, what is the available power gain of the cascade if $R_1 = 5 \, \Omega$, $R_2 = 3 \, \Omega$ and $R_s = 7 \, \Omega$?

8-65 Show that (8.8-4) is valid.

8-66 In a cascade of M network stages for which the mth stage has available power gain G_m and operating spot noise figure F_{opm} when driven by all previous stages as its source, show that the overall cascade's operating spot noise figure is

$$F_{op} = F_{op1} + \frac{T_{s1}(F_{op2} - 1)}{T_s G_1} + \cdots + \frac{T_{s(M-1)}(F_{opM} - 1)}{T_s G_1 G_2 \cdots G_{M-1}}$$

where $T_{s(m-1)}$ is the temperature of all stages prior to stage m treated as a source.

8-67 An amplifier has a standard spot noise figure $F_0 = 6.31$ (8.0 dB). An engineer uses the amplifier to amplify the output of an antenna that is known to have an antenna temperature of $T_a = 180$ K.

(a) What is the effective input noise temperature of the amplifier?

(b) What is the operating spot noise figure?

8-68 In a cascade of M stages for which F_{0m}, $m = 1, 2, \ldots, M$, is the standard spot noise figure of stage m which has available power gain G_m, show that the standard spot noise figure of the cascade of networks is

$$F_0 = F_{01} + \frac{F_{02} - 1}{G_1} + \frac{F_{03} - 1}{G_1 G_2} + \cdots + \frac{F_{0M} - 1}{G_1 G_2 \cdots G_{M-1}}$$

8-69 An amplifier has three stages for which $T_{e1} = 200$ K (first stage), $T_{e2} = 450$ K, and $T_{e3} = 1000$ K (last stage). If the available power gain of the second stage is 5, what gain must the first stage have to guarantee an effective input noise temperature of 250 K?

8-70 An amplifier has an operating spot noise figure of 10 dB when driven by a source of effective noise temperature 225 K.

(a) What is the standard spot noise figure of the amplifier?

(b) If a matched attenuator with a loss of 3.2 dB is placed between the source and amplifier's input, what is the operating spot noise figure of the attenuator-amplifier cascade if the attenuator's physical temperature is 290 K?

(c) What is the standard spot noise figure of the cascade in (b)?

8-71 One manufacturer sells a microwave receiver having an operating spot noise figure of 10 dB when driven by a source with effective noise temperature 130 K. Another sells a receiver with a standard spot noise figure of 6 dB.

(a) Find the effective input noise temperatures of the two receivers.

(b) All other parameters, such as gain, cost, etc., being the same, which receiver would be the best to purchase?

8-72 What is the maximum average effective input noise temperature that an amplifier can have if its average standard noise figure is to not exceed 1.7?

8-73 An amplifier has an average standard noise figure of 2.0 dB and an average operating noise figure of 6.5 dB when used with a source of average effective source temperature $\bar{T}_s$. What is $\bar{T}_s$?

8-74 An antenna with average noise temperature 60 K connects to a receiver through various microwave elements that can be modeled as an impedance-matched attenuator with an overall loss of 2.4 dB and a physical temperature of 275 K. The overall system noise temperature is $\bar{T}_{sys} = 820$ K.

(a) What is the average effective input noise temperature of the receiver?

(b) What is the average operating noise figure of the attenuator-receiver cascade?

(c) What is the available output noise power of the receiver if it has an available power gain of 110 dB and a noise bandwidth of 10 MHz?

8-75 If the antenna-attenuator cascade of Problem 8-74 is considered as a noise source, what is its average effective noise temperature?

8-76 The loss L in Figure 8.9-1a is replaced by two cascaded matched attenuators, one with loss L_1 at temperature T_1 attached to the antenna output, and one with loss L_2 at temperature T_2 that connects to the receiver. Derive a new expression for $\bar{T}_{sys}$ analogous to (8.9-24).

ADDITIONAL PROBLEMS

8-77 A network is driven by a resistive source as shown in Figure P8-77. Find: (a) Z_i, (b) Z_o, and (c) G_a. (d) Is the network a matched attenuator?

Figure P8-77

8-78 A network has the transfer function

$$H(\omega) = \frac{2e^{j\omega/20}}{(20 + j\omega)^3}$$

(a) Determine and sketch its impulse response. (*Hint:* Use Appendix E.)
(b) Is the network physically realizable?
(c) Determine if the network is stable by evaluating I in (8.1-16).

★8-79 Show that the impulse response of a cascade of N identical networks, each with transfer function

$$H_1(\omega) = 1/(\alpha + j\omega)$$

where $\alpha > 0$ is a constant, is given by

$$h_N(t) = u(t)\left[\frac{t^{N-1}}{(N-1)!}\right]\exp(-\alpha t)$$

8-80 A signal

$$x(t) = u(t)\exp(-\alpha t)$$

is applied to a network having an impulse response

$$h(t) = u(t)W^2 t\exp(-Wt)$$

Here $\alpha > 0$ and $W > 0$ are real constants. By use of (8.2-2) find the network's response $y(t)$.

8-81 Work Problem 8-80 assuming

$$h(t) = u(t)W^3t^2 \exp(-Wt)$$

8-82 A stationary random process $X(t)$ is applied to the input of a system for which

$$h(t) = 3u(t)t^2 \exp(-8t)$$

If $E[X(t)] = 2$ what is the mean value of the system's response $Y(t)$?

8-83 Work Problem 8-28 for the system of Problem 8-82.

8-84 White noise with power density $\mathcal{N}_0/2$ is applied to a network with impulse response

$$h(t) = u(t)Wt \exp(-Wt)$$

where $W > 0$ is a constant. Find the cross-correlations of the input and output.

8-85 Work Problem 8-84 for a network with impulse response.

$$h(t) = u(t)Wt \sin(\omega_0 t) \exp(-Wt)$$

where ω_0 is a constant.

8-86 A random process $X(t)$ is applied to a network with impulse response

$$h(t) = u(t)t \exp(-bt)$$

where $b > 0$ is a constant. The cross-correlation of $X(t)$ with the output $Y(t)$ is known to have the same form:

$$R_{XY}(\tau) = u(\tau)\tau \exp(-b\tau)$$

    (a) Find the autocorrelation of $Y(t)$.
    (b) What is the average power in $Y(t)$?

8-87 Work Problem 8-86 except assume

$$h(t) = u(t)t^2 \exp(-bt)$$

and

$$R_{XY}(\tau) = u(\tau)\tau^2 \exp(-b\tau)$$

8-88 Two identical networks are cascaded. Each has impulse response

$$h(t) = u(t)3t \exp(-4t)$$

A wide-sense stationary process $X(t)$ is applied to the cascade's input.
    (a) Find an expression for the response $Y(t)$ of the cascade.
    (b) If $E[X(t)] = \bar{X} = 6$, find $E[Y(t)]$.

8-89 A stationary random process $X(t)$, having an autocorrelation function

$$R_{XX}(\tau) = 2 \exp(-4|\tau|)$$

is applied to the network of Figure P8-89. Find: (a) $S_{XX}(\omega)$, (b) $|H(\omega)|^2$, and (c) $S_{YY}(\omega)$.

$C_1 = 4$ F

$Y(t)$

$X(t)$ $R = 1.5$ Ω $C_2 = 2$ F

Figure P8-89

8-90 A wide-sense stationary process $X(t)$, with mean value 5 and power spectrum

$$S_{XX}(\omega) = 50\pi\delta(\omega) + 3/[1 + (\omega/2)^2]$$

is applied to a network with impulse response

$$h(t) = 4 \exp(-4|t|)$$

(a) Find $H(\omega)$ for the network.
Determine: (b) the mean $\bar{Y}$, and (c) the power spectrum of the response $Y(t)$.

8-91 White noise, for which $R_{XX}(\tau) = 10^{-2}\delta(\tau)$, is applied to a network with impulse response

$$h(t) = u(t)3t \exp(-4t)$$

(a) Use (8.2-9) to obtain the network's output noise power (in a 1-ohm resistor).
(b) Obtain an expression for the output power spectrum.

8-92 White noise with power density $\mathcal{N}_0/2 = 6(10^{-6})$ W/Hz is applied to an ideal filter (gain = 1) with bandwidth W (rad/s). Find W so that the output's average noise power is 15 watts.

8-93 An ideal filter with a midband power gain of 8 and bandwidth of 4 rad/s has noise $X(t)$ at its input with power spectrum

$$S_{XX}(\omega) = (50/\sqrt{8\pi}) \exp(-\omega^2/8)$$

What is the noise power at the network's output?

8-94 White noise with power density $\mathcal{N}_0/2$, $\mathcal{N}_0 > 0$ a constant, is applied to a lowpass network for which $H(0) = 2$ and its noise bandwidth is 2 MHz. If average output noise power is 0.1 W in a 1-ohm resistor, what is $\mathcal{N}_0$?

8-95 A system's power transfer function is

$$|H(\omega)|^2 = 16/[256 + \omega^4]$$

(a) What is its noise bandwidth?

(b) If white noise with power density $6(10^{-3})$ W/Hz is applied to the input, find the noise power in the system's output.

*8-96 Assume a band-limited random process $N(t)$ has a power spectrum

$$S_{NN}(\omega) =$$

$$B[u(\omega - \omega_0 + W_1) - u(\omega - \omega_0 - W_2)] \exp[-a(\omega - \omega_0 + W_1)]$$

$$+ B[u(-\omega - \omega_0 + W_1) - u(-\omega - \omega_0 - W_2)] \exp[-a(-\omega - \omega_0 + W_1)]$$

where B, ω_0, W_1, and W_2 are positive constants, and a is a constant.

Assume $2\omega_0 > W_1 + W_2$ and find analytical expressions for (a) the power spectrum $S_{XX}(\omega)$ and (b) the cross-power spectrum $S_{XY}(\omega)$ for the processes $X(t)$ and $Y(t)$ involved in the representation of (8.6-2) for $N(t)$.

(c) Sketch $S_{XX}(\omega)$ and $S_{XY}(\omega)$ for $W_1 = W_2/2$ and $a = 1/W_1$.

(d) Repeat part (c) except with $a = -1/W_1$.

*8-97 Find the functions $R_{XX}(\tau)$ and $R_{XY}(\tau)$ applicable in Problem 8-96.

8-98 Determine the effective noise temperature of the network of Figure P8-98 if resistors R_1 and R_2 are at different physical temperatures T_1 and T_2, respectively.

Figure P8-98

8-99 Two resistors in series have different physical temperatures as in Example 8.7-1. Let R_1 and R_2 be independent random variables uniformly distributed on (1000, 1500) and (2200, 2700), respectively. Their average resistances are then $\bar{R}_1 = 1250 \ \Omega$ and $\bar{R}_2 = 2450 \ \Omega$.

(a) What is the effective noise temperature of the two resistors as a source if $T_1 = 250$ K and $T_2 = 330$ K and average resistors are used?

(b) What is the mean effective noise temperature of the source for the same values of T_1 and T_2?

8-100 An amplifier has three stages for which $T_{e1} = 150$ K (first stage), $T_{e2} = 350$ K, and $T_{e3} = 600$ K (output stage). Available power gain of the first stage is 10 and overall input effective noise temperature is 190 K.

(a) What is the available power gain of the second stage?

(b) What is the cascade's standard spot noise figure?

(c) What is the cascade's operating spot noise figure when used with a source of noise temperature $T_s = 50$ K?

8-101 Three networks are cascaded. Available power gains are $G_1 = 8$ (input stage), $G_2 = 6$, and $G_3 = 20$ (output stage). Respective input effective spot noise temperatures are $T_{e1} = 40$ K, $T_{e2} = 100$ K, and $T_{e3} = 280$ K.

(a) What is the input effective spot noise temperature of the cascade?

(b) If the cascade is used with a source of noise temperature $T_s = 30$ K, find the percentage of total available output noise power (in a band $d\omega$) due to each of the following: (1) source, and the excess noises of (2) network 1, (3) network 2, and (4) network 3.

8-102 An antenna with effective noise temperature $T_a = 90$ K is connected to an attenuator that is at a physical temperature of 270 K and has a loss of 1.9. What is the effective spot noise temperature of the antenna-attenuator cascade if its output is considered as a noise source?

8-103 An amplifier, when used with a source of average noise temperature 60 K, has an average operating noise figure of 5.

(a) What is $\bar{T}_e$?

(b) If the amplifier is sold to the engineering public, what noise figure would be quoted in a catalog (give a numerical answer)?

(c) What average operating noise figure results when the amplifier is used with an antenna of temperature 30 K?

8-104 An engineer purchases an amplifier with average operating noise figure of 1.8 when used with a 50-Ω broadband source having average source temperature of 80 K. When used with a different 50-Ω source the average operating noise figure is 1.25. What is the average noise temperature of the source?

8-105 An amplifier with a noise bandwidth of at least 1.8 MHz is needed by an engineer. Two units from which he can choose are: unit 1—average standard noise figure $= 3.98$, noise bandwidth $= 2.0$ MHz, and available power gain $= 10^6$; unit 2—average standard noise figure $= 2.82$, noise bandwidth $= 2.9$ MHz, and available power gain $= 10^6$.

Find: (a) $\bar{T}_e$ for unit 1, (b) $\bar{T}_e$ for unit 2, (c) excess noise power of unit 1, and (d) excess noise power of unit 2.

(e) If the source's noise temperature $\bar{T}_s$ is very small, which unit is the best to purchase and why?

(f) If $\bar{T}_s \gg \bar{T}_e$, which is best and why?

*8-106 A resistor is cooled to 75 K and serves as a noise source for a network with available power gain

$$G_a(\omega) = 10^{36}/(10^6 + \omega^2)^4$$

(a) Write an expression for the power spectrum of the network's output noise that is due to the source.

(b) Compute the available output noise power that is due to the source alone.

8-107 A broadband antenna, for which $T_a = 120$ K, connects through an attenuator with loss 2.5 to a receiver with average input effective noise temperature 80 K, available power gain 10^{12}, and noise bandwidth 20 MHz. The antenna and attenuator both have a physical temperature of 200 K.

(a) What is the attenuator's input effective noise temperature?

(b) What is the system's noise temperature?

(c) Find the average standard noise figure of the receiver by itself.

(d) What is the available noise power at the receiver's output (in system operation)?

(e) Determine the input effective noise temperature of the attenuator-receiver taken as a unit.

(f) What is the average operating noise figure of this system when the antenna is the source?

8-108 An antenna with average noise temperature 120 K connects to a receiver through an impedance-matched attenuator having a loss of 1.5 and physical temperature 75 K. For the overall system $\bar{T}_{sys} = 500$ K.

(a) What is the average effective input noise temperature of the receiver?

(b) What is the average operating noise figure of the attenuator-receiver cascade?

(c) What is the available output noise power of the receiver if its available power gain is 120 dB and its noise bandwidth is 20 MHz (system is connected)?

8-109 A receiving system consists of an antenna with noise temperature 80 K that feeds a matched attenuator with physical temperature 220 K and loss 2.6. The attenuator drives an amplifier with average effective noise temperature 170 K, noise bandwidth 4 MHz, and available power gain 10^8.

Find: (a) the overall system's average noise temperature $\bar{T}_{sys}$, (b) the available noise power N_{ao} at the system's output, (c) the total noise power available at the attenuator's output (within the noise bandwidth) and how much of the total (as a percentage) is due to the antenna alone, and (d) the average operating noise figure $\bar{F}_{op}$ of the system.

NINE

OPTIMUM LINEAR SYSTEMS

9.0 INTRODUCTION

The developments of the preceding chapter related entirely to the *analysis* of a linear system. In this chapter we do an about-face and concentrate only on the *synthesis* of a linear system. In particular, we choose the system in such a way that it satisfies certain rules that make it *optimum*.

In designing any optimum system we must consider three things: *input specification, system constraints*, and *criterion of optimality*.

Input specification means that at least some knowledge must be available about the input to the system. For example, we might specify the input to consist of the sum of a random signal and a noise. Alternatively, the input could be the sum of a deterministic signal and a noise. In addition, we may be able to specify signal and noise correlation functions, power spectrums, or probability densities. Thus, we may know a great deal about the inputs in some cases or little in others. Regardless, however, there is some minimum knowledge required of the characteristics of the input for any given problem.

System constraints define the form of the resulting system. For example, we might allow the system to be linear, nonlinear, time-invariant, realizable, etc. In our work we shall be exclusively concerned with linear time-invariant systems but will not necessarily require that they be realizable. By relaxing the realizability constraint, we shall be able to introduce the most important topics of interest without undue mathematical complexity.

In principle, there is great latitude available in choosing the criterion of optimality. In a practical sense, however, it should be a meaningful measure of "goodness" for the problem at hand and should correspond to equations that are mathematically tractable. We shall be concerned with only two criteria. One will involve the minimization of a suitably defined error quantity. The other will relate to maximization of the ratio of a signal power to a noise power. This last criterion leads us to an optimum system often called a *matched filter*.

9.1 SYSTEMS THAT MAXIMIZE SIGNAL-TO-NOISE RATIO

An important class of systems involves the transmission of a deterministic signal of known form in noise. A digital communication system is one example† where, during a time interval T, a known signal may arrive at the receiver in the presence of additive noise. The presence of the signal corresponds to transmission of a digital "1," while absence of the signal occurs when a digital "0" is transmitted (noise is always present). It would seem reasonable that some system (or filter‡) could be found that would enhance its output signal power at some instant in time while reducing its output average noise power. Indeed, such a filter that maximizes this output *signal-to-noise ratio* can be found and it is called a *matched filter*. It can be shown that decisions made as to whether the signal was present or not during time interval T have the smallest probability of being in error if they are based on samples taken at the times of maximum signal-to-noise ratio. Although our comments here are directed toward a digital communication system, we shall find as we progress that the matched filter concept is a broad one, applying to many situations.

In this section we shall consider the optimization of a linear time-invariant system when the input consists of the sum of a Fourier-transformable deterministic signal $x(t)$ of known form and continuous noise $n(t)$. If we denote by $x_o(t)$ and $n_o(t)$ the output signal and noise, the criterion of optimality we choose is the maximization of the ratio of the output signal power at some time t_o to the output average noise power. Thus, with $n_o(t)$ assumed to be a sample function of a wide-sense stationary random process§ $N_o(t)$, we maximize

$$\left(\frac{\hat{S}_o}{N_o}\right) = \frac{|x_o(t_o)|^2}{E[N_o^2(t)]} \tag{9.1-1}$$

where

$$\hat{S}_o = |x_o(t_o)|^2 \tag{9.1-2}$$

is the output signal power at time t_o and

$$N_o = E[N_o^2(t)] \tag{9.1-3}$$

is the output average noise power.

---

† Although we discuss only this example, many other systems such as radars, sonars, radio altimeters, ionospheric sounders, and automobile crash avoidance systems are other examples.

‡ We often use the words system, filter, or network in this chapter to convey the same meaning.

§ This assumption is equivalent to assuming the input noise is from a wide-sense stationary random process since the system is assumed to be linear and time-invariant (see Section 8.2).

Matched Filter for Colored Noise

Define $X(\omega)$ as the Fourier transform of $x(t)$, and $H(\omega)$ as the transfer function of the system. The output signal at any time t is

$$x_o(t) = \frac{1}{2\pi} \int_{-\infty}^{\infty} X(\omega)H(\omega)e^{j\omega t} \, d\omega \tag{9.1-4}$$

From (8.4-6), the output average noise power can be written in the form

$$N_o = E[N_o^2(t)] = \frac{1}{2\pi} \int_{-\infty}^{\infty} S_{NN}(\omega)|H(\omega)|^2 \, d\omega \tag{9.1-5}$$

where $S_{NN}(\omega)$ is the power density spectrum of the random process, denoted $N(t)$, that represents the input noise $n(t)$. By use of (9.1-4) at time t_o and (9.1-5), we can write (9.1-1) as

$$\left(\frac{\hat{S}_o}{N_o}\right) = \frac{\left|\dfrac{1}{2\pi} \displaystyle\int_{-\infty}^{\infty} X(\omega)H(\omega)e^{j\omega t_o} \, d\omega\right|^2}{\dfrac{1}{2\pi} \displaystyle\int_{-\infty}^{\infty} S_{NN}(\omega)|H(\omega)|^2 \, d\omega} \tag{9.1-6}$$

To find $H(\omega)$ that maximizes (9.1-6), we shall apply the *Schwarz†* inequality. If $A(\omega)$ and $B(\omega)$ are two possibly complex functions of the real variable ω, the inequality states that

$$\left|\int_{-\infty}^{\infty} A(\omega)B(\omega) \, d\omega\right|^2 \leq \int_{-\infty}^{\infty} |A(\omega)|^2 \, d\omega \int_{-\infty}^{\infty} |B(\omega)|^2 \, d\omega \tag{9.1-7}$$

The equality holds only when $B(\omega)$ is proportional to the complex conjugate of $A(\omega)$; that is, when

$$B(\omega) = CA^*(\omega) \tag{9.1-8}$$

where C is any *arbitrary* real constant.

By making the substitutions

$$A(\omega) = \sqrt{S_{NN}(\omega)}H(\omega) \tag{9.1-9}$$

$$B(\omega) = \frac{X(\omega)e^{j\omega t_o}}{2\pi\sqrt{S_{NN}(\omega)}} \tag{9.1-10}$$

in (9.1-7) we obtain

$$\left|\frac{1}{2\pi}\int_{-\infty}^{\infty} X(\omega)H(\omega)e^{j\omega t_o} \, d\omega\right|^2 \leq \int_{-\infty}^{\infty} S_{NN}(\omega)|H(\omega)|^2 \, d\omega \frac{1}{(2\pi)^2}\int_{-\infty}^{\infty} \frac{|X(\omega)|^2}{S_{NN}(\omega)} \, d\omega \tag{9.1-11}$$

† Named for the German mathematician Hermann Amandus Schwarz (1843–1921).

With this last result, we write (9.1-6) as

$$\left(\frac{\hat{S}_o}{N_o}\right) \leq \frac{1}{2\pi} \int_{-\infty}^{\infty} \frac{|X(\omega)|^2}{\mathcal{S}_{NN}(\omega)} \, d\omega \qquad (9.1\text{-}12)$$

The maximum value of $(\hat{S}_o/N_o)$ occurs when the equality holds in (9.1-12), which implies that (9.1-8) is true. Denote the optimum filter transfer function by $H_{opt}(\omega)$. We find this function by solving (9.1-8) using (9.1-9) and (9.1-10); the result is

$$H_{opt}(\omega) = \frac{1}{2\pi C} \frac{X^*(\omega)}{\mathcal{S}_{NN}(\omega)} e^{-j\omega t_o} \qquad (9.1\text{-}13)$$

From (9.1-13), we find that the optimum filter is proportional to the complex conjugate of the input signal's spectrum; we might say that the system is therefore *matched* to the specified signal since it depends so intimately on it. $H_{opt}(\omega)$ is also inversely proportional to the power spectrum of the input noise. In general, this noise has been assumed nonwhite; that is, colored. Because of these facts, an optimum filter given by (9.1-13) is called a *matched filter for colored noise*.

$H_{opt}(\omega)$ is also proportional to the inverse of the arbitrary constant C. In other words, $H_{opt}(\omega)$ has an arbitrary absolute magnitude. This fact allows the optimum system to have arbitrary gain. Intuitively, we feel that this should be true because gain affects both input signal and input noise in the same way, and, in the ratio of (9.1-1), gain cancels.

The time t_o at which the output ratio $(\hat{S}_o/N_o)$ is maximum enters into the optimum system transfer function only through the factor $\exp(-j\omega t_o)$. Such a factor only represents an ideal delay. Since t_o is a parameter that a designer may have some latitude in choosing, its value may be selected in some cases to make the optimum filter causal.

In general, the system defined by (9.1-13) may not be realizable. For certain forms of colored noise realizable filters may be found (Thomas, 1969, Chapter 5). In practice, one can always approximate (9.1-13) by a suitably chosen real filter.

Matched Filter for White Noise

If the input noise is white with power density $\mathcal{N}_0/2$, the optimum filter of (9.1-13) becomes

$$H_{opt}(\omega) = KX^*(\omega)e^{-j\omega t_o} \qquad (9.1\text{-}14)$$

where $K = 1/\pi C \mathcal{N}_0$ is an arbitrary constant. Here the optimum filter is related only to the input signal's spectrum and the time that $(\hat{S}_o/N_o)$ is maximum. Thus, the name *matched filter* is very appropriate. Indeed, the name was originally attached to the filter in white noise; we have liberalized the name to include the preceding colored noise case.

The impulse response denoted $h_{opt}(t)$ of the optimum filter is the inverse Fourier transform of $H_{opt}(\omega)$. From (9.1-14), it is easily found that

$$h_{opt}(t) = Kx^*(t_o - t) \qquad (9.1\text{-}15)$$

For real signals $x(t)$, (9.1-15) reduces to

$$h_{opt}(t) = Kx(t_o - t) \qquad (9.1\text{-}16)$$

Equation (9.1-16) indicates that the impulse response is equal to the input signal displaced to a new origin at $t = t_o$ and folded about this point so as to "run backward."

Example 9.1-1 We shall find the matched filter for the signal of Figure 9.1-1*a* when received in white noise. From (9.1-16), the matched filter's impulse response is as shown in (*b*). By Fourier transformation of the waveform in (*b*), we readily obtain

$$H_{opt}(\omega) = KA\tau \, \frac{\sin (\omega\tau/2)}{(\omega\tau/2)} \, e^{-j\omega[t_o + \tau_o - (\tau/2)]}$$

An alternative development consists of Fourier-transforming the input signal to get $X(\omega)$ and then using (9.1-14).

Whether or not any chance exists for the matched filter to be realizable may be determined from the impulse response of Figure 9.1-1*b*. Clearly, to be causal, and therefore realizable, the delay must be at least $\tau - \tau_o$; that is

$$t_o \geq \tau - \tau_o$$

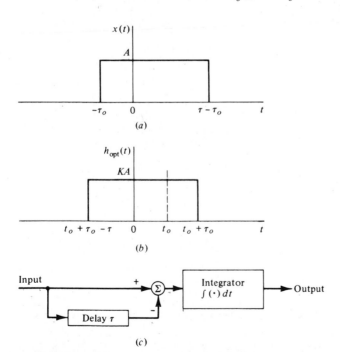

(a)

(b)

(c)

Figure 9.1-1 A matched filter and its related signals. (a) Input signal, (b) the filter's impulse response, and (c) the filter's block diagram [*Reproduced from Peebles (1976), with permission of publishers Addison-Wesley, Advanced Book Program.*]

If we assume this last condition is satisfied, the optimum filter is illustrated in (c) where the arbitrary constant K is set equal to $1/A$. This filter still requires that perfect integrators be possible. Of course, they are not. However, very good approximations are possible using modern operational amplifiers with feedback, so for all practical purposes matched filters for rectangular pulses in white noise may be constructed.†

9.2 SYSTEMS THAT MINIMIZE MEAN-SQUARED ERROR

A second class of optimum systems is concerned with causing the output to be a good estimate of some function of the input signal which arrives along with additive noise. One example corresponds to the output being a good estimate of the *derivative* of the input signal. In another case, the system could be designed so that its output is a good estimate of either the past, present, or future value of the input signal. We shall concern ourselves with only this last case. The optimum system or filter that results is called a *Wiener filter*.‡

Wiener Filters

The basic problem to be studied is depicted by Figure 9.2-1. The input signal $x(t)$ is now assumed to be *random*; it is therefore modeled as a sample function of a random process $X(t)$. It is applied to the input of the system along with additive noise $n(t)$ that is a sample function of a noise process $N(t)$. We assume $X(t)$ and

† Other techniques using *integrate-and-dump* methods exist. See Peebles (1976), pp. 361–362.

‡ After Norbert Wiener (1894–1964), a great American mathematician whose work has tremendously affected many areas of science and engineering.

Figure 9.2-1 Operations that define the Wiener filter problem.

$N(t)$ are jointly wide-sense stationary processes and that $N(t)$ has zero mean. The sum of signal and noise is denoted $W(t)$:

$$W(t) = X(t) + N(t) \tag{9.2-1}$$

The system is assumed to be linear and time-invariant with a real impulse response $h(t)$ and a transfer function $H(\omega)$. The output of the system is denoted $Y(t)$.

In general, we shall select $H(\omega)$ so that $Y(t)$ is the best possible estimate of the input signal $X(t)$ at a time $t + t_o$; that is, the best estimate of $X(t + t_o)$. If $t_o > 0$, $Y(t)$ is an estimate of a *future* value of $X(t)$ corresponding to a *prediction filter*. If $t_o < 0$, $Y(t)$ is an estimate of a *past* value of $X(t)$ and we have a *smoothing filter*. If $t_o = 0$, $Y(t)$ is an estimate of the current value of $X(t)$.

Now if $Y(t)$ differs from the desired true value of $X(t + t_o)$, we make an error of

$$\varepsilon(t) = X(t + t_o) - Y(t) \tag{9.2-2}$$

This error is illustrated conceptually in Figure 9.2-1 by dashed lines. The optimum filter will be chosen so as to minimize the mean-squared value of $\varepsilon(t)$.† We shall not be concerned with obtaining a system that is realizable. Some information is given by Thomas (1969) on the more difficult problem where $H(\omega)$ must be realizable. Thus, we seek to find $H(\omega)$ that minimizes

$$\begin{aligned} E[\varepsilon^2(t)] &= E[\{X(t + t_o) - Y(t)\}^2] \\ &= E[X^2(t + t_o) - 2Y(t)X(t + t_o) + Y^2(t)] \\ &= R_{XX}(0) - 2R_{YX}(t_o) + R_{YY}(0) \end{aligned} \tag{9.2-3}$$

From the Fourier transform relationship between an autocorrelation function and a power spectrum, we have

$$R_{XX}(0) = \frac{1}{2\pi} \int_{-\infty}^{\infty} \mathcal{S}_{XX}(\omega) \, d\omega \tag{9.2-4}$$

where $\mathcal{S}_{XX}(\omega)$ is the power density spectrum of $X(t)$. From a similar relationship and (8.4-1) we have

$$R_{YY}(0) = \frac{1}{2\pi} \int_{-\infty}^{\infty} \mathcal{S}_{WW}(\omega) |H(\omega)|^2 \, d\omega \tag{9.2-5}$$

where $\mathcal{S}_{WW}(\omega)$ is the power spectrum of $W(t)$. By substitution of (9.2-4) and (9.2-5) into (9.2-3), we have

$$E[\varepsilon^2(t)] = -2R_{YX}(t_o) + \frac{1}{2\pi} \int_{-\infty}^{\infty} \{\mathcal{S}_{XX}(\omega) + \mathcal{S}_{WW}(\omega) |H(\omega)|^2\} \, d\omega \tag{9.2-6}$$

† We could elect to minimize the average error, or even force such an error to be zero. This approach does not prevent large positive errors from being offset by large negative errors, however. By minimizing the squared error, we eliminate such possibilities.

To reduce (9.2-6) further, we develop the cross-correlation function:

$$R_{YX}(t_o) = E[Y(t)X(t + t_o)] = E\left[X(t + t_o)\int_{-\infty}^{\infty} h(\xi)W(t - \xi)\,d\xi\right]$$

$$= \int_{-\infty}^{\infty} R_{WX}(t_o + \xi)h(\xi)\,d\xi \qquad (9.2\text{-}7)$$

where $R_{WX}(\cdot)$ is the cross-correlation function of $W(t)$ and $X(t)$. After replacing $R_{WX}(t_o + \xi)$ by its equivalent, the inverse Fourier transform of the cross-power spectrum $S_{WX}(\omega)$, we obtain

$$R_{YX}(t_o) = \int_{-\infty}^{\infty} \frac{1}{2\pi}\int_{-\infty}^{\infty} S_{WX}(\omega)e^{j\omega(t_o + \xi)}\,d\omega\, h(\xi)\,d\xi$$

$$= \frac{1}{2\pi}\int_{-\infty}^{\infty} S_{WX}(\omega)e^{j\omega t_o}\left\{\int_{-\infty}^{\infty} h(\xi)e^{j\omega\xi}\,d\xi\right\}d\omega$$

$$= \frac{1}{2\pi}\int_{-\infty}^{\infty} S_{WX}(\omega)H(-\omega)e^{j\omega t_o}\,d\omega \qquad (9.2\text{-}8)$$

Substitution of this expression into (9.2-6) allows it to be written as

$$E[\varepsilon^2(t)] = \frac{1}{2\pi}\int_{-\infty}^{\infty} \{S_{XX}(\omega) - 2S_{WX}(\omega)H(-\omega)e^{j\omega t_o} + S_{WW}(\omega)|H(\omega)|^2\}\,d\omega$$

$$(9.2\text{-}9)$$

The transfer function that minimizes $E[\varepsilon^2(t)]$ is now found, We may write $H(\omega)$ in the form

$$H(\omega) = A(\omega)e^{jB(\omega)} \qquad (9.2\text{-}10)$$

where $A(\omega)$ is the magnitude of $H(\omega)$, and $B(\omega)$ is its phase. Next we observe that $S_{XX}(\omega)$ and $S_{WW}(\omega)$ are real nonnegative functions, since they are power spectrums, while the cross-power spectrum $S_{WX}(\omega)$ is complex in general and can be written as

$$S_{WX}(\omega) = C(\omega)e^{jD(\omega)} \qquad (9.2\text{-}11)$$

After using (9.2-10) and (9.2-11) in (9.2-9) and invoking the fact that

$$H(-\omega) = H^*(\omega) \qquad (9.2\text{-}12)$$

for filters having a real impulse response $h(t)$, we obtain

$$E[\varepsilon^2(t)] = \frac{1}{2\pi}\int_{-\infty}^{\infty} \{S_{XX}(\omega) + S_{WW}(\omega)A^2(\omega)\}\,d\omega$$

$$-\frac{1}{2\pi}\int_{-\infty}^{\infty} 2C(\omega)A(\omega)e^{j[\omega t_o + D(\omega) - B(\omega)]}\,d\omega \qquad (9.2\text{-}13)$$

We minimize $E[\varepsilon^2(t)]$ by first selecting the phase of $H(\omega)$ to maximize the second integral in (9.2-13) and then, with the optimum phase substituted, minimize the resulting expression by choice of $A(\omega)$. Clearly, choosing

$$B(\omega) = \omega t_o + D(\omega) \tag{9.2-14}$$

will maximize the second integral and give the expression

$$E[\varepsilon^2(t)] = \frac{1}{2\pi} \int_{-\infty}^{\infty} \{S_{XX}(\omega) - 2C(\omega)A(\omega) + S_{WW}(\omega)A^2(\omega)\}\, d\omega$$

$$= \frac{1}{2\pi} \int_{-\infty}^{\infty} \left\{ S_{XX}(\omega) - \frac{C^2(\omega)}{S_{WW}(\omega)} + S_{WW}(\omega)\left[A(\omega) - \frac{C(\omega)}{S_{WW}(\omega)} \right]^2 \right\} d\omega \tag{9.2-15}$$

In writing the last form of (9.2-15), we have completed the square in $A(\omega)$. Finally, it is clear that choosing

$$A(\omega) = \frac{C(\omega)}{S_{WW}(\omega)} \tag{9.2-16}$$

will minimize the right side of (9.2-15). By combining (9.2-16), (9.2-14), and (9.2-11) with (9.2-10) we have the optimum filter transfer function which we denote $H_{\mathrm{opt}}(\omega)$:

$$H_{\mathrm{opt}}(\omega) = \frac{S_{WX}(\omega)}{S_{WW}(\omega)}\, e^{j\omega t_o} \tag{9.2-17}$$

For the special case where input signal and noise are uncorrelated, it is easy to show that

$$S_{WW}(\omega) = S_{XX}(\omega) + S_{NN}(\omega) \tag{9.2-18}$$

$$S_{WX}(\omega) = S_{XX}(\omega) \tag{9.2-19}$$

where $S_{NN}(\omega)$ is the power spectrum of $N(t)$. Hence, for this special case

$$H_{\mathrm{opt}}(\omega) = \frac{S_{XX}(\omega)}{S_{XX}(\omega) + S_{NN}(\omega)}\, e^{j\omega t_o} \tag{9.2-20}$$

Example 9.2-1 We find the optimum filter for estimating $X(t + t_o)$ when there is no input noise. We let $S_{NN}(\omega) = 0$ in (9.2-20):

$$H_{\mathrm{opt}}(\omega) = e^{j\omega t_o}$$

This expression corresponds to an ideal delay line with delay $-t_o$. If $t_o > 0$, corresponding to prediction, we require an unrealizable negative delay line. If $t_o < 0$, corresponding to a smoothing filter, the required delay is positive and realizable. Of course, $t_o = 0$ results in $H_{\mathrm{opt}}(\omega) = 1$. In other words, the optimum filter for estimating $X(t)$ when no noise is present is just a direct connection from input to output, a result that is intuitively agreeable.

Minimum Mean-Squared Error

On substitution of (9.2-17) into (9.2-15), we readily find the mean-squared error of the optimum filter

$$E[\varepsilon^2(t)]_{min} = \frac{1}{2\pi} \int_{-\infty}^{\infty} \frac{S_{XX}(\omega)S_{WW}(\omega) - |S_{WX}(\omega)|^2}{S_{WW}(\omega)} \, d\omega \qquad (9.2\text{-}21)$$

For the special case where input signal and noise are uncorrelated, this equation reduces to

$$E[\varepsilon^2(t)]_{min} = \frac{1}{2\pi} \int_{-\infty}^{\infty} \frac{S_{XX}(\omega)S_{NN}(\omega)}{S_{XX}(\omega) + S_{NN}(\omega)} \, d\omega \qquad (9.2\text{-}22)$$

9.3 OPTIMIZATION BY PARAMETER SELECTION

We conclude our discussions of optimum linear systems by briefly considering a second approach that minimizes mean-squared error. The problem we undertake is identical to that of the last section up to (9.2-9), which defines the mean-squared error. Now, however, rather than seeking the filter that minimizes this error, we *specify* the *form* of the filter in terms of a number of unknown parameters and then determine the parameter values that minimize the mean-squared error. This procedure necessarily leads to a real filter so long as the form we choose corresponds to such a filter.

If we assume the special case where the input signal $X(t)$ and noise $N(t)$ are uncorrelated, (9.2-9) can be written as

$$E[\varepsilon^2(t)] = \frac{1}{2\pi} \int_{-\infty}^{\infty} S_{\varepsilon\varepsilon}(\omega) \, d\omega \qquad (9.3\text{-}1)$$

where

$$S_{\varepsilon\varepsilon}(\omega) = S_{XX}(\omega) - 2S_{XX}(\omega)H(-\omega)e^{j\omega t_o} + [S_{XX}(\omega) + S_{NN}(\omega)]|H(\omega)|^2 \qquad (9.3\text{-}2)$$

Since the imaginary part of $H(-\omega) \exp(j\omega t_o)$ is an odd function of ω when $h(t)$ is real (as assumed), the only contribution to the integral of (9.3-1) due to the middle term in (9.3-2) results from the real part of $H(-\omega) \exp(j\omega t_o)$. Thus, the error-contributing part of (9.3-2) can be written as†

$$S_{\varepsilon\varepsilon}(\omega) = S_{XX}(\omega)[1 - H(\omega)e^{-j\omega t_o} - H(-\omega)e^{j\omega t_o} + |H(\omega)|^2] + S_{NN}(\omega)|H(\omega)|^2$$

$$= S_{XX}(\omega)|1 - H(-\omega)e^{j\omega t_o}|^2 + S_{NN}(\omega)|H(\omega)|^2 \qquad (9.3\text{-}3)$$

because $H(-\omega) = H^*(\omega)$.

We summarize the synthesis procedure. First, a filter form is chosen for a real filter. The applicable transfer function $H(\omega)$ will depend on a number of unknown parameters. $H(\omega)$ is next substituted into (9.3-3), to obtain $S_{\varepsilon\varepsilon}(\omega)$, the power

---

† In writing (9.3-3), we also use the fact that $2 \operatorname{Re}(z) = z + z^*$ for any complex number z.

density spectrum of the error $\varepsilon(t)$. Finally, the error $E[\varepsilon^2(t)]$ is calculated from (9.3-1) and the parameters are then found by formally minimizing this error. Although this procedure is direct and conceptually simple to apply, the solution of the integral of (9.3-1) may be tedious. For the case where $S_{XX}(\omega)$ and $S_{NN}(\omega)$ are rational functions of ω and $H(\omega)$ corresponds to a real filter form, the resulting integral has been tabulated for a number of functions $S_{\varepsilon\varepsilon}(\omega)$ involving orders of ω up to 14 (Thomas, 1969, pp. 249 and 636, and James, et al., 1947, p. 369).

All the preceding discussion has related to the special case where the input signal and input zero-mean noise are jointly wide-sense stationary and uncorrelated. For the more general case of correlated signal and noise, the choice of form for $H(\omega)$ must be substituted into (9.2-9) and the integral solved. The unknown filter coefficients are then determined that minimize $E[\varepsilon^2(t)]$.

PROBLEMS

9-1 A matched filter is to be found for a signal defined by

$$x(t) = \begin{cases} A(\tau + t)/\tau & -\tau < t < 0 \\ A(\tau - t)/\tau & 0 < t < \tau \\ 0 & \text{elsewhere} \end{cases} \qquad \leftrightarrow \qquad X(\omega) = A\tau\left[\frac{\sin(\omega\tau/2)}{\omega\tau/2}\right]^2$$

when added to noise having a power density spectrum

$$S_{NN}(\omega) = \frac{W_2}{W_2^2 + \omega^2}$$

where A, τ, and W_2 are real positive constants.
  (a) Find the matched filter's transfer function $H_{opt}(\omega)$.
  (b) Find the filter's impulse response $h_{opt}(t)$. Plot $h_{opt}(t)$.
  (c) Is there a value of t_0 for which the filter is causal? If so, find it.
  (d) Sketch the block diagram of a network that has $H_{opt}(\omega)$ as its transfer function.

9-2 Work Problem 9-1 (a), (b), and (c) for the signal

$$x(t) = u(t)[e^{-W_2 t} - e^{-\alpha W_2 t}]$$

if $\alpha > 1$ is a real constant.

9-3 Work Problem 9-1 (a), (b), and (c) for the signal

$$x(t) = u(-t)[e^{W_2 t} - e^{\alpha W_2 t}]$$

if $\alpha > 1$ is a real constant.

*9-4 By proper inverse Fourier transformation of (9.1-13), show that the impulse response $h_{opt}(t)$ of the matched filter for signals in colored noise satisfies

$$\int_{-\infty}^{\infty} h_{opt}(\xi)R_{NN}(t - \xi)\,d\xi = x^*(t_0 - t)$$

9-5 A signal $x(t)$ and colored noise $N(t)$ are applied to the network of Figure P9-5. We select $|H_1(\omega)|^2 = 1/S_{NN}(\omega)$ so that the noise $N_1(t)$ is white. We also make $H_2(\omega)$ a matched filter for the signal $x_1(t)$ in the white noise $N_1(t)$. Show that the cascade is a matched filter for $x(t)$ in the noise $N(t)$.

Figure P9-5

9-6 For the matched filter of Example 9.1-1, find and sketch the output signal. [*Hint:* Fourier-transform $x(t)$ and use a transform pair from Appendix E to obtain $x_o(t)$.]

9-7 Assume the power density of the white noise at the input to the matched filter of Example 9.1-1 is $\mathcal{N}_0/2$ with $\mathcal{N}_0 > 0$ a real constant. Find the output signal-to-noise ratio of the filter at time t_o.

9-8 Show that the maximum output signal-to-noise ratio obtainable from a filter matched to a signal $x(t)$ in white noise with power density $\mathcal{N}_0/2$ is

$$\left(\frac{\hat{S}_o}{N_o}\right)_{max} = \frac{2}{\mathcal{N}_0} \int_{-\infty}^{\infty} |x(t)|^2 \, dt = \frac{2E}{\mathcal{N}_0}$$

where E is the energy in $x(t)$ and $\mathcal{N}_0 > 0$ is a real constant.

9-9 Let τ be a positive real constant. A pulse

$$x(t) = \begin{cases} A \cos(\pi t/\tau) & |t| < \tau/2 \\ 0 & |t| > \tau/2 \end{cases}$$

is added to white noise with a power density of $\mathcal{N}_0/2$. Find $(\hat{S}_o/N_o)_{max}$ for a filter matched to $x(t)$ by using the result of Problem 9-8.

9-10 Find the matched filter's transfer function applicable to Problem 9-9.

9-11 Show that the output signal $x_o(t)$ from a filter matched to a signal $x(t)$ in white noise is

$$x_o(t) = K \int_{-\infty}^{\infty} x^*(\xi) x(\xi + t - t_o) \, d\xi$$

That is, $x_o(t)$ is proportional to the *correlation integral* of $x(t)$.

9-12 Show that the output signal $x_o(t)$ from a filter matched to a signal in white noise reaches its maximum magnitude at $t = t_o$ if the filter impulse response is given by (9.1-15). (*Hint:* Use the result of Problem 9-11.)

9-13 Fourier-transform the signal of Figure 9.1-1a, and use (9.1-14) to verify the optimum system transfer function given in Example 9.1-1.

9-14 The signal

$$x(t) = u(t)e^{-Wt}$$

where $W > 0$ is a real constant, is applied to a filter along with white noise with power density $\mathcal{N}_0/2$, $\mathcal{N}_0 > 0$ being a real constant.

    (a) Find the transfer function of the filter matched to $x(t)$ at time t_o.

    (b) Find and sketch the filter's impulse response.

    (c) Is there any value of t_o that will make the filter causal?

    (d) Find the output maximum signal-to-noise ratio.

9-15 Work Problem 9-14 for the signal

$$x(t) = u(-t)e^{Wt}$$

9-16 Work Problem 9-14 for the signal

$$x(t) = u(t)te^{-Wt}$$

9-17 Work Problem 9-14 for the signal

$$x(t) = -u(-t)te^{Wt}$$

9-18 If a real signal $x(t)$ exists only in the interval $0 < t < T$, show that the *correlation receiver* of Figure P9-18 is a matched filter at time $t = T$; that is, show that the ratio of peak signal power to average noise power, both at time T, is the same as the ordinary matched filter. Assume white input noise.

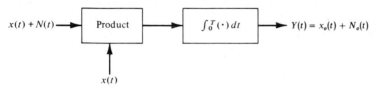

$x(t)$

Figure P9-18

9-19 Find the matched filter for the signal

$$x(t) = Ae^{-\alpha t^2}$$

in white noise with power density $\mathcal{N}_0/2$ where $\mathcal{N}_0 > 0$, $\alpha > 0$, and A are real constants.

9-20 A random signal $X(t)$ and uncorrelated white noise $N(t)$ have autocorrelation functions

$$R_{XX}(\tau) = \frac{WP}{2} e^{-W|\tau|}$$

$$R_{NN}(\tau) = (\mathcal{N}_0/2)\delta(\tau)$$

where $W > 0$, $P > 0$, and $\mathcal{N}_0 > 0$ are real constants.

    (a) Find the transfer function of the optimum Wiener filter.

    (b) Find and sketch the impulse response of the filter when $t_o < 0$, $t_o > 0$, and $t_o = 0$.

9-21 Find the minimum mean-squared error of the filter in Problem 9-20.

9-22 Work Problem 9-20 for colored noise defined by

$$R_{NN}(\tau) = W_N e^{-W_N|\tau|}$$

where $W_N > 0$ is a real constant.

9-23 Work Problem 9-21 for the noise defined in Problem 9-22.

9-24 A random signal $X(t)$ and additive uncorrelated noise $N(t)$ have respective power spectrums

$$S_{XX}(\omega) = \frac{9}{9 + \omega^4} \quad \text{and} \quad S_{NN}(\omega) = \frac{3}{6 + \omega^4}$$

(a) Find the transfer function of the Wiener filter for the given signal and noise.

(b) Find the minimum value of the error in predicting $X(t + t_o)$.

9-25 Work Problem 9-24 for signal and uncorrelated white noise defined by

$$S_{XX}(\omega) = \frac{A}{W^2 + \omega^4}$$

$$S_{NN}(\omega) = \mathcal{N}_0/2$$

where $A > 0$, $W > 0$, and $\mathcal{N}_0 > 0$ are real constants.

9-26 A deterministic signal $x(t) = A \cos(\omega_0 t)$ and white noise with power density $\mathcal{N}_0/2$ are applied to a one-section lowpass filter with transfer function $H(\omega) = W/(W + j\omega)$. Here $W > 0$, $\mathcal{N}_0 > 0$, ω_0, and A are all real constants. What value of W will cause the ratio of output *average* signal power to average noise power to be maximum?

9-27 Work Problem 9-26 if the network consists of two identical one-section filters in cascade.

9-28 Work Problem 9-26 if $x(t) = A \cos(\omega_0 t + \Theta)$, where Θ is a random variable uniformly distributed on the interval $(0, 2\pi)$.

9-29 A random signal $X(t)$ having the autocorrelation function

$$R_{XX}(\tau) = W_X e^{-W_X|\tau|}$$

and uncorrelated noise with power density $\mathcal{N}_0/2$ are applied to a lowpass filter with transfer function

$$H(\omega) = \frac{W}{W + j\omega}$$

Here $W > 0$ and $W_X > 0$ are real constants.

(a) What value of W will minimize the mean-squared error if the output is to be an estimate of $X(t)$?

(b) Calculate the minimum mean-squared error.

*9-30 Work Problem 9-29 by finding the real constants $G > 0$ and $W > 0$ for the filter defined by

$$H(\omega) = \frac{GW}{W + j\omega}$$

ADDITIONAL PROBLEMS

9-31 A signal $x(t) = u(t)5t^2 \exp(-2t)$ is added to white noise for which $\mathcal{N}_0/2 = 10^{-2}$ W/Hz. The sum is applied to a matched filter.
    (a) What is the filter's transfer function?
    (b) What is $(\hat{S}_o/N_o)$?
    (c) Sketch the impulse response of the filter.
    (d) Is the filter realizable?

9-32 A signal

$$x(t) = u(t)t^2 \exp(-Wt)$$

is added to noise with power spectrum

$$S_{NN}(\omega) = P/(W_N^2 + \omega^2)$$

where W, P, and W_N are positive constants. The sum is applied to a matched filter.
    (a) Find the filter's transfer function.
    (b) Find the filter's impulse response.
    (c) What is the signal-to-noise ratio at the output?

9-33 A pulse of amplitude $A > 0$ and duration $\tau > 0$ is $x(t) = A \, \mathrm{rect}\,(t/\tau)$. The pulse is added to white noise of power density $\mathcal{N}_0/2$ when it arrives at a receiver. For some practical reasons the receiver (filter) is not a matched filter but is a simple lowpass filter with transfer function

$$H(\omega) = W/(W + j\omega)$$

$W > 0$ a constant.
    (a) Find the ratio of instantaneous output signal power $x_o^2(t)$ at any time t to average noise power $E[N_o^2(t)]$ at the filter's output. At what time, denoted by t_o, is the ratio maximum?
    (b) At time t_o what bandwidth W will maximize signal-to-noise ratio?
    (c) Plot the loss in output signal-to-noise ratio that results, compared to a matched filter, for various values of $0 < W \le 5/\tau$. What is the minimum loss?

*9-34 Reconsider the system of Problem 9-33 except assume

$$H(\omega) = W^2/(W + j\omega)^2$$

    (a) Find the time t_o at which output signal-to-noise ratio is largest.
    (b) For the t_o found in (a) determine the output signal-to-noise ratio. Plot this result versus $W\tau$ for $0 < W\tau \le 6$ and determine what value of W gives the best performance.
    (c) What minimum loss in signal-to-noise ratio occurs compared to a matched filter?

9-35 A pulse

$$x(t) = A \, \mathrm{rect}\,(t/2\tau)[1 - (t/\tau)^2]$$

where A and $\tau > 0$ are constants, is added to white noise.

(a) Find the output signal $x_o(t)$ of a filter matched to the pulse.

(b) Sketch $x(t)$ and $x_o(t)$.

(c) What is the matched filter's output signal-to-noise ratio?

(d) What is its transfer function if K in (9.1-16) is chosen so that $|H_{opt}(0)| = 1$? Is there a value of t_o that makes the filter causal?

★9-36 A deterministic waveform $\psi(t)$ is defined by

$$\psi(t) = a(t)e^{j\phi(t) + j\omega_o t} = v(t)e^{j\omega_o t}$$

where $a(t)$ and $\phi(t)$ are "slowly" varying amplitude and phase "modulation" functions and $\omega_0 > 0$ is a large constant. The white-noise matched filter for $\psi(t)$ is defined by

$$h_{opt}(t) = \psi^*(t_o - t)$$

if $K = 1$ in (9.1-15). Now let $\psi(t)$ be offset in frequency by an amount ω_d before being applied to the "matched filter" so that

$$\psi_R(t) = \psi(t) \exp(-j\omega_d t)$$

is applied with noise to the filter.

(a) Show that the filter's response to $\psi_R(t)$ is

$$\chi(t_o - t, \omega_d) = \int_{-\infty}^{\infty} \psi(\xi)\psi^*(t_o - t + \xi)e^{-j\omega_d\xi}\,d\xi$$

The function $|\chi(\alpha, \omega_d)|^2$ is called the *ambiguity function* of the waveform $\psi(t)$.

(b) Show that the volume under the ambiguity function does not depend on the form of $\psi(t)$ but only on $|\chi(0, 0)|^2$.

(c) Show that

$$\chi(t_o - t, \omega_d) = e^{j\omega_0(t - t_o)} \int_{-\infty}^{\infty} v(\xi)v^*(t_o - t + \xi)e^{-j\omega_d\xi}\,d\xi$$

★9-37 Reconsider the ambiguity function of Problem 9-36.

(a) Show that $|\chi(\tau, \omega_d)|^2 \leq |\chi(0, 0)|^2$.

(b) Show that another form for $\chi(\tau, \omega_d)$ is

$$\chi(\tau, \omega_d) = \frac{1}{2\pi} \int_{-\infty}^{\infty} \Psi^*(\omega)\Psi(\omega + \omega_d)e^{-j\omega\tau}\,d\omega$$

where $\Psi(\omega)$ is the Fourier transform of $\psi(t)$.

(c) Show that

$$\chi(\tau, 0) = \int_{-\infty}^{\infty} \psi(\xi)\psi^*(\xi + \tau)\,d\xi$$

$$= \frac{1}{2\pi} \int_{-\infty}^{\infty} |\Psi(\omega)|^2 e^{-j\omega\tau}\,d\omega$$

$$\chi(0, \omega_d) = \int_{-\infty}^{\infty} |\psi(\xi)|^2 e^{-j\omega_d\xi}\,d\xi$$

$$= \frac{1}{2\pi} \int_{-\infty}^{\infty} \Psi^*(\omega)\Psi(\omega + \omega_d)\,d\omega$$

(d) Show that the symmetry of $\chi(\tau, \omega_d)$ is given by

$$\chi(\tau, \omega_d) = e^{j\omega_d\tau}\chi^*(-\tau, -\omega_d)$$

9-38 The deterministic signal

$$x(t) = \text{rect}\,(t/T)\exp\,(j\omega_0 t + j\mu t^2/2)$$

is a pulse having a linearly varying frequency with time during the pulse's duration T. The nominal frequency is ω_0(rad/s). The matched filter for white noise has the impulse response of (9.1-15) which, for $t_o = 0$, is

$$h_{opt}(t) = K\,\text{rect}\,(t/T)\exp\,(j\omega_0 t - j\mu t^2/2)$$

(a) If instantaneous frequency is to increase by a total amount $\Delta\omega$ (rad/s) during the pulse's duration T, how is the constant μ related to $\Delta\omega$ and T?

(b) Find the value of K such that $|H_{opt}(\omega_0)| = 1$ when μ is large. [*Hint:* Note that

$$C(x) = \int_0^x \cos\,(\pi\xi^2/2)\,d\xi$$

and

$$S(x) = \int_0^x \sin\,(\pi\xi^2/2)\,d\xi$$

called *Fresnel integrals*, approach $\frac{1}{2}$ as $x \to \infty$.]

(c) For the K found in (b), determine the output $x_o(t)$ of the filter. Sketch the envelopes of the signals $x(t)$ and $x_o(t)$ for $\Delta\omega T = 80\pi$ using the same time-voltage axes. What observations can you make about what has happended to $x(t)$ as it passes through the filter?

*9-39 (a) Find the transfer function $H_{opt}(\omega)$ of the matched filter of Problem 9-38. (*Hint:* Put the expression in terms of Fresnel integrals having arguments

$$x_1 = \sqrt{\Delta\omega T/2\pi}\,\{1 - [2(\omega - \omega_0)/\Delta\omega]\}/\sqrt{2}$$

and

$$x_2 = \sqrt{\Delta\omega T/2\pi}\,\{1 + [2(\omega - \omega_0)/\Delta\omega]\}/\sqrt{2}$$

where $\mu = \Delta\omega/T$.)

(b) Sketch the approximate form of $|H_{opt}(\omega)|$ that results when $\Delta\omega T$ is large.

9-40 A random signal $X(t)$ and uncorrelated white noise have respective power spectrums

$$S_{XX}(\omega) = 2\sqrt{2}\,P_{XX}W_X\,\omega^2/(W_X^4 + \omega^4)$$

and

$$S_{NN}(\omega) = \mathcal{N}_0/2$$

Here P_{XX} is the average power in $X(t)$, while W_X and $\mathcal{N}_0$ are positive constants.

(a) Find the transfer function of the Wiener filter for this signal and noise.

(b) What is the minimum mean-squared filter error?

(c) Evaluate the result of (b) for $P_{XX} = 2$ W, $W_X = 15$ rad/s, and $\mathcal{N}_0/2 = 0.1$ W/Hz. [*Hint:* Use the known integral (Thomas, 1969, p. 249)

$$I_2 = \frac{1}{2\pi} \int_{-\infty}^{\infty} \frac{(b_1 - b_0 \omega^2)\, d\omega}{a_2^2 + (a_1^2 - 2a_0 a_2)\omega^2 + a_0^2 \omega^4} = \frac{a_0 b_1 - a_2 b_0}{2a_0 a_1 a_2}$$

where b_0, b_1, a_0, a_1, and a_2 are constants and $a_0 \lambda^2 + a_1 \lambda + a_2$ has no roots in the lower half-plane when $\lambda = \omega + j\sigma$.]

9-41 Work Problem 9-40 for the signal with the power spectrum

$$S_{XX}(\omega) = A/(W_X^2 + \omega^2)^2$$

Put results in terms of the average power P_{XX} in $X(t)$.

9-42 The respective power spectrums of a random signal $X(t)$ and uncorrelated noise $N(t)$ are

$$S_{XX}(\omega) = (1/20)/(10^2 + \omega^2)$$

and

$$S_{NN}(\omega) = \omega^2/(16^2 + \omega^2)^2$$

(a) What is the transfer function of the Wiener filter?

(b) What is the minimum mean-squared prediction error? (*Hint:* Use results from Problem 9-40.)

*9-43 Generalize the random signal of Problem 9-42 by assuming its power spectrum is

$$S_{XX}(\omega) = (W_X^2/2000)/(W_X^2 + \omega^2)$$

where W_X is the signal's 3-dB bandwidth. Find the minimum mean-squared prediction error and plot the result for $W_X > 9.5$. What does an increase in W_X mean in a physical sense?

9-44 A random signal $X(t)$ plus uncorrelated noise $N(t)$, having respective power spectrums

$$S_{XX}(\omega) = 2P_{XX} W_X/(W_X^2 + \omega^2)$$

and

$$S_{NN}(\omega) = 4P_{NN} W_N^3/(W_N^2 + \omega^2)^2$$

is applied to a Wiener filter. Here P_{XX} and P_{NN} are the average signal and noise powers, respectively, while W_X and W_N are positive constants.

(a) Use (9.2-22) and find the filter's minimum mean-squared prediction error.

(b) Show that as $P_{XX} \to \infty$, $E[\varepsilon^2(t)]_{min} \to P_{NN}$, and that $E[\varepsilon^2(t)]_{min} \to P_{XX}$ if $P_{NN} \to \infty$.

(c) From a graphical plot of $E[\varepsilon^2(t)]_{min}/P_{NN}$ versus W_X/W_N, determine if there is a preferred bandwidth ratio when $P_{NN}/P_{XX} = 8$. Is there a ratio that should be avoided? Discuss. (*Hint:* Use the integral given in Problem 9-40.)

SOME PRACTICAL APPLICATIONS
OF THE THEORY

10.0 INTRODUCTION

The main purpose of this book has been to introduce the reader to the basic principles necessary to model random signals and noise. The principles were broad enough to include the descriptions of waveforms modified by passage through linear networks. In this chapter we shall apply the basic principles to a few practical problems that involve random signals, noise, and networks. Obviously, the list of practical applications is almost limitless and it is necessary to select only a finite few. Although the applications discussed here may not necessarily serve the main interests of all readers, they do represent important applications and do serve to illustrate the use of the book's theory.

In the following sections we shall describe two practical communication systems, two control systems (one with application to one of the communication systems), an application involving a computer-type signal, and two applications that relate to radar. In every case we are primarily interested in how these applications are affected by the presence of random noise. We begin by considering the common broadcast AM (amplitude modulation) communication system.

10.1 NOISE IN AN AMPLITUDE MODULATION COMMUNICATION SYSTEM

The communication system most familiar to the general public is probably the AM (amplitude modulation) system. In this system the amplitude of a high-frequency "carrier" is made to vary (be modulated) as a linear function of the message waveform, usually derived from music, speech, or other audio source. The carrier frequency assigned to a broadcast station in the United States is one of the values from 540 to 1600 kHz in 10-kHz steps. Each station must contain its radiated power to a 10-kHz band centered on its assigned frequency.

In this section we shall give a very brief introduction to the AM broadcast system and illustrate how the noise principles of the preceding chapters can be used to analyze the system's performance.

AM System and Waveforms

Figure 10.1-1 illustrates the basic *functions* that must be present in an AM system. In this figure we include only those functions necessary to the study of noise performance. A practical system would include many other devices such as amplifiers, mixers, oscillators, and antennas that do not directly affect our performance calculations.

The transmitted AM signal has the form

$$s_{AM}(t) = [A_0 + x(t)] \cos [\omega_0 t + \theta_0] \tag{10.1-1}$$

where $A_0 > 0$, ω_0, and θ_0 are constants, while $x(t)$ represents a message that we model as a sample function of a random process $X(t)$. Note that the amplitude $[A_0 + x(t)]$ of the carrier $\cos (\omega_0 t + \theta_0)$ is a linear function of $x(t)$. Now, in general, one has no control over θ_0 because the turn-on time of a transmitter is random and the channel itself may introduce a phase angle that is random (which we presume is absorbed in the value of θ_0). Thus, we may properly model θ_0 as a value of a random variable Θ_0 independent of $X(t)$ and uniformly distributed on $(0, 2\pi)$. These considerations allow $s_{AM}(t)$ to be modeled as a sample function of a transmitted random process $S_{AM}(t)$ given by

$$S_{AM}(t) = [A_0 + X(t)] \cos (\omega_0 t + \Theta_0) \tag{10.1-2}$$

The transmitted signal arrives at the receiver after passing through a channel with gain G_{ch}. The channel is assumed to add no signal distortion but does add zero-mean white gaussian noise of power density $\mathcal{N}_0/2$. A practical channel typically adds delay but this effect does not modify the noise performance. A receiver bandpass filter passes the received signal $s_R(t) = G_{ch} s_{AM}(t)$ with negligible distortion but has no wider bandwidth than necessary so as to not pass excessive noise.† The noise $n(t)$ at the filter's output is a bandpass noise so the theory of Section 8.6 applies.

We model waveforms $s_R(t)$ and $n(t)$ as sample functions of processes $S_R(t)$ and $N(t)$, respectively. Thus, we may write

$$S_R(t) = G_{ch} S_{AM}(t)$$

$$= G_{ch}[A_0 + X(t)] \cos (\omega_0 t + \Theta_0) \tag{10.1-3}$$

$$N(t) = N_c(t) \cos (\omega_0 t + \Theta_0) - N_s(t) \sin (\omega_0 t + \Theta_0) \tag{10.1-4}$$

where $N_c(t)$ and $N_s(t)$ are lowpass noises with average powers $\overline{N_c^2(t)} = \overline{N_s^2(t)} = \overline{N^2(t)}$ from Section 8.6.

---

† The required bandwidth W_{rec} must be at least twice the spectral extent W_X of $X(t)$.

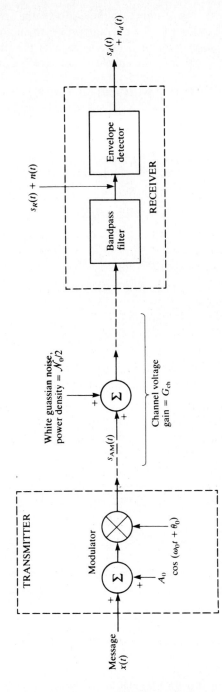

Figure 10.1-1 Functional block diagram of a broadcast AM system.

Noise Performance

A good measure of noise performance is the ratio of the average power in the output signal $s_d(t)$ of the system to the average power in the output noise $n_d(t)$. In the AM receiver an envelope detector is used to recover the transmitted message.

The total waveform applied to the envelope detector becomes

$$S_R(t) + N(t) = \{G_{ch}[A_0 + X(t)] + N_c(t)\} \cos(\omega_0 t + \Theta_0) - N_s(t) \sin(\omega_0 t + \Theta_0)$$

$$= A(t) \cos[\omega_0 t + \Theta_0 + \psi(t)] \tag{10.1-5}$$

where†

$$\psi(t) = \tan^{-1}\left\{\frac{N_s(t)}{G_{ch}[A_0 + X(t)] + N_c(t)}\right\} \tag{10.1-6}$$

$$A(t) = \langle\{G_{ch}[A_0 + X(t)] + N_c(t)\}^2 + N_s^2(t)\rangle^{1/2}$$

$$= G_{ch}[A_0 + X(t)]\left\langle 1 + \frac{2N_c(t)}{G_{ch}[A_0 + X(t)]} + \frac{N_c^2(t) + N_s^2(t)}{G_{ch}^2[A_0 + X(t)]^2}\right\rangle^{1/2} \tag{10.1-7}$$

Now only (10.1-7) is of interest because $A(t)$ is the envelope of $S_R(t) + N(t)$. The detector output is this envelope.

Since $N_c^2(t) + N_s^2(t)$ is the instantaneous envelope of the square of $N(t)$ (related to received noise power), while $G_{ch}^2[A_0 + X(t)]^2$ is the instantaneous envelope of the detector's input signal (related to received signal power), we make the assumption that input (received) signal-to-noise power ratio is large so that $[N_c^2(t) + N_s^2(t)]/G_{ch}^2[A_0 + X(t)]^2$ is small *most of the time*. The assumption allows

$$A(t) \approx G_{ch}[A_0 + X(t)] + N_c(t) \tag{10.1-8}$$

from (10.1-7). Only when this condition is true do we obtain quality performance anyway, so other situations are not usually of interest.

If we model $s_d(t)$ and $n_d(t)$ in Figure 10.1-1 as sample functions of processes $S_d(t)$ and $N_d(t)$, respectively, then (10.1-8) clearly gives

$$S_d(t) = G_{ch}[A_0 + X(t)] \tag{10.1-9}$$

$$N_d(t) = N_c(t) \tag{10.1-10}$$

The *useful* output signal average power, denoted by S_o, is that due to $X(t)$ in (10.1-9). If output average noise power is denoted by N_o then

$$S_o = G_{ch}^2 \overline{X^2(t)} \tag{10.1-11}$$

$$N_o = \overline{N_c^2(t)} = \overline{N^2(t)} \tag{10.1-12}$$

---

† Typically, *overmodulation* where $|X(t)|_{max}$, the maximum magnitude of $X(t)$, exceeds A_0 is undesirable in AM, so $[A_0 + X(t)] > 0$ is assumed in (10.1-7).

and performance is measured by

$$\left(\frac{S_o}{N_o}\right)_{AM} = \frac{G_{ch}^2 \overline{X^2(t)}}{\overline{N^2(t)}} \tag{10.1-13}$$

Next, we model the bandpass filter in Figure 10.1-1 as an ideal filter with bandwidth W_{rec} (rad/s). Noise power readily follows

$$\overline{N^2(t)} = \frac{1}{2\pi} 2 \int_{\omega_0 - (W_{rec}/2)}^{\omega_0 + (W_{rec}/2)} (\mathcal{N}_0/2) \, d\omega = \frac{\mathcal{N}_0 W_{rec}}{2\pi} \tag{10.1-14}$$

From (10.1-13) we have

$$\left(\frac{S_o}{N_o}\right)_{AM} = \frac{2\pi G_{ch}^2 \overline{X^2(t)}}{\mathcal{N}_0 W_{rec}} \tag{10.1-15}$$

Equation (10.1-15) is the principal result of this section. It describes the performance of the AM system. It is helpful to demonstrate the use of (10.1-15) by means of an example.

Example 10.1-1 Assume an AM system uses an unmodulated carrier of peak amplitude $A_0 = 10\sqrt{95}$ V and a message of power $\overline{X^2(t)} = 500$ W. Its channel has a gain $G_{ch} = \sqrt{32/100}$ with a noise density $\mathcal{N}_0/2 = (10^{-8})$ W/Hz. The receiver uses a filter with bandwidth $W_{rec} = 2\pi(10^4)$ rad/s. We compute various signal powers and system performance.

From Problem 10-1 the average power in the transmitted carrier is $A_0^2/2 = 4750$ W; the transmitted power due to message modulation is $R_{XX}(0)/2 = \overline{X^2(t)}/2 = 250$ W. Total average transmitted power is, therefore, 5000 W.

From (10.1-15) we compute

$$\left(\frac{S_o}{N_o}\right)_{AM} = \frac{2\pi(32)10^{-4}(500)}{2(10^{-8})2\pi(10^4)} = 8000 \qquad \text{(or 39.03 dB)}$$

This signal-to-noise ratio represents fairly good performance.

At the input to the envelope detector the received average signal power is 5000 W decreased by the loss incurred in passing over the channel: $5000(\sqrt{32}/100)^2 = 16$ W. From (10.1-14) and (10.1-12) the input average noise power is $10^{-8}2\pi(10^4)/\pi = 2(10^{-4})$ W. Input signal-to-noise ratio becomes $16/2(10^{-4}) = 80,000$ (or 49.03 dB). This value is well above the minimum for performance as required for (10.1-15) to be valid; in fact, if the performance of an AM system is satisfactory then (10.1-15) will always be valid (the reader should justify this fact by examining the *efficiency* of an AM system—see Problems 10-4 and 10-2.)

10.2 NOISE IN A FREQUENCY MODULATION COMMUNICATION SYSTEM

Another communication system with which the reader is familiar is the broadcast FM (frequency modulation) system. Here the instantaneous frequency of a sinusoidal "carrier" waveform is made to vary as a linear function of the message waveform. If $X(t)$ is a process representing the message, the FM transmitted waveform can be represented by the process

$$S_{FM}(t) = A \cos \left[\omega_0 t + \Theta_0 + k_{FM} \int X(t) \, dt \right] \tag{10.2-1}$$

where A, ω_0, and $k_{FM} > 0$ are constants† and Θ_0 is a random variable independent of $X(t)$ and uniformly distributed on $(0, 2\pi)$. In a practical station $\omega_0/2\pi$ is the station's assigned frequency and is one of 100 possible frequencies from 88.1 to 107.9 MHz. Each station transmits power in a 200-kHz "channel" centered on its assigned frequency.

The constant k_{FM} in (10.2-1) is the transmitter's modulation constant. Its unit is rad/second per volt when $X(t)$ is a voltage. Transmitted signal bandwidth is difficult to compute in FM because FM is a nonlinear modulation. If k_{FM} is large enough, this bandwidth can readily be much larger than the bandwidth of the message process $X(t)$. If $X(t)$ is presumed to be bounded at $|X(t)|_{max}$ and have a crest-factor defined by (Problem 10-3)

$$K_{cr}^2 = \frac{|X(t)|_{max}^2}{E[X^2(t)]} = \frac{|X(t)|_{max}^2}{\overline{X^2(t)}} \tag{10.2-2}$$

the bandwidth of $S_{FM}(t)$ for the broadband case is approximated by (Peebles, 1976)

$$W_{FM} \approx 2\Delta\omega = 2k_{FM} |X(t)|_{max}$$
$$= 2k_{FM} K_{cr} \sqrt{\overline{X^2(t)}} \tag{10.2-3}$$

Here

$$\Delta\omega = k_{FM} |X(t)|_{max} \tag{10.2-4}$$

is the peak frequency deviation that instantaneous frequency can make from ω_0 (on either side).

Although difficult to prove, the average transmitted waveform power is

$$P_{FM} = E[S_{FM}^2(t)] = \frac{A^2}{2} \tag{10.2-5}$$

which is independent of the modulation.

---

† If k_{FM} is negative its sign can be absorbed into the definition of $X(t)$.

FM System and Waveforms

Figure 10.2-1 illustrates the basic functions present in a typical FM system. The transmitted waveform passes over the channel modeled as a power gain G_{ch}^2 without distortion or delay (as also assumed in Section 10.1 above). The receiver's bandpass filter (BPF) is wide enough to pass $G_{ch} S_{FM}(t)$ with little distortion but not so wide as to pass excess noise. Its bandwidth is, therefore, $W_{FM} = 2\Delta\omega$.

The purpose of the limiter is to remove amplitude fluctuations in the received waveform. The limiter is necessary so that the receiver responds only to frequency variations (that contain the message) and not to amplitude variations that are mainly due to noise. The discriminator is the actual demodulation device; it produces a voltage proportional (constant of proportionality K_D) to instantaneous deviations of the frequency of its input waveform from a nominal value ω_0. Ideally, with no noise, the discriminator's output signal is $K_D k_{FM} X(t)$. The lowpass filter must pass this waveform with low distortion so that its output is proportional to $X(t)$

$$S_d(t) = K_D k_{FM} X(t) \tag{10.2-6}$$

It should have a bandwidth no wider than the spectral extent of $X(t)$, denoted by W_X, so as to not allow excessive output noise.

If the receiver's "input" is defined as the input to the limiter, the input signal's average power S_i is

$$S_i = G_{ch}^2 \frac{A^2}{2} \tag{10.2-7}$$

while the output signal power is

$$S_o = E[S_d^2(t)] = K_D^2 k_{FM}^2 \overline{X^2(t)} \tag{10.2-8}$$

By modeling the BPF in Figure 10.2-1 as an ideal filter the input noise power is readily found to be

$$N_i = \frac{1}{2\pi} 2 \int_{\omega_0 - \Delta\omega}^{\omega_0 + \Delta\omega} \frac{\mathcal{N}_0}{2} \, d\omega = \frac{\mathcal{N}_0 \Delta\omega}{\pi} \tag{10.2-9}$$

Input signal-to-noise power ratio is

$$\left(\frac{S_i}{N_i}\right)_{FM} = \frac{\pi G_{ch}^2 A^2}{2\mathcal{N}_0 \Delta\omega} \tag{10.2-10}$$

from (10.2-7) and (10.2-9).

Computation of output noise power is less straightforward than the preceding computations. However, its development forms the most interesting problem in computing system performance.

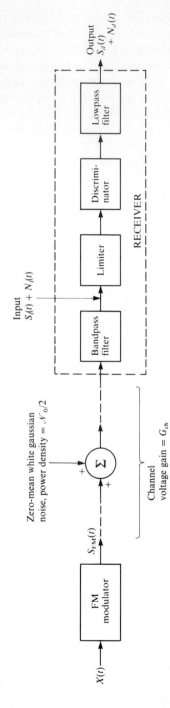

Figure 10.2-1 Functional block diagram of an FM communication system.

FM System Performance

Care must be exercised in finding output noise power because FM is a nonlinear operation. For relatively large $(S_i/N_i)_{FM}$ and wideband operation (developed above), signal and noise powers may be independently found. Signal power is found assuming noise zero (above). Noise power is found assuming the message is zero but carrier is still transmitted. In this latter case the waveform at the limiter is

$$G_{ch} A \cos [\omega_0 t + \Theta_0] + N_c(t) \cos (\omega_0 t + \Theta_0) - N_s(t) \sin (\omega_0 t + \Theta_0)$$
$$= A(t) \cos [\omega_0 t + \Theta_0 + \psi(t)] \quad (10.2\text{-}11)$$

where the bandpass noise $N_i(t)$ is modeled as in (10.1-4) (see also Section 8.6) and

$$A(t) = \{[G_{ch} A + N_c(t)]^2 + N_s^2(t)\}^{1/2} \quad (10.2\text{-}12)$$

$$\psi(t) = \tan^{-1} \left\{ \frac{N_s(t)}{G_{ch} A + N_c(t)} \right\} \quad (10.2\text{-}13)$$

For large input signal-to-noise ratio we have $|G_{ch} A| \gg |N_c(t)|$ and $|G_{ch} A| \gg |N_s(t)|$ *most of the time*, so (10.2-13) becomes

$$\psi(t) \approx \tan^{-1} \left[\frac{N_s(t)}{G_{ch} A} \right] \approx \frac{N_s(t)}{G_{ch} A} \quad (10.2\text{-}14)$$

Equation (10.2-11) is now approximated by

$$A(t) \cos [\omega_0 t + \Theta_0 + \psi(t)] \approx A(t) \cos \left[\omega_0 t + \Theta_0 + \frac{N_s(t)}{G_{ch} A} \right] \quad (10.2\text{-}15)$$

Because the limiter removes $A(t)$ and the discriminator responds only to instantaneous frequency deviations from ω_0, the input to the lowpass filter is

$$\left(\frac{K_D}{G_{ch} A} \right) \frac{dN_s(t)}{dt} \quad (10.2\text{-}16)$$

If $S_{N_s N_s}(\omega)$ is the power spectrum of $N_s(t)$ the power spectrum of (10.2-16) is

$$\left(\frac{K_D}{G_{ch} A} \right)^2 \omega^2 S_{N_s N_s}(\omega) \quad (10.2\text{-}17)$$

However, we may use (8.6-17) and (8.6-16) to write this power spectrum as

$$\left(\frac{K_D}{G_{ch} A} \right)^2 \omega^2 [S_{N_i N_i}(\omega - \omega_0) + S_{N_i N_i}(\omega + \omega_0)] \quad |\omega| < \Delta\omega \quad (10.2\text{-}18)$$

where $S_{N_i N_i}(\omega)$ is the power spectrum of $N_i(t)$; it is constant at $\mathcal{N}_0/2$ over bands of width $2\Delta\omega$ centered at ω_0 and $-\omega_0$.

Final output noise power results from the action of the lowpass filter on (10.2-18). We have

$$
N_o = E[N_d^2(t)] = \frac{1}{2\pi} \int_{-W_X}^{W_X} \left(\frac{K_D}{G_{ch}A}\right)^2 \omega^2 [\mathcal{S}_{N_iN_i}(\omega - \omega_0) + \mathcal{S}_{N_iN_i}(\omega + \omega_0)] \, d\omega
$$

$$
= \frac{K_D^2}{2\pi G_{ch}^2 A^2} \int_{-W_X}^{W_X} \omega^2 \left[\frac{\mathcal{N}_0}{2} + \frac{\mathcal{N}_0}{2}\right] d\omega = \frac{K_D^2 \mathcal{N}_0 W_X^3}{3\pi G_{ch}^2 A^2} \tag{10.2-19}
$$

Output performance is determined by

$$
\left(\frac{S_o}{N_o}\right)_{FM} = \frac{3\pi G_{ch}^2 A^2 k_{FM}^2 \overline{X^2(t)}}{\mathcal{N}_0 W_X^3} \tag{10.2-20}
$$

from (10.2-8) and (10.2-19). An alternative form of (10.2-20) is

$$
\left(\frac{S_o}{N_o}\right)_{FM} = \frac{6}{K_{cr}^2} \left(\frac{\Delta\omega}{W_X}\right)^3 \left(\frac{S_i}{N_i}\right)_{FM} \tag{10.2-21}
$$

An important observation derives from (10.2-21). Since FM bandwidth is $2\Delta\omega$, we see that performance increases as the *cube* of bandwidth relative to $(S_i/N_i)_{FM}$. However, $(S_i/N_i)_{FM}$ decreases as the reciprocal of bandwidth from (10.2-10), so the *net* performance increases as the *square* of bandwidth. By simply increasing bandwidth at the transmitter, system performance rapidly increases. There is a limit to this procedure, unfortunately, that occurs when conditions under which the performance equations were derived are no longer valid. The break point, or *threshold*, occurs approximately where $(S_i/N_i)_{FM}$ drops below about 10 (or 10 dB). For a more detailed discussion of FM threshold the reader is referred to Peebles (1976). We shall emphasize FM system performance through an example.

Example 10.2-1 An FM system uses a message with crest factor 3 and bandwidth $W_X/2\pi = 3$ kHz. The FM modulator's bandwidth is $2\Delta\omega/2\pi = 20$ kHz and the receiver's input signal-to-noise ratio is 81. From (10.2-21) $(S_o/N_o)_{FM} = 2000$ (or 33.01 dB). We determine how much performance can be increased by raising $\Delta\omega$.

From (10.2-10) $(S_i/N_i)_{FM}$ decreases to 10 from 81 if $\Delta\omega$ increases by a factor of 8.1. Next, we again use (10.2-21) but now with $\Delta\omega/2\pi = 8.1(10)$ kHz and $(S_i/N_i)_{FM} = 10$:

$$
\left(\frac{S_o}{N_o}\right)_{FM} = \frac{6}{9} \left(\frac{81}{3}\right)^3 (10) = 131,220
$$

(or 51.18 dB). The bandwidth increase of 8.1 times has improved $(S_o/N_o)_{FM}$ by 65.61 times.

10.3 NOISE IN A SIMPLE CONTROL SYSTEM

In this section we shall briefly consider the noise response of a simple control system modeled by the block diagram shown in Figure 10.3-1. The following section will then illustrate how a very practical network can be analyzed by applying the results developed here.

Transfer Function

Typical loop behavior in Figure 10.3-1 is to force the feedback signal F to approximate the command C so that the error C-F is small. The control loop's response R may be conveniently chosen. For example, if R in the time domain is to be the derivative of the command then $H_2(\omega) = 1/j\omega$, the transfer function of an integrator. If R is to approximate C then $H_2(\omega) = 1$.

From Figure 10.3-1 it is clear that

$$R(\omega) = H_1(\omega)[C(\omega) - H_2(\omega)R(\omega)] \tag{10.3-1}$$

so

$$R(\omega) = C(\omega)\left[\frac{H_1(\omega)}{1 + H_1(\omega)H_2(\omega)}\right] \tag{10.3-2}$$

We define the *transfer function* of the control loop as

$$H(\omega) = \frac{R(\omega)}{C(\omega)} = \frac{H_1(\omega)}{1 + H_1(\omega)H_2(\omega)} \tag{10.3-3}$$

The transfer function (10.3-3) is not always stable. There are combinations of $H_1(\omega)$ and $H_2(\omega)$ that can cause instability. In general, if $H_1(\omega)$ and $H_2(\omega)$ are stable and $|H_1(\omega)H_2(\omega)|$ falls below unity, as a function of ω, before the phase of $H_1(\omega)H_2(\omega)$ becomes $-\pi$, and if the phase of $H_1(\omega)H_2(\omega)$ equals $-\pi$ at only one frequency, the transfer function $H(\omega)$ is stable. The product $H_1(\omega)H_2(\omega)$ is called the *open-loop transfer function* of the control system. Stability is a deep subject in control systems and we shall not develop it further because it detracts from the simple points to be made here.

Now suppose the command waveform in Figure 10.3-1 is the sum of a signal $S_c(t)$ and noise $N_c(t)$. Because the system is linear its responses to signal and noise may be computed separately. If $S_{N_c N_c}(\omega)$ is the power spectrum of $N_c(t)$ then the power spectrum of the response noise $N_R(t)$ is

$$S_{N_R N_R}(\omega) = S_{N_c N_c}(\omega)\left|\frac{H_1(\omega)}{1 + H_1(\omega)H_2(\omega)}\right|^2 \tag{10.3-4}$$

whenever the network is stable.

An example serves to illustrate the use of (10.3-4).

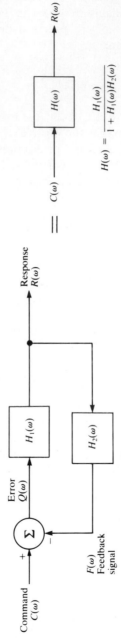

Figure 10.3-1 Block diagram of a simple control system.

Example 10.3-1 Let a signal

$$S_c(t) = Au(t)e^{-Wt}$$

plus white noise of power density $\mathcal{N}_0/2$ be applied to the control network
where

$$H_1(\omega) = \frac{K_1 W_1}{W_1 + j\omega} \qquad K_1 \gg 1$$

$$H_2(\omega) = 1$$

This choice means that we desire the response to equal the command. We
find the output signal and the output noise power.

From (10.3-3)

$$H(\omega) = \frac{K_1 W_1/(W_1 + j\omega)}{1 + [K_1 W_1/(W_1 + j\omega)]} = \frac{K_1 W_1}{(1 + K_1)W_1 + j\omega}$$

From pair 15 of Appendix E the inverse transform of $H(\omega)$ is

$$h(t) = K_1 W_1 u(t)e^{-(1 + K_1)W_1 t}$$

The response signal becomes

$$S_R(t) = \int_{-\infty}^{\infty} h(\xi)S_c(t - \xi)\,d\xi$$

$$= K_1 W_1 A \int_{-\infty}^{\infty} u(\xi)u(t - \xi)e^{-[(1 + K_1)W_1 - W]\xi}\,d\xi e^{-Wt}$$

$$= K_1 W_1 Au(t)e^{-Wt} \int_0^t e^{-[(1 + K_1)W_1 - W]\xi}\,d\xi$$

$$= \frac{K_1 W_1}{(1 + K_1)W_1 - W}\langle 1 - \exp\{-[(1 + K_1)W_1 - W]t\}\rangle S_c(t)$$

For $K_1 \gg 1$ so that $(1 + K_1)W_1 \gg W$ this result becomes

$$S_R(t) \approx S_c(t)$$

The approximation is more accurate as t becomes large.

From (10.3-4) the output noise power spectrum is

$$\mathcal{S}_{N_R N_R}(\omega) = \frac{\mathcal{N}_0(K_1 W_1)^2/2}{[(1 + K_1)W_1]^2 + \omega^2}$$

Output noise power is found using (C-25):

$$P_{N_R N_R} = \frac{1}{2\pi}\int_{-\infty}^{\infty} \mathcal{S}_{N_R N_R}(\omega)\,d\omega$$

$$= \frac{\mathcal{N}_0 K_1^2 W_1}{4(1 + K_1)} \approx \frac{\mathcal{N}_0 K_1 W_1}{4}$$

We observe in passing that this control loop is stable and its transfer function is equivalent to a simple lowpass filter of gain $K_1/(1 + K_1) \approx 1$ and 3-dB bandwidth $(1 + K_1)W_1 \approx K_1W_1$. This unity-gain large-bandwidth filter resulted from a narrowband (bandwidth W_1) high gain filter (gain K_1) inside the loop.

Error Function

The error $Q = C - F$ in Figure 10.3-1 is readily found. From

$$Q(\omega) = C(\omega) - F(\omega) = C(\omega) - H_2(\omega)H_1(\omega)Q(\omega) \tag{10.3-5}$$

we have

$$Q(\omega) = \frac{C(\omega)}{1 + H_1(\omega)H_2(\omega)} \tag{10.3-6}$$

Wiener Filter Application

By comparing (10.3-3) with the transfer function of a Wiener filter for uncorrelated signal and noise as given by (9.2-20) we see that the Wiener filter can be implemented as a loop. From (9.2-20)

$$H_{opt}(\omega) = \frac{e^{j\omega t_o}}{1 + [S_{NN}(\omega)/S_{XX}(\omega)]} \tag{10.3-7}$$

Thus

$$H(\omega) = H_{opt}(\omega) \tag{10.3-8}$$

if

$$H_1(\omega) = e^{j\omega t_o} \tag{10.3-9}$$

$$H_2(\omega) = [S_{NN}(\omega)/S_{XX}(\omega)]e^{-j\omega t_o} \tag{10.3-10}$$

Of course these functions $H_1(\omega)$ and $H_2(\omega)$ may not be realizable even for realizable signal and noise power spectrums. Other choices for $H_1(\omega)$ and $H_2(\omega)$ are also possible (Problem 10-10).

10.4 NOISE IN A PHASE-LOCKED LOOP

The phase-locked loop (PLL) is a practical system to which the noise theory of this book can be applied as a good example. The PLL is also an example of the control system of the preceding section.

Figure 10.4-1 depicts the block diagram of a PLL. Broadly, the action of the loop is to force the phase of the output of the voltage-controlled oscillator (VCO)

Figure 10.4-1 Block diagram of a phase-locked loop (PLL).

Input signal
$A_i \cos[\omega_0 t + \theta_0 + \theta_i(t)]$
$= A_i \cos[\theta_1(t)]$

Feedback signal
$A_V \cos[\omega_0 t + \theta_0 + \theta_V(t)]$
$= A_V \cos[\theta_2(t)]$

K_P

Phase
detector

Error signal
$e_L(t)$

Filter
$H_L(\omega)$

k_V

Voltage
controlled
oscillator

Output signal
$s_R(t)$

to closely follow the phase of the input signal. This action leads to one of the most important uses of the PLL, that of demodulating a frequency-modulated signal. If there is no input noise $N_i(t)$ and the VCOs phase follows that of the input FM signal, then the VCOs signal has the same FM as that transmitted. Since the VCO is just a frequency modulator, its input waveform (loop's output waveform) has to be proportional to the original message used at the transmitter. When input noise is present there is noise on the output signal. In this section we shall develop this output noise power and find the available output signal-to-noise power ratio.

Phase Detector

Consider first the phase detector. Although there are many forms of phase detector [Blanchard (1976) and Klapper et al. (1972)] they all provide an output response proportional to the difference between the phases of the two input waveforms for small difference phases. Thus

$$e_L(t) \approx K_P[\theta_1(t) - \theta_2(t)] \tag{10.4-1}$$

if the two input waveform's phases are defined as $\theta_1(t)$ and $\theta_2(t)$. The constant K_P is the phase detector's sensitivity constant; its unit is volts per radian for $e_L(t)$ a voltage. In some phase detectors the response is also proportional to the amplitudes of the two input waveforms. Others depend only on one input amplitude because the other is large enough to saturate the device giving a type of limiting. Another type allows both inputs to limit in the detector and the output is not a function of either input waveform's level. We shall assume either this last form of detector or that an actual limiter is in the path of the signal's input when a detector is used with limiting in the feedback path's input. Thus our phase detector is described by (10.4-1).

Loop Transfer Function

Since the VCO in Figure 10-4-1 acts like a frequency modulator for the "message" $s_R(t)$, its output can be written as

$$\text{VCO output} = A_V \cos [\omega_0 t + \theta_0 + \theta_V(t)]$$

$$= A_V \cos \left[\omega_0 t + \theta_0 + k_V \int s_R(t) \, dt \right]$$

$$= A_V \cos [\theta_2(t)] \tag{10.4-2}$$

where k_V is the VCO's modulation constant,

$$\theta_2(t) = \omega_0 t + \theta_0 + k_V \int s_R(t) \, dt \tag{10.4-3}$$

and

$$\theta_V(t) = k_V \int s_R(t)\, dt \qquad (10.4\text{-}4)$$

The other phase detector input signal, from Figure 10.4-1, is the input waveform. If we define its phase as

$$\theta_1(t) = \omega_0 t + \theta_0 + \theta_i(t) \qquad (10.4\text{-}5)$$

then the phase detector's response (10.4-1) becomes

$$e_L(t) = K_P\left[\omega_0 t + \theta_0 + \theta_i(t) - \omega_0 t - \theta_0 - k_V \int s_R(t)\, dt\right]$$

$$= K_P\left[\theta_i(t) - k_V \int s_R(t)\, dt\right] \qquad (10.4\text{-}6)$$

Next, if we define Fourier transforms as follows

$$e_L(t) \leftrightarrow E_L(\omega) \qquad (10.4\text{-}7)$$

$$\theta_i(t) \leftrightarrow \Theta_i(\omega) \qquad (10.4\text{-}8)$$

$$s_R(t) \leftrightarrow S_R(\omega) \qquad (10.4\text{-}9)$$

we may write (10.4-6) as

$$E_L(\omega) = K_P\left[\Theta_i(\omega) - \frac{k_V S_R(\omega)}{j\omega}\right] \qquad (10.4\text{-}10)$$

From Figure 10.4-1

$$E_L(\omega) = \frac{S_R(\omega)}{H_L(\omega)} \qquad (10.4\text{-}11)$$

On equating (10.4-10) and (10.4-11) we find the PLL's transfer function, denoted by $H_T(\omega)$, to be

$$H_T(\omega) = \frac{S_R(\omega)}{\Theta_i(\omega)} = \frac{K_P j\omega H_L(\omega)}{j\omega + K_P k_V H_L(\omega)} = \frac{j\omega}{k_V} H(\omega) \qquad (10.4\text{-}12)$$

where we also define†

$$H(\omega) = \frac{K_P k_V H_L(\omega)}{j\omega + K_P k_V H_L(\omega)} \qquad (10.4\text{-}13)$$

---

† In many texts $H(\omega)$ is called the PLL transfer function but the loop's output is defined at a different point. (Where would it be?)

Figure 10.4-2 (*a*) Equivalent block diagram of the linear PLL of Figure 10.4-1, and (*b*) the transfer function equivalent of the loop in (*a*).

It should be noted that the above definition of transfer function relates the output *signal* to the input signal's *phase modulation* $\theta_i(t)$ according to

$$S_R(\omega) = H_T(\omega)\Theta_i(\omega) \tag{10.4-14}$$

or

$$s_R(t) = \int_{-\infty}^{\infty} h_T(t - \xi)\theta_i(\xi)\, d\xi \tag{10.4-15}$$

where $h(t)$ denotes the inverse transform of $H_T(\omega)$

$$h_T(t) \leftrightarrow H_T(\omega) \tag{10.4-16}$$

The above developments show, in effect, that Figure 10.4-2 is an equivalent form for the loop of Figure 10.4-1.

Loop Noise Performance

We shall apply the preceding results to the case where the input to the PLL is the sum of an FM signal plus bandpass noise $N_i(t)$ modeled as

$$N_i(t) = N_c(t)\cos(\omega_0 t) - N_s(t)\sin(\omega_0 t) \tag{10.4-17}$$

The representation (10.4-17) follows developments of Section 8.6 where $N_c(t)$ and $N_s(t)$ are lowpass random processes having the properties defined in (8.6-7) through (8.6-19). The actual input to the PLL is, therefore,

$$A_i \cos\left[\omega_0 t + \theta_0 + k_{FM}\int X(t)\, dt\right] + N_c(t)\cos(\omega_0 t) - N_s(t)\sin(\omega_0 t) \tag{10.4-18}$$

where k_{FM} is the FM modulator's constant, $X(t)$ is the message process, and A_i, ω_0, and θ_0 are the input FM signal's peak amplitude, frequency, and phase, respectively.

The exact analysis of the PLL's response to the waveform of (10.4-18) is very involved. However, it can be shown that the waveform of (10.4-18) can be put in the form (Problem 10-11)

$$R(t) \cos \left[\omega_0 t + \theta_0 + \theta_{FM}(t) + \theta_N(t) \right] \qquad (10.4\text{-}19)$$

where

$$\theta_{FM}(t) = k_{FM} \int X(t) \, dt \qquad (10.4\text{-}20)$$

and $\theta_N(t)$ is a phase angle caused by noise. For large-input signal-to-noise ratio $(A_i^2/2)/E[N_i^2(t)]$ and input noise $N_i(t)$ broadband relative to the FM signal, the autocorrelation function of $\theta_N(t)$ is approximately $1/A_i^2$ times the autocorrelation function of $N_c(t)$ (Problem 10-12). This fact means that, within a reasonable approximation, $\theta_N(t)$ can be replaced by the *equivalent* angle $N_c(t)/A_i$.

With the above noise equivalence used, the input phase modulation to the PLL from (10.4-19) is

$$\theta_i(t) = \theta_{FM}(t) + \theta_N(t)$$

$$= \theta_{FM}(t) + \frac{N_c(t)}{A_i} \qquad (10.4\text{-}21)$$

The component $\theta_{FM}(t)$ is due to the signal. If $X(t)$ is a random process with power spectrum $S_{XX}(\omega)$, we use (10.4-20) in (10.4-21) and find that the power spectrum of $\theta_i(t)$ is

$$S_{\theta_i\theta_i}(\omega) = \frac{k_{FM}^2 \, S_{XX}(\omega)}{\omega^2} + \frac{S_{N_c N_c}(\omega)}{A_i^2} \qquad (10.4\text{-}22)$$

After using the PLL's transfer function (10.4-12), the output waveform's power spectrum becomes

$$S_{S_R S_R}(\omega) = S_{\theta_i\theta_i}(\omega) | H_T(\omega) |^2$$

$$= S_{XX}(\omega) \left(\frac{k_{FM}}{k_V} \right)^2 | H(\omega) |^2 + S_{N_c N_c}(\omega) \frac{\omega^2}{A_i^2 k_V^2} | H(\omega) |^2 \qquad (10.4\text{-}23)$$

The first right-side term in (10.4-23) is due to the desired message while the second is due to noise. Loop design is typically chosen so that $| H(\omega) |^2 \approx 1$ for all frequencies of interest in $S_{XX}(\omega)$. In fact, if the message is to be preserved with very small distortion the bandwidth of the transfer function $H(\omega)$ may be significantly *larger* than the frequencies of interest in $S_{XX}(\omega)$. Thus, if W_X is the spectral extent of the message $X(t)$ then the power in the output signal component is

$$
S_o = \frac{1}{2\pi} \int_{-\infty}^{\infty} S_{XX}(\omega) \left(\frac{k_{FM}}{k_V} \right)^2 |H(\omega)|^2 \, d\omega \approx \left(\frac{k_{FM}}{k_V} \right)^2 \frac{1}{2\pi} \int_{-\infty}^{\infty} S_{XX}(\omega) \, d\omega
$$

$$
= \left(\frac{k_{FM}}{k_V} \right)^2 \overline{X^2(t)} \tag{10.4-24}
$$

In some loops (see example to follow) $|H(\omega)|^2$ does not decrease rapidly enough to remove high-frequency noise due to the factor ω^2 in $\omega^2 |H(\omega)|^2$ in (10.4-23). In these cases it may be necessary to follow the loop with a separate filter to better remove noise spectral components at frequencies $|\omega| > W_X$. As long as either the loop or a separate filter removes these components, the overall output noise power is approximately

$$
N_o \approx \frac{1}{2\pi} \int_{-W_X}^{W_X} S_{N_c N_c}(\omega) \frac{\omega^2}{A_i^2 k_V^2} |H(\omega)|^2 \, d\omega
$$

$$
\approx \frac{\mathcal{N}_0}{2\pi A_i^2 k_V^2} \int_{-W_X}^{W_X} \omega^2 \, d\omega = \frac{\mathcal{N}_0 W_X^3}{3\pi A_i^2 k_V^2} \tag{10.4-25}
$$

Finally, we determine output signal-to-noise power ratio from (10.4-25) and (10.4-24). As in Section 10.2, we let A be the peak amplitude of the transmitted FM signal and let G_{ch} be the gain of the channel, so that

$$
A_i = A G_{ch} \tag{10.4-26}
$$

Thus,

$$
\left(\frac{S_o}{N_o} \right)_{FM} = \frac{3\pi G_{ch}^2 A^2 k_{FM}^2 \overline{X^2(t)}}{\mathcal{N}_0 W_X^3} \tag{10.4-27}
$$

On comparing (10.4-27) with (10.2-20) we find that both the discriminator and PLL forms of FM receiver have the same performance when the received (input) signal-to-noise ratio is large.

Example 10.4-1 As an example of a practical PLL's transfer function let the loop filter be a simple low-pass function with 3-dB bandwidth W_L where

$$
H_L(\omega) = \frac{W_L}{W_L + j\omega}
$$

The function $H(\omega)$, from (10.4-13), becomes

$$
H(\omega) = \frac{1}{1 - \left(\dfrac{\omega}{\omega_n} \right)^2 + j2\zeta \left(\dfrac{\omega}{\omega_n} \right)}
$$

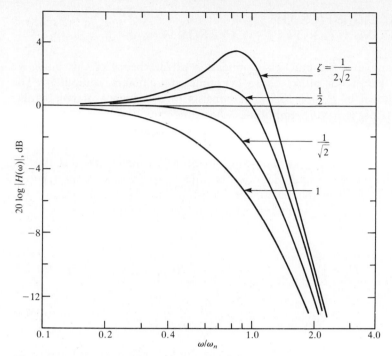

Figure 10.4-3 $|H(\omega)|$ for the loop of Example 10.4-1.

where the quantities defined by

$$\omega_n = (K_P k_V W_L)^{1/2}$$

$$\zeta = \frac{1}{2}\sqrt{\frac{W_L}{K_P k_V}}$$

are called the *natural frequency* and *damping* factor, respectively, of the loop. Figure 10.4-3 illustrates how $|H(\omega)|$ behaves with ω/ω_n for ζ as a parameter. The curve for $\zeta = 1/\sqrt{2}$ is most flat in the sense that the largest number of derivatives of $|H(\omega)|$ are zero at $\omega = 0$.

For $\zeta = 1/\sqrt{2}$ and $\omega_n = W_X$, the signal's spectral extent, we have

$$|H(\omega)|^2 = \frac{W_X^4}{W_X^4 + \omega^4}$$

The more exact power in the noise term of (10.4-23) becomes

$$N_o = \frac{\mathcal{N}_0 W_X^4}{2\pi A_i^2 k_V^2} \int_{-\infty}^{\infty} \frac{\omega^2 \, d\omega}{W_X^4 + \omega^4} = \frac{\mathcal{N}_0 W_X^3}{2\sqrt{2} \, A_i^2 \, k_V^2}$$

after using (C-38). On comparing this result with (10.4-25) we see the noise in the loop output is $3\pi/2\sqrt{2} \approx 3.33$ times that of a broadband loop followed by an abrupt-cutoff filter of bandwidth W_X.

10.5 CHARACTERISTICS OF RANDOM COMPUTER-TYPE WAVEFORM

As another example of the practical application of the theory of this book we examine a waveform not unlike those encountered in binary computers. The waveform is shown in Figure 10.5-1; it consists of a sequence of rectangular pulses of durations T_b having amplitudes that randomly may equal A or $-A$. Amplitudes A and $-A$ are assumed to occur with equal probability and the amplitude of any pulse interval is assumed to be statistically independent of the amplitudes of all other intervals. The random process from which this type of waveform is modeled as a sample function is called a *semirandom binary process* (see also Problem 6-4); in the remainder of this section we shall examine the description, power spectrum, and autocorrelation function of this process.

Process Description

The semirandom binary process $X(t)$ can be described by

$$X(t) = \sum_{k=-\infty}^{\infty} A_k \, \text{rect} \left[\frac{t - kT_b}{T_b} \right] \tag{10.5-1}$$

where $\{A_k\}$ is a set of statistically independent random variables and rect $(\cdot)$ is defined by (E-2). The A_k satisfy

$$E[A_K] = 0 \qquad k = 0, \pm 1, \pm 2, \ldots \tag{10.5-2}$$

$$E[A_k A_m] = \begin{cases} A^2 & k = m \\ 0 & k \neq m \end{cases} \tag{10.5-3}$$

The truncated version of $X(t)$ is needed in calculating power spectrum. We truncate to a time interval $2T$ centered on $t = 0$ that is a discrete multiple of T_b according to

$$2T = (2K + 1)T_b \tag{10.5-4}$$

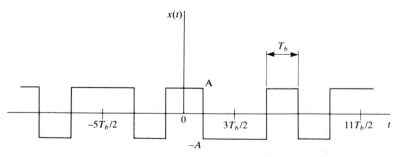

Figure 10.5-1 Typical waveform of a semirandom binary random process.

Thus, the truncated process $X_T(t)$ is

$$X_T(t) = \sum_{k=-K}^{K} A_k \text{ rect}\left[\frac{t - kT_b}{T_b}\right] \tag{10.5-5}$$

Power Spectrum

We compute the power spectrum $S_{XX}(\omega)$ of $X(t)$ by use of (7.1-11). The Fourier transform of $X_T(t)$, denoted by $X_T(\omega)$, is

$$X_T(\omega) = T_b \sum_{k=-K}^{K} A_k \text{ Sa}(\omega T_b/2)e^{-jk\omega T_b}$$

$$= T_b \text{ Sa}(\omega T_b/2) \sum_{k=-K}^{K} A_k e^{-jk\omega T_b} \tag{10.5-6}$$

from (10.5-5) and pair 5 of Table E-1. Next,

$$\frac{E[|X_T(\omega)|^2]}{2T} = \frac{T_b \text{ Sa}^2(\omega T_b/2)}{(2K+1)} \sum_{k=-K}^{K} \sum_{m=-K}^{K} E[A_k A_m]e^{-j(k-m)\omega T_b}$$

$$= A^2 T_b \text{ Sa}^2(\omega T_b/2) \tag{10.5-7}$$

Now because (10.5-7) does not depend on K, and therefore not on T through (10.5-4), we have

$$S_{XX}(\omega) = \lim_{T\to\infty} \frac{E[|X_T(\omega)|^2]}{2T} = A^2 T_b \text{ Sa}^2(\omega T_b/2) \tag{10.5-8}$$

The bandwidth of this power spectrum at its -3-dB point is $0.4429(2\pi/T_b) = 0.4429\omega_b$.

Autocorrelation Function

It follows from (10.5-1) through (10.5-3) that $E[X(t)X(t + \tau)]$ is zero unless both t and $t + \tau$ fall in the same pulse interval. The autocorrelation function is, therefore,

$$R_{XX}(t, t + \tau) = E[X(t)X(t + \tau)]$$

$$= \begin{cases} A^2 & (k - \tfrac{1}{2})T_b < (t \text{ and } t + \tau) < (k + \tfrac{1}{2})T_b \\ 0 & \text{elsewhere} \end{cases} \tag{10.5-9}$$

Thus, the process $X(t)$ is not even wide-sense stationary since (10.5-9) depends on absolute time t.

The time-averaged autocorrelation function is readily obtained by inverse Fourier transforming (10.5-8) according to (7.2-9). After using pair 7 of Table E-1 we obtain

$$R_{XX}(\tau) = \lim_{T\to\infty} \frac{1}{2T} \int_{-T}^{T} R_{XX}(t, t + \tau)\, dt = A^2 \text{ tri}\left(\frac{t}{T_b}\right) \tag{10.5-10}$$

The direct computation of $R_{XX}(\tau)$ by time-averaging $R_{XX}(t, t + \tau)$ is possible, but a bit more complicated than the inverse transform procedure used here (see Thomas, 1969, p. 107).

10.6 ENVELOPE AND PHASE OF A SINUSOIDAL SIGNAL PLUS NOISE

Many practical problems involve the probability density function of the envelope of the sum of a sinusoidal signal and noise. A radar, for example, may be interested in determining if a short segment (pulse) of a sinusoidal waveform is being received at some time or if only noise is being received. This problem is one of detection based on observing the received waveform's envelope; if the envelope is large enough (because of the signal's presence) the radar decides both the signal and the noise are present. We examine radar detection further in Section 10.7.

In this section we discuss probability densities involved in describing the envelope and phase of the sum of the sinusoidal signal and noise.

Waveforms

Let the signal be

$$s(t) = A_0 \cos (\omega_0 t + \theta_0) = A_0 \cos (\theta_0) \cos (\omega_0 t) - A_0 \sin (\theta_0) \sin (\omega_0 t) \quad (10.6\text{-}1)$$

where A_0, ω_0, and θ_0 are constants. We assume the noise $n(t)$ to be added to $s(t)$ is a sample function of a zero-mean, wide-sense stationary gaussian bandpass process $N(t)$ with power $E[N^2(t)] = \sigma^2$. From (8.6-2), the sum can be written as

$$s(t) + N(t) = [A_0 \cos (\theta_0) + X(t)] \cos (\omega_0 t) - [A_0 \sin (\theta_0) + Y(t)] \sin (\omega_0 t)$$

$$= R(t) \cos [\omega_0 t + \Theta(t)] \quad (10.6\text{-}2)$$

where $X(t)$ and $Y(t)$ are zero-mean, gaussian, lowpass processes having the same powers $E[X^2(t)] = E[Y^2(t)] = E[N^2(t)] = \sigma^2$. Other properties of $X(t)$ and $Y(t)$ are given in (8.6-7) through (8.6-19). The envelope and phase of the sum are $R(t)$ and $\Theta(t)$, respectively. We may think of $R(t)$ and $\Theta(t)$ as transformations of $X(t)$ and $Y(t)$ as follows:

$$R = T_1(X, Y) = \{[A_0 \cos (\theta_0) + X]^2 + [A_0 \sin (\theta_0) + Y]^2\}^{1/2} \quad (10.6\text{-}3a)$$

$$\Theta = T_2(X, Y) = \tan^{-1}\left[\frac{A_0 \sin (\theta_0) + Y}{A_0 \cos (\theta_0) + X}\right] \quad (10.6\text{-}3b)$$

Inverse transformations are:

$$X = T_1^{-1}(R, \Theta) = R \cos (\Theta) - A_0 \cos (\theta_0) \quad (10.6\text{-}4a)$$

$$Y = T_2^{-1}(R, \Theta) = R \sin (\Theta) - A_0 \sin (\theta_0) \quad (10.6\text{-}4b)$$

The functional dependence on t has been suppressed in writing (10.6-3) and (10.6-4) with the implied understanding that the quantities X, Y, R, and Θ are random variables defined from the respective processes at time t.

Probability Density of The Envelope

From (8.6-15), processes $X(t)$ and $Y(t)$ are statistically independent (at the same times t) because they are gaussian and uncorrelated. The joint density of random variables X and Y is, therefore,

$$f_{X,Y}(x, y) = \frac{e^{-(x^2+y^2)/2\sigma^2}}{2\pi\sigma^2} \tag{10.6-5}$$

From (5.4-4) the jacobian of the transformations (10.6-4) is readily found to be R. We next apply (5.4-6) to obtain the joint density of random variables R and Θ:

$$f_{R,\Theta}(r, \theta) = \frac{u(r)r}{2\pi\sigma^2} \exp\left\{-\frac{1}{2\sigma^2}\left[r^2 - 2rA_0 \cos(\theta - \theta_0) + A_0^2\right]\right\} \tag{10.6-6}$$

The density of R alone is obtained by integrating over all values of Θ:

$$f_R(r) = \int_0^{2\pi} f_{R,\Theta}(r, \theta)\, d\theta$$

$$= \frac{u(r)r}{\sigma^2} e^{-(r^2+A_0^2)/2\sigma^2} \frac{1}{2\pi} \int_0^{2\pi} e^{rA_0 \cos(\theta-\theta_0)/\sigma^2}\, d\theta \tag{10.6-7}$$

The integral is known to equal the modified Bessel function of order zero

$$I_0(\beta) = \frac{1}{2\pi} \int_0^{2\pi} e^{\beta \cos(\theta)}\, d\theta \tag{10.6-8}$$

Thus,

$$f_R(r) = \frac{u(r)}{\sigma^2}\, rI_0\left(\frac{rA_0}{\sigma^2}\right) e^{-(r^2+A_0^2)/2\sigma^2} \tag{10.6-9}$$

which is known as the *Rice* probability density.

Equation (10.6-9) is our principal result; it is the density of the envelope $R(t)$ at any time t. Figure 10.6-1 illustrates the behavior of (10.6-9). For $A_0/\sigma = 0$, the case of no signal, the density is Rayleigh. For A_0/σ large the density becomes gaussian. To show this last fact we note that

$$I_0(\beta) \approx \frac{e^\beta}{\sqrt{2\pi\beta}} \qquad \beta \gg 1 \tag{10.6-10}$$

so for rA_0/σ^2 large

$$f_R(r) \approx u(r)\sqrt{\frac{r}{2\pi A_0 \sigma^2}} \exp\left[\frac{-(r-A_0)^2}{2\sigma^2}\right] \tag{10.6-11}$$

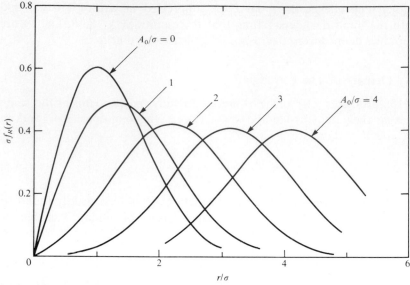

Figure 10.6-1 Probability densities of the envelope of a sinusoidal signal (amplitude A_0) plus noise (power σ^2) for various ratios A_0/σ.

This function peaks for r near A_0, and since $A_0 \gg \sigma$, the most significant values of r exist only near A_0. Therefore, with $r \approx A_0$ (10.6-11) becomes

$$f_R(r) \approx \frac{e^{-(r-A_0)^2/2\sigma^2}}{\sqrt{2\pi\sigma^2}} \tag{10.6-12}$$

which is a gaussian function with mean A_0 and variance σ^2.

Although difficult to derive, the mean and variance of R as found from (10.6-9) are known (Appendix F).

Probability Density of Phase

The density of the phase Θ of (10.6-2) derives by integrating (10.6-6) over all values of R. We shall leave the detailed steps for the reader as an exercise (Problem 10-16). The procedure is to first complete the square in r in the exponent, and, after a suitable variable change, integrate the sum of two terms. The result becomes (Middleton, 1960, p. 417)

$$\begin{aligned}
f_\Theta(\theta) = (1/2\pi) \exp\left(-A_0^2/2\sigma^2\right) \\
+ \frac{A_0 \cos(\theta - \theta_0)}{\sqrt{2\pi}\,\sigma} \exp\left[\frac{-A_0^2 \sin^2(\theta - \theta_0)}{2\sigma^2}\right] \\
\cdot F\left[\frac{A_0 \cos(\theta - \theta_0)}{\sigma}\right]
\end{aligned} \tag{10.6-13}$$

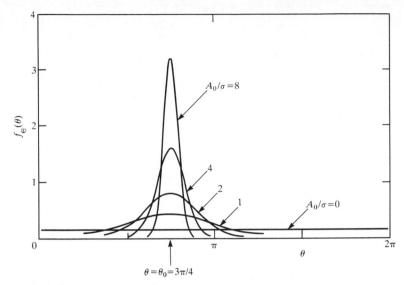

Figure 10.6-2 Probability density function of the phase of the sum of a sinusoidal signal and gaussian noise. Curves are plotted for a signal phase of $\theta_0 = 3\pi/4$.

where the function $F(\cdot)$ is given by (B-3). Figure 10.6-2 illustrates the behavior of $f_\Theta(\theta)$ for various values of A_0/σ when $\theta_0 = 3\pi/4$.

For noise only, which is the case of $A_0/\sigma = 0$, Figure 10.6-2 shows that the density of Θ is uniform on $(0, 2\pi)$. As A_0/σ becomes large the density approaches an impulse function located at the signal's phase (at $\theta = \theta_0$). Thus,

$$\lim_{A_0/\sigma \to \infty} [f_\Theta(\theta)] = \delta(\theta - \theta_0) \tag{10.6-14}$$

(Problem 10-17).

10.7 RADAR DETECTION USING A SINGLE OBSERVATION

Radar can be used to detect the presence (and distance) of a nearby object (called the radar *target*). A representative problem might be to detect the presence of an aircraft approaching an airport. Here the airport's radar radiates a pulse of radio frequency (RF) energy. The pulse propagates outward until it strikes the target (aircraft), whereupon some of the energy is reflected back toward the radar. The target's presence can be detected at the radar simply by detecting the presence of the reflected RF pulse. Once the received pulse is detected the delay between the time of the radiated pulse and the received pulse is proportional to the target's distance from the radar. After a sufficient time interval (called the *pulse repetition frequency*, or PRF, *interval*, chosen for the most distant detection of interest) the radar transmits another RF pulse and the entire "detection" process is repeated.

Figure 10.7-1 Simple radar detection network.

A straightforward implementation within the radar receiver to achieve detection is depicted in Figure 10.7-1. During any PRF interval noise is always being received (mainly due to the radar's own self-generated noise). A reflected pulse is received with this noise only when a target is present. The envelope detector produces an output $W(t)$ that is some monotonic function $g(\cdot)$ of the envelope $R(t)$ of the received signal-plus-noise waveform. The first-order probability density function of $R(t)$ was developed in the preceding Section 10.6. On the average $R(t)$, and therefore $W(t)$, with a target present will be larger than $R(t)$ when only noise is being received. A suitable detection logic compares $W(t)$ to a *threshold* W_T; if $W(t) > W_T$ the receiver decides that a target is present; if $W(t) \le W_T$ it assumes only noise is being received. These tests amount to determining when $D > 0$ in Figure 10.7-1; when $D > 0$ a target is declared to be present.

On the *average* the detection logic is valid. On any one PRF interval, however, it is possible for the receiver to make mistakes. For example, if no target is truly present it may occur that noise could become large enough at some time to make $W(t)$ exceed W_T and cause a false detection; this type of detection is called a *false alarm*. The probability of a false alarm, denoted by P_{fa}, is

$$P_{fa} = \int_{W_T}^{\infty} f_0(w) \, dw \tag{10.7-1}$$

where $f_0(w)$ is the probability density of $W(t)$ given that there is no target present. Generally, a radar wants P_{fa} to be small.

Another type of error occurs when a target is actually present but noise is such as to cancel its effect during the signal's duration and force $W(t) < W_T$. The radar usually is designed such that the probability of this event, called the *probability of a miss*, is small; it equals one minus the *detection* probability, denoted by P_d, given by

$$P_d = \int_{W_T}^{\infty} f_1(w) \, dw \tag{10.7-2}$$

Here $f_1(w)$ is the probability density of $W(t)$ when a target is present.

In most radars P_d and P_{fa} are parameters of greatest importance. W_T is usually chosen to give a prescribed value of P_{fa}. P_d then depends on the amplitude of the target's returned signal. In this section we shall develop expressions for P_{fa} and P_d when the radar makes detection decisions based on a single observation (uses only one PRF interval). Our results can be extended to multiple observations but the details are complicated and we only refer the reader to the literature (Difranco and Rubin, 1968).

False Alarm Probability and Threshold

When there is no target only noise is present at the input to the envelope detector. From (10.6-9) the density of the envelope of the noise is

$$f_R(r) = \frac{u(r)}{\sigma^2} \, r e^{-r^2/2\sigma^2} \tag{10.7-3}$$

where σ^2 is the power in the input noise. Because the detector characteristic $g(R)$ is assumed monotonic, there is an *equivalent threshold* R_T on R that is related to W_T by

$$W_T = g(R_T) \tag{10.7-4}$$

$$R_T = g^{-1}(W_T) \tag{10.7-5}$$

where $g^{-1}(\cdot)$ is the inverse function of $g(\cdot)$. We may then compute P_{fa} from the envelope as follows:

$$P_{fa} = \int_{W_T}^{\infty} f_0(w) \, dw = \int_{R_T}^{\infty} f_R(r) \, dr$$

$$= \int_{R_T}^{\infty} \frac{r}{\sigma^2} e^{-r^2/2\sigma^2} \, dr = e^{-R_T^2/2\sigma^2} \tag{10.7-6}$$

Thus,

$$R_T = \left\{ 2\sigma^2 \ln \left(\frac{1}{P_{fa}} \right) \right\}^{1/2} \tag{10.7-7}$$

and

$$W_T = g\left[\left\{ 2\sigma^2 \ln \left(\frac{1}{P_{fa}} \right) \right\}^{1/2} \right] \tag{10.7-8}$$

where $\ln (\cdot)$ represents the natural logarithm.

Equation (10.7-8) gives the threshold W_T that is to be used to realize a specified value of P_{fa} when the noise power level is σ^2 at the detector's input.

Example 10.7-1 A radar receiver uses a square-law envelope detector defined by $W = 3R^2$. We find what threshold is required when noise power at the detector's input is $\sigma^2 = 0.025$ W and $P_{fa} = 10^{-6}$ is required. From (10.7-8)

$$W_T = 3\left[2(0.025) \ln \left(\frac{1}{10^{-6}} \right) \right] \approx 2.07 \text{ V}$$

Detection Probability

When a target signal is present the density of the received waveform's envelope is given by (10.6-9). Again using the idea of an equivalent threshold R_T on the envelope R we expand (10.7-2) to get

$$P_d = \int_{W_T}^{\infty} f_1(w)\,dw = \int_{R_T}^{\infty} f_R(r)\,dr$$

$$= \int_{\sqrt{2\sigma^2 \ln (1/P_{fa})}}^{\infty} \frac{r}{\sigma^2} I_0\!\left(\frac{rA_0}{\sigma^2}\right) e^{-(r^2 + A_0{}^2)/2\sigma^2}\,dr$$

$$= Q\!\left[\sqrt{\frac{A_0^2}{\sigma^2}},\ \sqrt{2 \ln\left(\frac{1}{P_{fa}}\right)}\right] \tag{10.7-9}$$

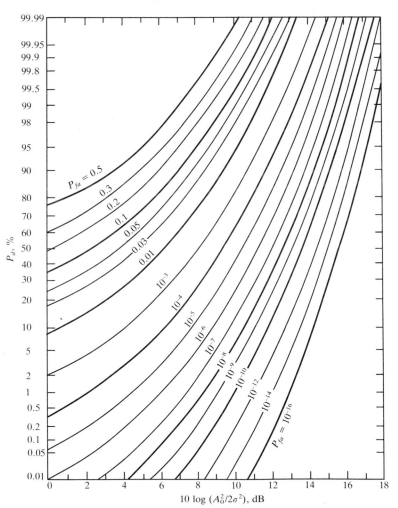

Figure 10.7-2 Radar detection probabilities for various false alarm probabilities when detection is based on a single observation. [*Adapted from Barton (1964) with permission.*]

where

$$Q(\alpha, \beta) = \int_{\beta}^{\infty} \xi I_0(\alpha\xi) e^{-(\xi^2 + \alpha^2)/2} \, d\xi \qquad (10.7\text{-}10)$$

is called Marcum's Q-function (Marcum, 1950, 1960). Figure 10.7-2 illustrates P_d for various values of $A_0^2/2\sigma^2$ with P_{fa} as a parameter. Generally, the smaller P_{fa} is required to be the larger is the necessary signal strength to achieve a given value of P_d.

When P_{fa} is small while P_d is relatively large so that the threshold W_T is large and signal strength is relatively large, the approximation of (10.6-12) can be used in (10.7-9) to obtain

$$P_d \approx F\left[\frac{A_0}{\sigma} - \sqrt{2 \ln\left(\frac{1}{P_{fa}}\right)}\right] \qquad (10.7\text{-}11)$$

where $F(\cdot)$ is given by (B-3).

Example 10.7-2 We find the value of P_d in a receiver having $P_{fa} = 10^{-10}$ when the received signal-to-noise power ratio at the detector's input is 16.0 dB. Here $A_0^2/2\sigma^2 = 39.811$ (16 dB). Thus, $(A_0/\sigma) - \sqrt{2 \ln (1/P_{fa})} \approx 2.137$. From Table B-1 and (10.7-11), $P_d \approx F(2.137) \approx 0.9837$ or 98.37%, which is in agreement with Figure 10.7-2.

PROBLEMS

10-1 Show that (a) the time-averaged autocorrelation function of $S_{AM}(t)$, as given by (10.1-2) is

$$R_{AM}(\tau) = \frac{1}{2}[A_0^2 + R_{XX}(\tau)] \cos (\omega_0 \tau)$$

if $X(t)$ is a zero-mean process, and (b) the power spectrum is

$$S_{AM}(\omega) = \frac{A_0^2 \pi}{2} [\delta(\omega - \omega_0) + \delta(\omega + \omega_0)] + \frac{1}{4}[S_{XX}(\omega - \omega_0) + S_{XX}(\omega + \omega_0)]$$

where $S_{XX}(\omega)$ is the power spectrum of $X(t)$.

10-2 Define transmitter efficiency η_{AM} in an amplitude modulation communication system as the ratio of transmitted power due to the message to the total power. For a zero-mean stationary random message show that

$$\eta_{AM} = \frac{R_{XX}(0)}{A_0^2 + R_{XX}(0)} = \frac{\int_{-\infty}^{\infty} S_{XX}(\omega) \, d\omega}{2\pi A_0^2 + \int_{-\infty}^{\infty} S_{XX}(\omega) \, d\omega} = \frac{\overline{X^2(t)}}{A_0^2 + \overline{X^2(t)}}$$

where $R_{XX}(\tau)$ and $S_{XX}(\omega)$ are the autocorrelation function and power spectrum, respectively, of the message $X(t)$.

10-3 Define *crest factor* K_{cr} for a zero-mean, bounded, random signal by

$$K_{cr}^2 = |X(t)|_{max}^2 / \overline{X^2(t)}$$

If no *overmodulation* is to occur, such that $|X(t)|_{max} \le A_0$ in the transmitted signal of an amplitude modulation system, show that the transmitter efficiency (Problem 10-2) is

$$\eta_{AM} \le \frac{1}{1 + K_{cr}^2}$$

What is the maximum efficiency for a message $X(t) = A_m \cos(\omega_m t + \Theta_m)$, where A_m and ω_m are constants while Θ_m is a random variable uniform on $(0, 2\pi)$?

10-4 Use (10.1-3), (10.1-4), and (10.1-14) to show that the input signal-to-noise power ratio at the envelope detector of Figure 10.1-1 is

$$\left(\frac{S_i}{N_i}\right)_{AM} = \frac{\overline{S_R^2(t)}}{\overline{N^2(t)}} = \frac{\pi G_{ch}^2 [A_0^2 + \overline{X^2(t)}]}{\mathcal{N}_0 W_{rec}}$$

Use this result to show that (10.1-15) can be written in the form

$$\left(\frac{S_o}{N_o}\right)_{AM} = 2\eta_{AM}\left(\frac{S_i}{N_i}\right)_{AM}$$

where η_{AM} is defined in Problem 10-2.

10-5 In an AM broadcast system the total average transmitted power is 1 kW. The channel gain is $G_{ch} = 3\sqrt{2}(10^{-3})$. Average noise power at the envelope detector's input is 10^{-5} W and the output signal-to-noise power ratio of the receiver is 180 (or 22.55 dB).

(a) What is the average signal power at the input to the envelope detector?

(b) Find $(S_i/N_i)_{AM}$.

(c) What is the transmitter's efficiency?

(*Hint:* Use results of Problem 10-4.)

10-6 When the message in an FM system is a sinusoid, such as $x(t) = A_m \cos(\omega_m t)$ where $A_m > 0$ and ω_m are constants, *modulation index* β_{FM} is defined by $\beta_{FM} = \Delta\omega/\omega_m$.

(a) Write an expression for the instantaneous frequency (rad/s) of the FM waveform in terms of β_{FM}.

(b) What is the approximate bandwidth of the FM signal in terms of β_{FM} if $\Delta\omega$ is large relative to ω_m?

(c) For the specific waveform $x(t) = 0.1 \cos(10^3 t)$, what are β_{FM} and the transmitter's constant k_{FM} if the approximate bandwidth is to be 200 kHz?

10-7 Find an expression for the autocorrelation function of $S_{FM}(t)$, as given by (10.2-1), when $X(t)$ is a gaussian, zero-mean process. Formulate the expression in terms of the correlation coefficient and variance of the process

$$\Gamma(t) = k_{FM} \int X(t) \, dt$$

[*Hint:* Note that the expectation involving $X(t)$ leads to a characteristic function.]

10-8 In an FM system the transmitted signal has 10 kW of average power and a bandwidth of approximately 150 kHz when a random message with a crest factor of 4 is used (Problem 10-3). The signal passes over a channel for which $G_{ch} = 10^{-6}$ and $\mathcal{N}_0/2 = 5(10^{-15})/3$.

(*a*) Find the signal and noise average powers and the signal-to-noise ratio at the receiver's input.

(*b*) What is the message's spectral extent if the output signal-to-noise power ratio of the receiver is found to be 25,000?

10-9 Let $H_1(\omega) = K_1 W_1/(W_1 + j\omega)$ and $H_2(\omega) = 1/j\omega$ in (10.3-3) where $K_1 > 0$ and $W_1 > 0$ are constants.

(*a*) Are there any values of K_1 and/or W_1 that will make the loop of Figure 10.3-1 unstable?

(*b*) If $W_1 = 200$ and $K_1 = 40$ find the loop's output noise power if white noise of power density $\mathcal{N}_0/2 = 10^{-4}$ W/Hz is applied at the input. (*Hint:* Use the integral given in Problem 9-40.)

10-10 Show that the transfer function of the control system of Figure 10.3-1 is the same as the Wiener filter of (9.2-20) if

$$H_1(\omega) = \left[\frac{S_{XX}(\omega)}{S_{NN}(\omega)}\right] e^{j\omega t_o}$$

and

$$H_2(\omega) = e^{-j\omega t_o}$$

*__10-11__ Show that the sum of an FM waveform plus noise as given by (10.4-18) can be written in the form

$$R(t) \cos\left[\omega_0 t + \theta_0 + \theta_{FM}(t) + \theta_N(t)\right]$$

where

$$\theta_{FM}(t) = k_{FM} \int X(t)\, dt$$

and

$$R(t) = \langle\{N_c(t) + A_i \cos\left[\theta_0 + \theta_{FM}(t)\right]\}^2 + \{N_s(t) + A_i \sin\left[\theta_0 + \theta_{FM}(t)\right]\}^2\rangle^{1/2}$$

$$\theta_N(t) = \tan^{-1}\left\{\frac{N_s(t)\cos\left[\theta_0 + \theta_{FM}(t)\right] - N_c(t)\sin\left[\theta_0 + \theta_{FM}(t)\right]}{A_i + N_c(t)\cos\left[\theta_0 + \theta_{FM}(t)\right] + N_s(t)\sin\left[\theta_0 + \theta_{FM}(t)\right]}\right\}$$

*__10-12__ Assume the bandpass noise $N_i(t)$ in Problem 10-11 is wide-sense stationary and gaussian and note that if $|A_i| \gg |N_c(t)|$ and $|A_i| \gg |N_s(t)|$ *most of the time,* then

$$\theta_N(t) \approx \frac{N_s(t)}{A_i} \cos\left[\theta_0 + \theta_{FM}(t)\right] - \frac{N_c(t)}{A_i} \sin\left[\theta_0 + \theta_{FM}(t)\right]$$

(a) Show that the autocorrelation function of the process $\Theta_N(t)$, for which $\theta_N(t)$ is a sample function, is

$$R_{\Theta_N\Theta_N}(t, t + \tau) = \frac{1}{A_i^2} R_{N_cN_c}(\tau)E\left[\cos\left\{k_{FM}\int_t^{t+\tau} X(\xi)\,d\xi\right\}\right]$$
$$+ \frac{1}{A_i^2} R_{N_cN_s}(\tau)E\left[\sin\left\{k_{FM}\int_t^{t+\tau} X(\xi)\,d\xi\right\}\right]$$

where $R_{N_cN_c}(\tau)$ and $R_{N_cN_s}(\tau)$ are the correlation functions of $N_c(t)$ and $N_s(t)$, and the expectations are with respect to the message process $X(t)$ assumed statistically independent of the noises. (*Hint*: Use the results of Section 8.6.)

(b) If noises $N_c(t)$ and $N_s(t)$ are broadband relative to the FM signal, justify that

$$R_{\Theta_N\Theta_N}(t, t + \tau) \approx \frac{1}{A_i^2} R_{N_cN_c}(\tau) = R_{\Theta_N\Theta_N}(\tau)$$

(c) If the message process varies slowly enough for values of τ that are important to $R_{N_cN_c}(\tau)$ such that

$$k_{FM}\int_t^{t+\tau} X(\xi)\,d\xi \approx k_{FM}X(t)\tau$$

is valid, show that the expression of part (a) reduces to

$$R_{\Theta_N\Theta_N}(t, t + \tau) \approx \frac{1}{A_i^2}\exp\left[\frac{-\sigma_X^2 k_{FM}^2 \tau^2}{2}\right] R_{N_cN_c}(\tau)$$

if $X(t)$ is a zero-mean, wide-sense stationary gaussian message of power σ_X^2. (*Hint*: Make use of characteristic functions.)

10-13 In Example 10.4-1 let $\zeta = \frac{1}{2}$ instead of $1/\sqrt{2}$ and recompute the loop's output noise power N_o. Compare the result with that of (10.4-25). Is there any improvement over the case where $\zeta = 1/\sqrt{2}$? (*Hint*: Make use of the integral given in Problem 9-40.)

10-14 Assume white noise is added to an FM signal and the sum is applied to a phase-locked loop for message recovery. Thus, $S_{N_cN_c}(\omega) = \mathcal{N}_0$ in (10.4-23).

(a) If

$$H_L(\omega) = \frac{W_1 W_3(W_2 + j\omega)}{W_2(W_1 + j\omega)(W_3 + j\omega)}$$

where W_1, W_2, and W_3 are positive constants, find an expression for the power contained in the noise part of (10.4-23).

(b) Assume the loop is designed so that $W_3 = 2\omega_0$, $W_2 = \omega_0/5$, and $W_1 = \omega_0^2/5K$, where $K = K_P k_V$ and ω_0 is called the loop's *crossover frequency* (rad/s); it is the frequency where $|KH_L(\omega)/j\omega| = 1$. Evaluate the result found in part (a) when ω_0 equals the message's spectral extent W_X.

(c) If K is very large, to what does the evaluation of part (b) approach? [*Hint*: Use the known integral

$$I_3 = \frac{1}{2\pi} \int_{-\infty}^{\infty} \frac{(b_0 \omega^4 - b_1 \omega^2 + b_2)\, d\omega}{a_0^2 \omega^6 + (a_1^2 - 2a_0 a_2)\omega^4 + (a_2^2 - 2a_1 a_3)\omega^2 + a_3^2}$$

$$= \frac{a_0 b_1 - a_2 b_0 - (a_0 a_1 b_2 / a_3)}{2a_0(a_0 a_3 - a_1 a_2)}$$

where a_0, a_1, a_2, a_3, b_0, b_1, and b_2 are constants and $a_0 \lambda^3 + a_1 \lambda^2 + a_2 \lambda + a_3$ has no roots in the lower half-plane when $\lambda = \omega + j\sigma$ (Thomas, 1969, p. 249).]

10-15 A sample function of a semirandom binary process is to be passed through a lowpass filter with transfer function $H(\omega) = W_L/(W_L + j\omega)$ where W_L is its 3-dB bandwidth. If the rise and fall times of the pulses in the output waveform are not to exceed 5% of the pulse duration T_b what minimum value of W_L is required? (*Hint*: Assume the input waveform has been at level $-A$ for many pulse intervals and suddenly makes a transition to level A; determine rise time as that required for the output to rise from $-A$ to $0.9A$.)

*__10-16__ Carry out the steps suggested in the text and show that (10.6-13) derives from (10.6-6).

*__10-17__ If $A_0/\sigma \to \infty$ in (10.6-13) show that (10.6-14) is true.

10-18 A radar receiver uses a linear envelope detector where $W = R$. Find an expression for false alarm probability P_{fa} in terms of W_T, the threshold voltage level.

10-19 Work Problem 10-18 for a square-law detector defined by $W = KR^2$, where $K > 0$ is a constant.

10-20 A radar uses a linear envelope detector defined by $W = R/4$. The threshold voltage is $W_T = 0.7$ volt. Measurements show that $P_{fa} = 4(10^{-7})$. What is the noise power at the envelope detector's input?

10-21 Work Problem 10-20 for a square-law detector with characteristic $W = R^2/4$.

10-22 False alarm probability is 10^{-8} in a radar that must have a detection probability of 0.9901. When target is present what signal-to-noise power ratio is necessary at the envelope detector's input? [*Hint*: Assume (10.7-11) applies.]

*__10-23__ A radar receiver as shown in Figure 10.7-1 uses a square-law detector defined by $W = KR^2$ where $K > 0$ is a constant. Find an expression for the probability density of W.

*__10-24__ A radar receiver uses a binary detection logic based on observing N PRF intervals (multiple observations). If the observations in the N intervals are statistically independent and the detection and false alarm probabilities on any one observation are P_{d1} and P_{fa1}, respectively, find P_d and P_{fa} that correspond to an overall detection logic based on obtaining at least n detections in N intervals.

A

REVIEW OF THE IMPULSE FUNCTION

There are several ways of defining what is known as the impulse function (Papoulis, 1962) denoted $\delta(x)$. The most mathematically sound approach is to define $\delta(x)$ on the basis of its integral property. If $\phi(x)$ is any arbitrary function of x,† $x_1 < x_2$ are two values of x, and x_0 is the point of "occurrence" of the impulse, then $\delta(x)$ satisfies (Korn and Korn, 1961, p. 742)

$$\int_{x_1}^{x_2} \phi(x)\delta(x - x_0)\,dx = \begin{cases} 0 & x_2 < x_0 \quad \text{or} \quad x_0 < x_1 \\[2mm] \dfrac{1}{2}\left[\phi(x_0^+) + \phi(x_0^-)\right] & x_1 < x_0 < x_2 \\[2mm] \dfrac{1}{2}\,\phi(x_0^+) & x_0 = x_1 \\[2mm] \dfrac{1}{2}\,\phi(x_0^-) & x_0 = x_2 \end{cases} \qquad \text{(A-1)}$$

It can be shown, using (A-1), that $\delta(x)$ behaves as a function having even symmetry, an area of unity, a vanishingly small "duration," and an infinite "amplitude" (Peebles, 1976, pp. 34–35).

---

† The function is also assumed to have *bounded variation* in the neighborhood of $x = x_0$ (see footnote, page 321).

A simpler form of (A-1) is often applicable to many practical situations. If $x_1 = -\infty$, $x_2 = \infty$, and $\phi(x)$ is arbitrary except that it is continuous at $x = x_0$, then

$$\int_{-\infty}^{\infty} \phi(x)\delta(x - x_0) \, dx = \phi(x_0) \tag{A-2}$$

A useful fact that is easily obtained from (A-1) is

$$\int_{-\infty}^{x} \delta(\xi - x_0) \, d\xi = u(x - x_0) \tag{A-3}$$

or, equivalently

$$\frac{du(x)}{dx} = \delta(x) \tag{A-4}$$

where $u(x)$ is the unit-step function defined by

$$u(x) = \begin{cases} 1 & 0 < x \\ 0 & x < 0 \end{cases} \tag{A-5}$$

The impulse function can be generalized to N-dimensional space (Korn and Korn, 1961, p. 745). If we assume a cartesian coordinate system with axes ξ_1, ξ_2, ..., ξ_N, and a function $\phi(\xi_1, \xi_2, \ldots, \xi_N)$ that is continuous at the point ($\xi_1 = x_1$, $\xi_2 = x_2, \ldots, \xi_N = x_N$), then an *N-dimensional impulse function* $\delta(\xi_1, \xi_2, \ldots, \xi_N)$ is defined by

$$\int_{-\infty}^{\infty} \cdots \int_{-\infty}^{\infty} \phi(\xi_1, \xi_2, \ldots, \xi_N)\delta(\xi_1 - x_1, \xi_2 - x_2, \ldots, \xi_N - x_N) \, d\xi_1 \cdots d\xi_N$$

$$= \phi(x_1, x_2, \ldots, x_N) \tag{A-6}$$

Of special interest is the two-dimensional case; it is known that $\delta(\xi_1, \xi_2)$ can be written as (Bracewell, 1965, p. 85)

$$\delta(\xi_1, \xi_2) = \delta(\xi_1)\delta(\xi_2) \tag{A-7}$$

so (A-6) becomes

$$\int_{-\infty}^{\infty} \int_{-\infty}^{\infty} \phi(\xi_1, \xi_2)\delta(\xi_1 - x_1)\delta(\xi_2 - x_2) \, d\xi_1 \, d\xi_2 = \phi(x_1, x_2) \tag{A-8}$$

By using (A-7) with an appropriate choice of $\phi(\xi_1, \xi_2)$ we readily find that, for $N = 2$, (A-6) can be written as

$$\int_{-\infty}^{y} \int_{-\infty}^{x} \delta(\xi_1 - x_0, \xi_2 - y_0) \, d\xi_1 \, d\xi_2$$

$$= \int_{-\infty}^{y} \delta(\xi_2 - y_0) \, d\xi_2 \int_{-\infty}^{x} \delta(\xi_1 - x_0) \, d\xi_1$$

$$= u(x - x_0)u(y - y_0) \tag{A-9}$$

If $u(x - x_0)u(y - y_0)$ is interpreted as a *two-dimensional unit-step function* $u(x - x_0, y - y_0)$, we have

$$\frac{\partial^2 u(x - x_0, y - y_0)}{\partial x \, \partial y} = \delta(x - x_0, y - y_0) \tag{A-10}$$

where

$$\delta(x - x_0, y - y_0) = \delta(x - x_0)\delta(y - y_0) \tag{A-11}$$

$$u(x - x_0, y - y_0) = u(x - x_0)u(y - y_0) \tag{A-12}$$

GAUSSIAN DISTRIBUTION FUNCTION

The general gaussian or normal probability density and distribution functions are:

$$f_X(x) = \frac{1}{\sqrt{2\pi\sigma_X^2}} \, e^{-(x-a_X)^2/2\sigma_X^2} \tag{B-1}$$

$$F_X(x) = \int_{-\infty}^{x} f_X(\xi) \, d\xi = F\left(\frac{x - a_X}{\sigma_X}\right) \tag{B-2}$$

where $-\infty < a_X < \infty$, $0 < \sigma_X$ are constants and $F(\cdot)$ is the "normalized" distribution function for $a_X = 0$ and $\sigma_X = 1$; that is

$$F(x) = \int_{-\infty}^{x} \frac{1}{\sqrt{2\pi}} \, e^{-\xi^2/2} \, d\xi \tag{B-3}$$

$F(x)$ is listed in the following table. When $a_X \neq 0$ and $\sigma_X \neq 1$, $F_X(x)$ can be found from $F(x)$ by use of (B-2). For negative values of x, use

$$F(-x) = 1 - F(x) \tag{B-4}$$

Table B-1 Values of $F(x)$ for $0 \leq x \leq 3.89$ in steps of 0.01

x	.00	.01	.02	.03	.04	.05	.06	.07	.08	.09
0.0	.5000	.5040	.5080	.5120	.5160	.5199	.5239	.5279	.5319	.5359
0.1	.5398	.5438	.5478	.5517	.5557	.5596	.5636	.5675	.5714	.5753
0.2	.5793	.5832	.5871	.5910	.5948	.5987	.6026	.6064	.6103	.6141
0.3	.6179	.6217	.6255	.6293	.6331	.6368	.6406	.6443	.6480	.6517
0.4	.6554	.6591	.6628	.6664	.6700	.6736	.6772	.6808	.6844	.6879
0.5	.6915	.6950	.6985	.7019	.7054	.7088	.7123	.7157	.7190	.7224
0.6	.7257	.7291	.7324	.7357	.7389	.7422	.7454	.7486	.7517	.7549
0.7	.7580	.7611	.7642	.7673	.7704	.7734	.7764	.7794	.7823	.7852
0.8	.7881	.7910	.7939	.7967	.7995	.8023	.8051	.8078	.8106	.8133
0.9	.8159	.8186	.8212	.8238	.8264	.8289	.8315	.8340	.8365	.8389
1.0	.8413	.8438	.8461	.8485	.8508	.8531	.8554	.8577	.8599	.8621
1.1	.8643	.8665	.8686	.8708	.8729	.8749	.8770	.8790	.8810	.8830
1.2	.8849	.8869	.8888	.8907	.8925	.8944	.8962	.8980	.8997	.9015
1.3	.9032	.9049	.9066	.9082	.9099	.9115	.9131	.9147	.9162	.9177
1.4	.9192	.9207	.9222	.9236	.9251	.9265	.9279	.9292	.9306	.9319
1.5	.9332	.9345	.9357	.9370	.9382	.9394	.9406	.9418	.9429	.9441
1.6	.9452	.9463	.9474	.9484	.9495	.9505	.9515	.9525	.9535	.9545
1.7	.9554	.9564	.9573	.9582	.9591	.9599	.9608	.9616	.9625	.9633
1.8	.9641	.9649	.9656	.9664	.9671	.9678	.9686	.9693	.9699	.9706
1.9	.9713	.9719	.9726	.9732	.9738	.9744	.9750	.9756	.9761	.9767
2.0	.9772	.9778	.9783	.9788	.9793	.9798	.9803	.9808	.9812	.9817
2.1	.9821	.9826	.9830	.9834	.9838	.9842	.9846	.9850	.9854	.9857
2.2	.9861	.9864	.9868	.9871	.9875	.9878	.9881	.9884	.9887	.9890
2.3	.9893	.9896	.9898	.9901	.9904	.9906	.9909	.9911	.9913	.9916
2.4	.9918	.9920	.9922	.9925	.9927	.9929	.9931	.9932	.9934	.9936
2.5	.9938	.9940	.9941	.9943	.9945	.9946	.9948	.9949	.9951	.9952
2.6	.9953	.9955	.9956	.9957	.9959	.9960	.9961	.9962	.9963	.9964
2.7	.9965	.9966	.9967	.9968	.9969	.9970	.9971	.9972	.9973	.9974
2.8	.9974	.9975	.9976	.9977	.9977	.9978	.9979	.9979	.9980	.9981
2.9	.9981	.9982	.9982	.9983	.9984	.9984	.9985	.9985	.9986	.9986
3.0	.9987	.9987	.9987	.9988	.9988	.9989	.9989	.9989	.9990	.9990
3.1	.9990	.9991	.9991	.9991	.9992	.9992	.9992	.9992	.9993	.9993
3.2	.9993	.9993	.9994	.9994	.9994	.9994	.9994	.9995	.9995	.9995
3.3	.9995	.9995	.9996	.9996	.9996	.9996	.9996	.9996	.9996	.9997
3.4	.9997	.9997	.9997	.9997	.9997	.9997	.9997	.9997	.9998	.9998
3.5	.9998	.9998	.9998	.9998	.9998	.9998	.9998	.9998	.9998	.9998
3.6	.9998	.9999	.9999	.9999	.9999	.9999	.9999	.9999	.9999	.9999
3.7	.9999	.9999	.9999	.9999	.9999	.9999	.9999	.9999	.9999	.9999
3.8	.9999	.9999	.9999	.9999	.9999	.9999	.9999	1.0000	1.0000	1.0000

USEFUL MATHEMATICAL QUANTITIES

TRIGONOMETRIC IDENTITIES

$$\cos (x \pm y) = \cos (x) \cos (y) \mp \sin (x) \sin (y) \qquad \text{(C-1)}$$

$$\sin (x \pm y) = \sin (x) \cos (y) \pm \cos (x) \sin (y) \qquad \text{(C-2)}$$

$$\cos \left(x \pm \frac{\pi}{2} \right) = \mp \sin (x) \qquad \text{(C-3)}$$

$$\sin \left(x \pm \frac{\pi}{2} \right) = \pm \cos (x) \qquad \text{(C-4)}$$

$$\cos (2x) = \cos^2 (x) - \sin^2 (x) \qquad \text{(C-5)}$$

$$\sin (2x) = 2 \sin (x) \cos (x) \qquad \text{(C-6)}$$

$$2 \cos (x) = e^{jx} + e^{-jx} \qquad \text{(C-7)}$$

$$2j \sin (x) = e^{jx} - e^{-jx} \qquad \text{(C-8)}$$

$$2 \cos (x) \cos (y) = \cos (x - y) + \cos (x + y) \qquad \text{(C-9)}$$

$$2 \sin (x) \sin (y) = \cos (x - y) - \cos (x + y) \qquad \text{(C-10)}$$

$$2 \sin (x) \cos (y) = \sin (x - y) + \sin (x + y) \qquad \text{(C-11)}$$

$$2 \cos^2 (x) = 1 + \cos (2x) \qquad \text{(C-12)}$$

$$2 \sin^2 (x) = 1 - \cos (2x) \qquad \text{(C-13)}$$

$$4 \cos^3 (x) = 3 \cos (x) + \cos (3x) \qquad \text{(C-14)}$$

$$4 \sin^3 (x) = 3 \sin (x) - \sin (3x) \qquad \text{(C-15)}$$

$$8 \cos^4 (x) = 3 + 4 \cos (2x) + \cos (4x) \qquad \text{(C-16)}$$

$$8 \sin^4 (x) = 3 - 4 \cos (2x) + \cos (4x) \qquad \text{(C-17)}$$

$$A \cos (x) - B \sin (x) = R \cos (x + \theta) \qquad \text{(C-18)}$$

where

$$R = \sqrt{A^2 + B^2} \qquad \text{(C-19a)}$$

$$\theta = \tan^{-1} (B/A) \qquad \text{(C-19b)}$$

$$A = R \cos (\theta) \qquad \text{(C-19c)}$$

$$B = R \sin (\theta) \qquad \text{(C-19d)}$$

INDEFINITE INTEGRALS

Rational Algebraic Functions

$$\int (a + bx)^n \, dx = \frac{(a + bx)^{n+1}}{b(n + 1)} \qquad 0 < n \qquad \text{(C-20)}$$

$$\int \frac{dx}{a + bx} = \frac{1}{b} \ln |a + bx| \qquad \text{(C-21)}$$

$$\int \frac{dx}{(a + bx)^n} = \frac{-1}{(n - 1)b(a + bx)^{n-1}} \qquad 1 < n \qquad \text{(C-22)}$$

$$\int \frac{dx}{c + bx + ax^2} = \frac{2}{\sqrt{4ac - b^2}} \tan^{-1} \left(\frac{2ax + b}{\sqrt{4ac - b^2}} \right) \qquad b^2 < 4ac$$

$$= \frac{1}{\sqrt{b^2 - 4ac}} \ln \left| \frac{2ax + b - \sqrt{b^2 - 4ac}}{2ax + b + \sqrt{b^2 - 4ac}} \right| \qquad b^2 > 4ac$$

$$= \frac{-2}{2ax + b} \qquad b^2 = 4ac \qquad \text{(C-23)}$$

$$\int \frac{x \, dx}{c + bx + ax^2} = \frac{1}{2a} \ln |ax^2 + bx + c| - \frac{b}{2a} \int \frac{dx}{c + bx + ax^2} \qquad \text{(C-24)}$$

$$\int \frac{dx}{a^2 + b^2 x^2} = \frac{1}{ab} \tan^{-1} \left(\frac{bx}{a} \right) \qquad \text{(C-25)}$$

$$\int \frac{x \, dx}{a^2 + x^2} = \frac{1}{2} \ln (a^2 + x^2) \qquad \text{(C-26)}$$

$$\int \frac{x^2 \, dx}{a^2 + x^2} = x - a \tan^{-1} \left(\frac{x}{a} \right) \qquad \text{(C-27)}$$

$$\int \frac{dx}{(a^2 + x^2)^2} = \frac{x}{2a^2(a^2 + x^2)} + \frac{1}{2a^3} \tan^{-1}\left(\frac{x}{a}\right) \tag{C-28}$$

$$\int \frac{x\,dx}{(a^2 + x^2)^2} = \frac{-1}{2(a^2 + x^2)} \tag{C-29}$$

$$\int \frac{x^2\,dx}{(a^2 + x^2)^2} = \frac{-x}{2(a^2 + x^2)} + \frac{1}{2a} \tan^{-1}\left(\frac{x}{a}\right) \tag{C-30}$$

$$\int \frac{dx}{(a^2 + x^2)^3} = \frac{x}{4a^2(a^2 + x^2)^2} + \frac{3x}{8a^4(a^2 + x^2)} + \frac{3}{8a^5} \tan^{-1}\left(\frac{x}{a}\right) \tag{C-31}$$

$$\int \frac{x^2\,dx}{(a^2 + x^2)^3} = \frac{-x}{4(a^2 + x^2)^2} + \frac{x}{8a^2(a^2 + x^2)} + \frac{1}{8a^3} \tan^{-1}\left(\frac{x}{a}\right) \tag{C-32}$$

$$\int \frac{x^4\,dx}{(a^2 + x^2)^3} = \frac{a^2 x}{4(a^2 + x^2)^2} - \frac{5x}{8(a^2 + x^2)} + \frac{3}{8a} \tan^{-1}\left(\frac{x}{a}\right) \tag{C-33}$$

$$\int \frac{dx}{(a^2 + x^2)^4} = \frac{x}{6a^2(a^2 + x^2)^3} + \frac{5x}{24a^4(a^2 + x^2)^2} + \frac{5x}{16a^6(a^2 + x^2)} + \frac{5}{16a^7} \tan^{-1}\left(\frac{x}{a}\right) \tag{C-34}$$

$$\int \frac{x^2\,dx}{(a^2 + x^2)^4} = \frac{-x}{6(a^2 + x^2)^3} + \frac{x}{24a^2(a^2 + x^2)^2} + \frac{x}{16a^4(a^2 + x^2)} + \frac{1}{16a^5} \tan^{-1}\left(\frac{x}{a}\right) \tag{C-35}$$

$$\int \frac{x^4\,dx}{(a^2 + x^2)^4} = \frac{a^2 x}{6(a^2 + x^2)^3} - \frac{7x}{24(a^2 + x^2)^2} + \frac{x}{16a^2(a^2 + x^2)} + \frac{1}{16a^3} \tan^{-1}\left(\frac{x}{a}\right) \tag{C-36}$$

$$\int \frac{dx}{a^4 + x^4} = \frac{1}{4a^3\sqrt{2}} \ln\left(\frac{x^2 + ax\sqrt{2} + a^2}{x^2 - ax\sqrt{2} + a^2}\right) + \frac{1}{2a^3\sqrt{2}} \tan^{-1}\left(\frac{ax\sqrt{2}}{a^2 - x^2}\right) \tag{C-37}$$

$$\int \frac{x^2\,dx}{a^4 + x^4} = -\frac{1}{4a\sqrt{2}} \ln\left(\frac{x^2 + ax\sqrt{2} + a^2}{x^2 - ax\sqrt{2} + a^2}\right) + \frac{1}{2a\sqrt{2}} \tan^{-1}\left(\frac{ax\sqrt{2}}{a^2 - x^2}\right) \tag{C-38}$$

Trigonometric Functions

$$\int \cos(x)\,dx = \sin(x) \tag{C-39}$$

$$\int x\cos(x)\,dx = \cos(x) + x\sin(x) \tag{C-40}$$

$$\int x^2 \cos(x)\, dx = 2x \cos(x) + (x^2 - 2) \sin(x) \tag{C-41}$$

$$\int \sin(x)\, dx = -\cos(x) \tag{C-42}$$

$$\int x \sin(x)\, dx = \sin(x) - x \cos(x) \tag{C-43}$$

$$\int x^2 \sin(x)\, dx = 2x \sin(x) - (x^2 - 2) \cos(x) \tag{C-44}$$

Exponential Functions

$$\int e^{ax}\, dx = \frac{e^{ax}}{a} \qquad a \text{ real or complex} \tag{C-45}$$

$$\int x e^{ax}\, dx = e^{ax}\left[\frac{x}{a} - \frac{1}{a^2}\right] \qquad a \text{ real or complex} \tag{C-46}$$

$$\int x^2 e^{ax}\, dx = e^{ax}\left[\frac{x^2}{a} - \frac{2x}{a^2} + \frac{2}{a^3}\right] \qquad a \text{ real or complex} \tag{C-47}$$

$$\int x^3 e^{ax}\, dx = e^{ax}\left[\frac{x^3}{a} - \frac{3x^2}{a^2} + \frac{6x}{a^3} - \frac{6}{a^4}\right] \qquad a \text{ real or complex} \tag{C-48}$$

$$\int e^{ax} \sin(x)\, dx = \frac{e^{ax}}{a^2 + 1}[a \sin(x) - \cos(x)] \tag{C-49}$$

$$\int e^{ax} \cos(x)\, dx = \frac{e^{ax}}{a^2 + 1}[a \cos(x) + \sin(x)] \tag{C-50}$$

DEFINITE INTEGRALS

$$\int_{-\infty}^{\infty} e^{-a^2 x^2 + bx}\, dx = \frac{\sqrt{\pi}}{a}\, e^{b^2/(4a^2)} \qquad a > 0 \tag{C-51}$$

$$\int_0^{\infty} x^2 e^{-x^2}\, dx = \sqrt{\pi}/4 \tag{C-52}$$

$$\int_0^{\infty} \text{Sa}(x)\, dx = \int_0^{\infty} \frac{\sin(x)}{x}\, dx = \frac{\pi}{2} \tag{C-53}$$

$$\int_0^{\infty} \text{Sa}^2(x)\, dx = \pi/2 \tag{C-54}$$

FINITE SERIES

$$\sum_{n=1}^{N} n = \frac{N(N+1)}{2} \tag{C-55}$$

$$\sum_{n=1}^{N} n^2 = \frac{N(N+1)(2N+1)}{6} \tag{C-56}$$

$$\sum_{n=1}^{N} n^3 = \frac{N^2(N+1)^2}{4} \tag{C-57}$$

$$\sum_{n=0}^{N} x^n = \frac{x^{N+1}-1}{x-1} \tag{C-58}$$

$$\sum_{n=0}^{N} \frac{N!}{n!(N-n)!} x^n y^{N-n} = (x+y)^N \tag{C-59}$$

$$\sum_{n=0}^{N} e^{j(\theta+n\phi)} = \frac{\sin\,[(N+1)\phi/2]}{\sin\,(\phi/2)} e^{j[\theta+(N\phi/2)]} \tag{C-60}$$

$$\sum_{n=0}^{N} \binom{N}{n} = \sum_{n=0}^{N} \frac{N!}{n!(N-n)!} = 2^N \tag{C-61}$$

REVIEW OF FOURIER TRANSFORMS

The *Fourier transform*† or spectrum $X(\omega)$ of a signal $x(t)$ is given by

$$X(\omega) = \int_{-\infty}^{\infty} x(t)e^{-j\omega t} \, dt \qquad \text{(D-1)}$$

The *inverse Fourier transform* allows the recovery of $x(t)$ from its spectrum $X(\omega)$. It is given by

$$x(t) = \frac{1}{2\pi} \int_{-\infty}^{\infty} X(\omega)e^{j\omega t} \, d\omega \qquad \text{(D-2)}$$

Together, (D-1) and (D-2) form a *Fourier transform pair*. Extensive tables of transform pairs exist (Campbell and Foster, 1948). A transform pair is often symbolized by use of a double-ended arrow:

$$x(t) \leftrightarrow X(\omega) \qquad \text{(D-3)}$$

The Fourier transform $X(\omega)$ is valid for real or complex signals, and, in general, is a complex function of ω even for real signals $x(t)$. $X(\omega)$ describes the relative complex voltages (amplitudes and phases) as a function of ω that are present in a waveform $x(t)$. From (D-1), we see that the unit of $X(\omega)$ is volts per hertz if $x(t)$ is a voltage-time waveform. Thus, $X(\omega)$ can be considered as the *density of voltage* in $x(t)$ as a function of angular frequency ω.

---

† Named for the great French mathematician and physicist Baron Jean Baptiste Joseph Fourier (1768–1830).

EXISTENCE

Conditions that guarantee the existence of the Fourier transform of a waveform $x(t)$ are:

1. that $x(t)$ be bounded with at most a finite number of maxima and minima and a finite number of discontinuities in any finite time interval,† and

2.
$$\int_{-\infty}^{\infty} |x(t)|\, dt < \infty \tag{D-4}$$

These conditions are only *sufficient* for $X(\omega)$ to exist; they are *not necessary*. Many signals of practical interest do not satisfy these conditions but do have transforms. Examples are: the unit-impulse function $\delta(t)$ that has the transform $X(\omega) = 1$; and the unit-step function $u(t)$, defined by $u(t) = 1$ for $0 < t$ and $u(t) = 0$ for $t < 0$, that has the transform $X(\omega) = \pi\delta(\omega) + (1/j\omega)$.

PROPERTIES

A number of extremely useful properties of Fourier transforms may be stated. We give these without proofs since the proofs may readily be found in the literature (Peebles, 1976, p. 29; Papoulis, 1962, p. 14). In these properties, we assume the Fourier transform of some signal $x(t)$ is $X(\omega)$, while the notation $X_n(\omega)$ implies the transform of a signal $x_n(t)$ with $n = 1, 2, \ldots, N$.

Linearity

For constants α_n (that may be complex):

$$x(t) = \sum_{n=1}^{N} \alpha_n x_n(t) \leftrightarrow \sum_{n=1}^{N} \alpha_n X_n(\omega) = X(\omega) \tag{D-5}$$

Time and Frequency Shifting

With t_0 and ω_0 real constants:

$$x(t - t_0) \leftrightarrow X(\omega)e^{-j\omega t_0} \tag{D-6}$$

$$x(t)e^{j\omega_0 t} \leftrightarrow X(\omega - \omega_0) \tag{D-7}$$

† These are known as the *Dirichlet conditions*, after the German mathematician Peter Gustov Lejeune Dirichlet (1805–1859). A signal satisfying them is said to have *bounded variation* (Thomas, 1969, p. 579).

Scaling

With α a real constant:

$$x(\alpha t) \leftrightarrow \frac{1}{|\alpha|} X\left(\frac{\omega}{\alpha}\right) \qquad \text{(D-8)}$$

Duality

$$X(t) \leftrightarrow 2\pi x(-\omega) \qquad \text{(D-9)}$$

Differentiation

$$\frac{d^n x(t)}{dt^n} \leftrightarrow (j\omega)^n X(\omega) \qquad \text{(D-10)}$$

$$(-jt)^n x(t) \leftrightarrow \frac{d^n X(\omega)}{d\omega^n} \qquad \text{(D-11)}$$

Integration

$$\int_{-\infty}^{t} x(\tau)\,d\tau \leftrightarrow \pi X(0)\delta(\omega) + \frac{X(\omega)}{j\omega} \qquad \text{(D-12)}$$

$$\pi x(0)\delta(t) - \frac{x(t)}{jt} \leftrightarrow \int_{-\infty}^{\omega} X(\xi)\,d\xi \qquad \text{(D-13)}$$

Conjugation

$$x^*(t) \leftrightarrow X^*(-\omega) \qquad \text{(D-14)}$$

$$x^*(-t) \leftrightarrow X^*(\omega) \qquad \text{(D-15)}$$

Convolution

$$x(t) = \int_{-\infty}^{\infty} x_1(\tau)x_2(t-\tau)\,d\tau \leftrightarrow X_1(\omega)X_2(\omega) = X(\omega) \qquad \text{(D-16)}$$

$$x(t) = x_1(t)x_2(t) \leftrightarrow \frac{1}{2\pi} \int_{-\infty}^{\infty} X_1(\xi)X_2(\omega-\xi)\,d\xi = X(\omega) \qquad \text{(D-17)}$$

Correlation

$$x(t) = \int_{-\infty}^{\infty} x_1^*(\tau)x_2(\tau+t)\,d\tau \leftrightarrow X_1^*(\omega)X_2(\omega) = X(\omega) \qquad \text{(D-18)}$$

$$x(t) = x_1^*(t)x_2(t) \leftrightarrow \frac{1}{2\pi} \int_{-\infty}^{\infty} X_1^*(\xi)X_2(\xi+\omega)\,d\xi = X(\omega) \qquad \text{(D-19)}$$

Parseval's† Theorem

$$\int_{-\infty}^{\infty} x_1^*(\tau)x_2(\tau) \, d\tau = \frac{1}{2\pi} \int_{-\infty}^{\infty} X_1^*(\omega)X_2(\omega) \, d\omega \tag{D-20}$$

An alternative form occurs when $x_1(t) = x_2(t) = x(t)$:

$$\int_{-\infty}^{\infty} |x(t)|^2 \, dt = \frac{1}{2\pi} \int_{-\infty}^{\infty} |X(\omega)|^2 \, d\omega \tag{D-21}$$

MULTIDIMENSIONAL FOURIER TRANSFORMS

The Fourier transform $X(\omega_1, \omega_2)$ of a function $x(t_1, t_2)$ of two "time" variables t_1 and t_2 is defined as the iterated double transform. Upon Fourier transforming $x(t_1, t_2)$ first with respect to t_1 we have

$$X(\omega_1, t_2) = \int_{-\infty}^{\infty} x(t_1, t_2)e^{-j\omega_1 t_1} \, dt_1 \tag{D-22}$$

$X(\omega_1, \omega_2)$ results from Fourier transformation of $X(\omega_1, t_2)$ with respect to t_2:

$$X(\omega_1, \omega_2) = \int_{-\infty}^{\infty} X(\omega_1, t_2)e^{-j\omega_2 t_2} \, dt_2 \tag{D-23}$$

or

$$X(\omega_1, \omega_2) = \int_{-\infty}^{\infty} \int_{-\infty}^{\infty} x(t_1, t_2)e^{-j\omega_1 t_1 - j\omega_2 t_2} \, dt_1 \, dt_2 \tag{D-24}$$

By use of similar logic, the two-dimensional inverse Fourier transform is

$$x(t_1, t_2) = \frac{1}{(2\pi)^2} \int_{-\infty}^{\infty} \int_{-\infty}^{\infty} X(\omega_1, \omega_2)e^{j\omega_1 t_1 + j\omega_2 t_2} \, d\omega_1 \, d\omega_2 \tag{D-25}$$

The extension of the above procedures to an N-dimensional function is direct; we obtain the Fourier transform pair

$$X(\omega_1, \ldots, \omega_N) = \int_{-\infty}^{\infty} \cdots \int_{-\infty}^{\infty} x(t_1, \ldots, t_N)e^{-j\omega_1 t_1 - \cdots - j\omega_N t_N} \, dt_1 \cdots dt_N \tag{D-26}$$

$$x(t_1, \ldots, t_N) = \frac{1}{(2\pi)^N} \int_{-\infty}^{\infty} \cdots \int_{-\infty}^{\infty} X(\omega_1, \ldots, \omega_N)e^{j\omega_1 t_1 + \cdots + j\omega_N t_N} \, d\omega_1 \cdots d\omega_N \tag{D-27}$$

† Named for M. A. Parseval.

PROBLEMS

D-1 Find the Fourier transform of a pulse $x(t)$ defined by

$$x(t) = \begin{cases} A & -\tau/2 < t < \tau/2 \\ 0 & \text{elsewhere} \end{cases}$$

where $\tau > 0$ and A are real constants.

D-2 If a signal $y(t)$ is the product of $x(t)$ of Problem D-1 with a cosine wave, that is, if

$$y(t) = x(t) \cos (\omega_0 t + \theta_0)$$

where ω_0 and θ_0 are real constants, what is the Fourier transform of $y(t)$?

D-3 Find the Fourier transform of the waveform

$$x(t) = \begin{cases} A\left(1 - \dfrac{|t|}{\tau}\right) & |t| \leq \tau \\ 0 & |t| > \tau \end{cases}$$

where $\tau > 0$ and A are real constants.

D-4 By direct use of (D-1), find the Fourier transform of the waveform

$$x(t) = \begin{cases} A \cos (\pi t/2\tau) & |t| \leq \tau \\ 0 & |t| > \tau \end{cases}$$

where $\tau > 0$ and A are real constants.

D-5 The waveform of Problem D-4 can be written in the form

$$x(t) = A \text{ rect } (t/2\tau) \cos (\pi t/2\tau)$$

where rect $(t/2\tau)$ is defined by (E-2). By using (D-19), find the Fourier transform of $x(t)$.

D-6 The complex form of the *Fourier series* of an arbitrary periodic signal $y(t)$ of period T is

$$y(t) = \sum_{n=-\infty}^{\infty} C_n e^{jn2\pi t/T}$$

where the *Fourier series coefficients* are given by

$$C_n = \frac{1}{T} \int_{-T/2}^{T/2} y(t) e^{-jn2\pi t/T} \, dt$$

for $n = 0, \pm 1, \pm 2, \dots$. Show that the Fourier transform of this arbitrary periodic signal is

$$Y(\omega) = 2\pi \sum_{n=-\infty}^{\infty} C_n \delta\left(\omega - \frac{n2\pi}{T}\right)$$

where $\delta(\cdot)$ is the unit-impulse function of Appendix A.

*D-7 Prove the Fourier transform pair

$$\sum_{n=-\infty}^{\infty} \delta(t - nT) \leftrightarrow \frac{2\pi}{T} \sum_{n=-\infty}^{\infty} \delta\left(\omega - \frac{n2\pi}{T}\right)$$

where $T > 0$ is a real constant and $\delta(\,\cdot\,)$ is the impulse function of Appendix A. (*Hint:* Represent the time function by a complex Fourier series as in Problem D-6, find the Fourier coefficients of the series, and then Fourier-transform the series).

*D-8 From the expression in Problem D-7, it is readily shown that

$$\sum_{n=-\infty}^{\infty} e^{-jn\omega T} = \frac{2\pi}{T} \sum_{n=-\infty}^{\infty} \delta\left(\omega - \frac{n2\pi}{T}\right)$$

Use this result to prove that the periodic signal

$$y(t) = \sum_{n=-\infty}^{\infty} x(t - nT)$$

comprised of repetitions in each period T of a basic waveform $x(t)$, has the Fourier transform $Y(\omega)$ given by

$$Y(\omega) = \frac{2\pi}{T} \sum_{n=-\infty}^{\infty} X\left(\frac{n2\pi}{T}\right) \delta\left(\omega - \frac{n2\pi}{T}\right)$$

where $X(\omega)$ is the Fourier transform of $x(t)$. By using the result of Problem D-6, we see that the coefficient C_n of the Fourier series of $y(t)$ is related to the Fourier transform of its component waveform $x(t)$ by

$$C_n = \frac{1}{T} X\left(\frac{n2\pi}{T}\right)$$

D-9 Find the Fourier transform of the waveform

$$x(t) = u(t)e^{j\omega_0 t}$$

where $u(\,\cdot\,)$ is the unit-step function of (A-5) and ω_0 is a real constant.

D-10 Find the Fourier transform of a sequence of $2N + 1$ pulses of the form given in Problem D-1, where $N = 0, 1, 2, \ldots$. That is, find the transform of

$$y(t) = \sum_{n=-N}^{N} x(t - nT)$$

with $T > 0$ a real constant and $\tau < T$.

D-11 Determine the Fourier transform of the signal

$$x(t) = \begin{cases} At^2 & 0 < t < \tau \\ 0 & \text{elsewhere} \end{cases}$$

where $\tau > 0$ and A are real constants.

D-12 Show that the inverse Fourier transform of the function

$$X(\omega) = \begin{cases} K & -W < \omega < W \\ 0 & \text{elsewhere} \end{cases}$$

is

$$x(t) = (KW/\pi) \, \text{Sa} \, (Wt)$$

where $W > 0$ and K are real constants and Sa $(\cdot)$ is the *sampling function* defined by (E-3).

D-13 The transfer function $H(\omega)$ of a lowpass filter can be approximated by

$$H(\omega) = \begin{cases} K_0 + 2 \sum_{n=1}^{N} K_n \cos(n\pi\omega/W) & -W < \omega < W \\ 0 & \text{elsewhere} \end{cases}$$

Here $W > 0$, K_0, K_1, ..., K_N are real constants and $N \geq 0$ is an integer. Find the inverse Fourier transform $h(t)$ of $H(\omega)$ which is the *impulse response* of the network, in terms of sampling functions (see Problem D-12).

D-14 Let $x(t)$ have the Fourier transform $X(\omega)$. Find the transforms of the following functions in terms of $X(\omega)$:

(a) $x(t-2) \exp(j\omega_0 t)$ (b) $\dfrac{dx(t)}{dt} \exp[j\omega_0(t-3)]$ (c) $x(t-3) - 3x(2t)$

Here ω_0 is a real constant.

D-15 If $x(t) \leftrightarrow X(\omega)$, find the inverse transforms of the following functions in terms of $x(t)$:

(a) $X(\omega)X^*(\omega + \omega_0)$ (b) $X(\omega - \omega_0) \dfrac{dX(\omega)}{d\omega}$ (c) $X^*(-\omega) + X(\omega)$

Here * represents complex conjugation and ω_0 is a real constant.

D-16 A voltage $x(t)$ exists across a resistor of resistance R. Show that the real energy E expended in the resistance is

$$E = \frac{1}{2\pi R} \int_{-\infty}^{\infty} |X(\omega)|^2 \, d\omega$$

where $X(\omega)$ is the Fourier transform of $x(t)$.

D-17 It is known that

$$x(t) = e^{-\alpha|t|} \leftrightarrow \frac{2\alpha}{\alpha^2 + \omega^2} = X(\omega)$$

where $\alpha > 0$ is a real constant. Find the Fourier transform $Y(\omega)$ of

$$y(t) = \frac{6}{4 + t^2}$$

D-18 Use the definition (A-2) of an impulse function to prove that the impulse has the Fourier transform 1. That is, show that

$$\delta(t) \leftrightarrow 1$$

D-19 By use of various Fourier transform properties, show that the following are true:

(a) $A \leftrightarrow A(2\pi)\delta(\omega)$ where A is a constant

(b) $\cos(\omega_0 t) \leftrightarrow \pi[\delta(\omega - \omega_0) + \delta(\omega + \omega_0)]$ where ω_0 is a real constant

D-20 Use the facts that

$$u(t)e^{-\alpha t} \leftrightarrow \frac{1}{\alpha + j\omega}$$

and

$$\cos(\omega_0 t) \leftrightarrow \pi[\delta(\omega - \omega_0) + \delta(\omega + \omega_0)]$$

where $\alpha > 0$ and ω_0 are real constants, to prove that

$$u(t)e^{-\alpha t} \cos(\omega_0 t) \leftrightarrow \frac{\alpha + j\omega}{(\alpha^2 + \omega_0^2 - \omega^2) + j(2\alpha\omega)}$$

D-21 Prove (D-6) and (D-10).

*★***D-22** Prove (D-12). [*Hint:* Use (D-16).]

D-23 Prove (D-18).

D-24 Find the Fourier transform of the signal

$$x(t_1, t_2) = \begin{cases} A & -\tau_1 < t_1 < \tau_1 \quad \text{and} \quad -\tau_2 < t_2 < \tau_2 \\ 0 & \text{elsewhere} \end{cases}$$

where $\tau_1 > 0$, $\tau_2 > 0$, and A are real constants.

E

TABLE OF USEFUL FOURIER TRANSFORMS

In the following table of Fourier transform pairs, we define

$$u(\xi) = \begin{cases} 1 & \xi > 0 \\ 0 & \xi < 0 \end{cases} \tag{E-1}$$

$$\text{rect}\,(\xi) = \begin{cases} 1 & |\xi| < \frac{1}{2} \\ 0 & |\xi| > \frac{1}{2} \end{cases} \tag{E-2}$$

$$\text{Sa}\,(\xi) = \frac{\sin\,(\xi)}{\xi} \tag{E-3}$$

$$\text{tri}\,(\xi) = \begin{cases} 1 - |\xi| & |\xi| < 1 \\ 0 & |\xi| > 1 \end{cases} \tag{E-4}$$

$$x(t) \leftrightarrow X(\omega) \tag{E-5}$$

and let α, τ, σ, ω_0, and W be real constants.

Table E-1 Fourier Transform Pairs

Pair	$x(t)$	$X(\omega)$	Notes
1	$\alpha\delta(t)$	α	
2	$\alpha/2\pi$	$\alpha\delta(\omega)$	
3	$u(t)$	$\pi\delta(\omega) + (1/j\omega)$	
4	$\frac{1}{2}\delta(t) - \dfrac{1}{j2\pi t}$	$u(\omega)$	
5	rect (t/τ)	τ Sa $(\omega\tau/2)$	$\tau > 0$
6	(W/π) Sa (Wt)	rect $(\omega/2W)$	$W > 0$
7	tri (t/τ)	τ Sa^2 $(\omega\tau/2)$	$\tau > 0$
8	(W/π) Sa^2 (Wt)	tri $(\omega/2W)$	$W > 0$
9	$e^{j\omega_0 t}$	$2\pi\delta(\omega - \omega_0)$	
10	$\delta(t - \tau)$	$e^{-j\omega\tau}$	
11	$\cos(\omega_0 t)$	$\pi[\delta(\omega - \omega_0) + \delta(\omega + \omega_0)]$	
12	$\sin(\omega_0 t)$	$-j\pi[\delta(\omega - \omega_0) - \delta(\omega + \omega_0)]$	
13	$u(t)\cos(\omega_0 t)$	$\dfrac{\pi}{2}[\delta(\omega - \omega_0) + \delta(\omega + \omega_0)] + \dfrac{j\omega}{\omega_0^2 - \omega^2}$	
14	$u(t)\sin(\omega_0 t)$	$-j\dfrac{\pi}{2}[\delta(\omega - \omega_0) - \delta(\omega + \omega_0)] + \dfrac{\omega_0}{\omega_0^2 - \omega^2}$	
15	$u(t)e^{-\alpha t}$	$\dfrac{1}{\alpha + j\omega}$	$\alpha > 0$
16	$u(t)te^{-\alpha t}$	$\dfrac{1}{(\alpha + j\omega)^2}$	$\alpha > 0$
17	$u(t)t^2 e^{-\alpha t}$	$\dfrac{2}{(\alpha + j\omega)^3}$	$\alpha > 0$
18	$u(t)t^3 e^{-\alpha t}$	$\dfrac{6}{(\alpha + j\omega)^4}$	$\alpha > 0$
19	$e^{-\alpha|t|}$	$\dfrac{2\alpha}{\alpha^2 + \omega^2}$	$\alpha > 0$
20	$e^{-t^2/(2\sigma^2)}$	$\sigma\sqrt{2\pi}\, e^{-\sigma^2\omega^2/2}$	$\sigma > 0$

SOME PROBABILITY DENSITIES
AND DISTRIBUTIONS

For convenience of reference we list below the probability density $f_X(x)$ and distribution function $F_X(x)$ for some well-known distributions. Where appropriate, we also give the mean $\bar{X}$, variance σ_X^2, and characteristic function $\Phi_X(\omega)$.

A number of constants and functions are used as defined below:†

$$a, a_1, a_2, b, b_1, b_2, \sigma, \text{ and } p \text{ are real constants} \qquad \text{(F-1}a)$$

$$N \text{ is a positive integer} \qquad \text{(F-1}b)$$

$$\delta(\xi) = \text{impulse function of (2.3-2)} \qquad \text{(F-1}c)$$

$$u(\xi) = \text{unit-step function of (2.2-4)} \qquad \text{(F-1}d)$$

$$\text{rect } (\xi) = \text{rectangular function of (E-2)} \qquad \text{(F-1}e)$$

$$\Gamma(x) = \int_0^\infty \xi^{x-1} e^{-\xi} \, d\xi \qquad \text{Re } (x) > 0$$

$$= \text{gamma function} \qquad \text{(F-1}f)$$

$$P(\alpha, \beta) = \frac{1}{\Gamma(\alpha)} \int_0^\beta \xi^{\alpha-1} e^{-\xi} \, d\xi \qquad \text{Re } (\alpha) > 0$$

$$= \text{incomplete gamma function} \qquad \text{(F-1}g)$$

---

† Re (z) denotes the real part of z.

$$P(x \mid N) = \frac{1}{2^{N/2}\Gamma(N/2)} \int_0^x \xi^{(N/2)-1} e^{-\xi/2} \, d\xi$$

$$= \text{chi-square probability function}$$

$$= P\!\left(\frac{N}{2}, \frac{x}{2}\right) \qquad\qquad \text{(F-1}h\text{)}$$

$$I(u, p) = \frac{1}{\Gamma(p+1)} \int_0^{u\sqrt{p+1}} \xi^p e^{-\xi} \, d\xi$$

$$= \text{Pearson's form of incomplete gamma function (Pearson, 1934)}$$

$$= P(p + 1, u\sqrt{p+1}) \qquad\qquad \text{(F-1}i\text{)}$$

$$I_x(a, b) = \frac{\Gamma(a+b)}{\Gamma(a)\Gamma(b)} \int_0^x \xi^{a-1}(1 - \xi)^{b-1} \, d\xi$$

$$= \text{incomplete beta function} \qquad\qquad \text{(F-1}j\text{)}$$

$$F(x) = \text{gaussian distribution of (B-3)} \qquad\qquad \text{(F-1}k\text{)}$$

$$I_n(x) = (x/2)^n \sum_{k=0}^{\infty} \frac{(x/2)^{2k}}{k!\,(n+k)!}$$

$$= \frac{1}{\pi} \int_0^{\pi} e^{x \cos(\theta)} \cos(n\theta) \, d\theta$$

$$= \text{modified Bessel function of first kind of order } n = 0, 1, 2, \ldots \quad \text{(F-1}l\text{)}$$

$$Q(\alpha, \beta) = \int_{\beta}^{\infty} \xi I_0(\alpha\xi) \exp\left[\frac{-(\xi^2 + \alpha^2)}{2}\right] d\xi \qquad\qquad \text{(F-1}m\text{)}$$

The functions of (F-1f) through (F-1j) and that of (F-1l) are discussed in detail in Abramowitz and Stegun, editors (1964). $Q(\alpha, \beta)$ is Marcum's Q-function; it is tabulated in Marcum (1950).

DISCRETE FUNCTIONS

Bernoulli

For $0 < p < 1$

$$f_X(x) = (1 - p)\delta(x) + p\delta(x - 1) \qquad\qquad \text{(F-2)}$$

$$F_X(x) = (1 - p)u(x) + pu(x - 1) \qquad\qquad \text{(F-3)}$$

$$\bar{X} = p \qquad\qquad \text{(F-4)}$$

$$\sigma_X^2 = p(1 - p) \qquad\qquad \text{(F-5)}$$

$$\Phi_X(\omega) = 1 - p + pe^{j\omega} \qquad\qquad \text{(F-6)}$$

Binomial

For $0 < p < 1$ and $N = 1, 2, \ldots$

$$f_X(x) = \sum_{k=0}^{N} \binom{N}{k} p^k (1-p)^{N-k} \delta(x-k) \tag{F-7}$$

$$F_X(x) = \sum_{k=0}^{N} \binom{N}{k} p^k (1-p)^{N-k} u(x-k) \tag{F-8}$$

$$\bar{X} = Np \tag{F-9}$$

$$\sigma_X^2 = Np(1-p) \tag{F-10}$$

$$\Phi_X(\omega) = [1 - p + pe^{j\omega}]^N \tag{F-11}$$

Pascal†

For $0 < p < 1$ and $N = 1, 2, \ldots$

$$f_X(x) = \sum_{k=N}^{\infty} \binom{k-1}{N-1} p^N (1-p)^{k-N} \delta(x-k) \tag{F-12}$$

$$F_X(x) = \sum_{k=N}^{\infty} \binom{k-1}{N-1} p^N (1-p)^{k-N} u(x-k) \tag{F-13}$$

$$\bar{X} = \frac{N}{p} \tag{F-14}$$

$$\sigma_X^2 = \frac{N(1-p)}{p^2} \tag{F-15}$$

$$\Phi_X(\omega) = p^N e^{jN\omega} [1 - (1-p)e^{j\omega}]^{-N} \tag{F-16}$$

Poisson

For $b > 0$

$$f_X(x) = e^{-b} \sum_{k=0}^{\infty} \frac{b^k}{k!} \delta(x-k) \tag{F-17}$$

$$F_X(x) = e^{-b} \sum_{k=0}^{\infty} \frac{b^k}{k!} u(x-k) \tag{F-18}$$

$$\bar{X} = b \tag{F-19}$$

$$\sigma_X^2 = b \tag{F-20}$$

$$\Phi_X(\omega) = \exp\,[b(e^{j\omega} - 1)] \tag{F-21}$$

† Blaise Pascal (1623–1662) was a French mathematician.

CONTINUOUS FUNCTIONS

Arcsine

For $a > 0$

$$f_X(x) = \frac{\text{rect } (x/2a)}{\pi\sqrt{a^2 - x^2}} \tag{F-22}$$

$$F_X(x) = \begin{cases} 0 & -\infty < x < -a \\ \dfrac{1}{2} + \dfrac{1}{\pi}\sin^{-1}\left(\dfrac{x}{a}\right) & -a \le x < a \\ 1 & a \le x < \infty \end{cases} \tag{F-23}$$

$$\bar{X} = 0 \tag{F-24}$$

$$\sigma_X^2 = \frac{a^2}{2} \tag{F-25}$$

Beta

For $a > 0$ and $b > 0$

$$f_X(x) = \frac{\Gamma(a + b)}{\Gamma(a)\Gamma(b)}\left[u(x) - u(x - 1)\right]x^{a-1}(1 - x)^{b-1} \tag{F-26}$$

$$F_X(x) = \begin{cases} I_x(a, b)u(x) & x < 1 \\ 1 & x \ge 1 \end{cases} \tag{F-27}$$

$$\bar{X} = \frac{a}{a + b} \tag{F-28}$$

$$\sigma_X^2 = \frac{ab}{(a + b)^2(a + b + 1)} \tag{F-29}$$

Cauchy

For $b > 0$ and $-\infty < a < \infty$

$$f_X(x) = \frac{(b/\pi)}{b^2 + (x - a)^2} \tag{F-30}$$

$$F_X(x) = \frac{1}{2} + \frac{1}{\pi}\tan^{-1}\left(\frac{x - a}{b}\right) \tag{F-31}$$

$$\bar{X} = \text{is undefined} \tag{F-32}$$

$$\sigma_X^2 = \text{is undefined} \tag{F-33}$$

$$\Phi_X(\omega) = e^{ja\omega - b|\omega|} \tag{F-34}$$

Chi-Square with N Degrees of Freedom

For $N = 1, 2, \ldots$

$$f_X(x) = \frac{x^{(N/2)-1}}{2^{N/2}\Gamma(N/2)}\, e^{-x/2} u(x) \tag{F-35}$$

$$F_X(x) = P(x\,|\,N) = P\!\left(\frac{N}{2}, \frac{x}{2}\right) \tag{F-36}$$

$$\bar{X} = N \tag{F-37}$$

$$\sigma_X^2 = 2N \tag{F-38}$$

$$\Phi_X(\omega) = (1 - j2\omega)^{-N/2} \tag{F-39}$$

Erlang

For $N = 1, 2, \ldots$ and $a > 0$

$$f_X(x) = \frac{a^N x^{N-1} e^{-ax}}{(N-1)!}\, u(x) \tag{F-40}$$

$$F_X(x) = \left[1 - e^{-ax}\sum_{n=0}^{N-1}\frac{(ax)^n}{n!}\right] u(x) \tag{F-41}$$

$$\bar{X} = \frac{N}{a} \tag{F-42}$$

$$\sigma_X^2 = \frac{N}{a^2} \tag{F-43}$$

$$\Phi_X(\omega) = \left(\frac{a}{a - j\omega}\right)^N \tag{F-44}$$

Exponential

For $a > 0$

$$f_X(x) = a e^{-ax} u(x) \tag{F-45}$$

$$F_X(x) = [1 - e^{-ax}] u(x) \tag{F-46}$$

$$\bar{X} = \frac{1}{a} \tag{F-47}$$

$$\sigma_X^2 = \frac{1}{a^2} \tag{F-48}$$

$$\Phi_X(\omega) = \frac{a}{a - j\omega} \tag{F-49}$$

Gamma

For $a > 0$ and $b > 0$

$$f_X(x) = \frac{a^b x^{b-1} e^{-ax}}{\Gamma(b)} u(x) \tag{F-50}$$

$$F_X(x) = I\left(\frac{ax}{\sqrt{b}}, b-1\right) u(x) \tag{F-51}$$

$$\bar{X} = \frac{b}{a} \tag{F-52}$$

$$\sigma_X^2 = \frac{b}{a^2} \tag{F-53}$$

$$\Phi_X(\omega) = \left(\frac{a}{a - j\omega}\right)^b \tag{F-54}$$

Note that if b is a positive integer the gamma density becomes the Erlang density. Also if $b = N/2$, for $N = 1, 2, \ldots,$ and $a = \frac{1}{2}$ the gamma density becomes the chi-square density.

Gaussian-Univariate

For $b > 0$ and $-\infty < a < \infty$

$$f_X(x) = (\pi b)^{-1/2} e^{-(x-a)^2/b} \tag{F-55}$$

$$F_X(x) = F\left(\frac{x - a}{\sqrt{b/2}}\right) \tag{F-56}$$

$$\bar{X} = a \tag{F-57}$$

$$\sigma_X^2 = \frac{b}{2} \tag{F-58}$$

$$\Phi_X(\omega) = e^{j\omega a - (\omega^2 b/4)} \tag{F-59}$$

Gaussian-Bivariate

For $-\infty < a_1 < \infty$, $-\infty < a_2 < \infty$, $b_1 > 0$, $b_2 > 0$ and $-1 \leq \rho \leq 1$

$$f_{X_1, X_2}(x_1, x_2) = [\pi^2 b_1 b_2 (1 - \rho^2)]^{-1/2}$$

$$\cdot \exp\left\{\frac{-1}{(1 - \rho^2)}\left[\frac{(x_1 - a_1)^2}{b_1}\right.\right.$$

$$\left.\left. - \frac{2\rho(x_1 - a_1)(x_2 - a_2)}{\sqrt{b_1 b_2}} + \frac{(x_2 - a_2)^2}{b_2}\right]\right\} \tag{F-60}$$

$$F_{X_1, X_2}(x_1, x_2) = L\left(-\left[\frac{x_1 - a_1}{\sqrt{b_1/2}}\right], -\left[\frac{x_2 - a_2}{\sqrt{b_2/2}}\right], \rho\right) \tag{F-61}$$

where $L(x_1, x_2, \rho)$ is a probability function discussed extensively and graphed in Abramowitz and Stegun, editors (1964), p. 936. Also

$$\bar{X}_1 = a_1 \tag{F-62}$$

$$\bar{X}_2 = a_2 \tag{F-63}$$

$$\sigma_{X_1}^2 = b_1/2 \tag{F-64}$$

$$\sigma_{X_2}^2 = b_2/2 \tag{F-65}$$

$$\Phi_{X_1, X_2}(\omega_1, \omega_2) = \exp\left\{j\omega_1 a_1 + j\omega_2 a_2 - \tfrac{1}{4}[\omega_1^2 b_1 + 2\rho\omega_1\omega_2\sqrt{b_1 b_2} + \omega_2^2 b_2]\right\} \tag{F-66}$$

Laplace

For $b > 0$ and $-\infty < a < \infty$

$$f_X(x) = \frac{b}{2} e^{-b|x-a|} \tag{F-67}$$

$$F_X(x) = \begin{cases} \tfrac{1}{2}e^{b(x-a)} & -\infty < x < a \\ 1 - \tfrac{1}{2}e^{-b(x-a)} & a \le x < \infty \end{cases} \tag{F-68}$$

$$\bar{X} = a \tag{F-69}$$

$$\sigma_X^2 = \frac{2}{b^2} \tag{F-70}$$

$$\Phi_X(\omega) = b^2 \frac{e^{ja\omega}}{b^2 + \omega^2} \tag{F-71}$$

Log-Normal

For $-\infty < a < \infty$, $-\infty < b < \infty$, and $\sigma > 0$

$$f_X(x) = \frac{u(x-b)e^{-[\ln(x-b)-a]^2/2\sigma^2}}{\sqrt{2\pi}(x-b)\sigma} \tag{F-72}$$

$$F_X(x) = u(x-b)F\{\sigma^{-1}[\ln(x-b)-a]\} \tag{F-73}$$

$$\bar{X} = b + \exp\left(a + \frac{\sigma^2}{2}\right) \tag{F-74}$$

$$\sigma_X^2 = [\exp(\sigma^2) - 1]\exp(2a + \sigma^2) \tag{F-75}$$

Rayleigh

For $-\infty < a < \infty$ and $b > 0$

$$f_X(x) = \frac{2}{b}(x-a)e^{-(x-a)^2/b}u(x-a) \tag{F-76}$$

$$F_X(x) = [1 - e^{-(x-a)^2/b}]u(x-a) \tag{F-77}$$

$$\bar{X} = a + \sqrt{\frac{\pi b}{4}} \tag{F-78}$$

$$\sigma_X^2 = \frac{b(4-\pi)}{4} \tag{F-79}$$

Rice [Thomas (1969), Middleton (1960)]

For $a > 0$ and $b > 0$

$$f_X(x) = \frac{x}{b^2} e^{-(a^2+x^2)/2b^2} I_0\left(\frac{ax}{b^2}\right) u(x) \tag{F-80}$$

$$F_X(x) = \left[1 - Q\left(\frac{a}{b}, \frac{x}{b}\right)\right] u(x) \tag{F-81}$$

$$\bar{X} = b\sqrt{\frac{\pi}{2}} e^{-k^2/4}\left[\left(1+\frac{k^2}{2}\right)I_0\left(\frac{k^2}{4}\right) + \frac{k^2}{2} I_1\left(\frac{k^2}{4}\right)\right] \tag{F-82}$$

$$\sigma_X^2 = b^2(2+k^2) - (\bar{X})^2 \tag{F-83}$$

$$k^2 = \frac{a^2}{b^2} \tag{F-84}$$

Uniform

For $-\infty < a < b < \infty$

$$f_X(x) = \frac{u(x-a) - u(x-b)}{b-a} \tag{F-85}$$

$$F_X(x) = \begin{cases} \dfrac{(x-a)u(x-a)}{b-a} & x < b \\ 1 & x \geq b \end{cases} \tag{F-86}$$

$$\bar{X} = \frac{a+b}{2} \tag{F-87}$$

$$\sigma_X^2 = \frac{(b-a)^2}{12} \tag{F-88}$$

$$\Phi_X(\omega) = \frac{e^{j\omega b} - e^{j\omega a}}{j\omega(b-a)} \tag{F-89}$$

Weibull

For $a > 0$ and $b > 0$

$$f_X(x) = abx^{b-1}e^{-ax^b}u(x) \tag{F-90}$$

$$F_X(x) = [1 - e^{-ax^b}]u(x) \tag{F-91}$$

$$\bar{X} = \frac{\Gamma(1 + b^{-1})}{a^{1/b}} \tag{F-92}$$

$$\sigma_X^2 = \frac{\Gamma(1 + 2b^{-1}) - [\Gamma(1 + b^{-1})]^2}{a^{2/b}} \tag{F-93}$$

Note that if $b = 2$ the Weibull density becomes a Rayleigh density.

BIBLIOGRAPHY

Abramowitz, M., and I. Stegun, editors (1964): *Handbook of Mathematical Functions with Formulas, Graphs, and Mathematical Tables*, National Bureau of Standards Applied Mathematics Series 55, U.S. Government Printing Office, Washington, D.C.

Barton, D. K. (1964): *Radar System Analysis*, Prentice-Hall, Englewood Cliffs, New Jersey.

Blanchard, A. (1976): *Phase-Locked Loops, Application to Coherent Receiver Design*, John Wiley & Sons, New York.

Bracewell, R. (1965): *The Fourier Transform and Its Applications*, McGraw-Hill, New York.

Campbell, G. A., and R. M. Foster (1948): *Fourier Integrals for Practical Applications*, Van Nostrand, Princeton, New Jersey.

Carlson, A. B. (1975): *Communication Systems, An Introduction to Signals and Noise in Electrical Communication*, 2d ed., McGraw-Hill, New York.

Clarke, A. B., and R. L. Disney (1970): *Probability and Random Processes for Engineers and Scientists*, Wiley, New York.

Cooper, G. R., and C. D. McGillem (1971): *Probabilistic Methods of Signal and System Analysis*, Holt, Rinehart and Winston, New York.

Cramér, H. (1946): *Mathematical Methods of Statistics*, Princeton University Press, Princeton, New Jersey.

Davenport, W. B., Jr., and W. L. Root (1958): *An Introduction to the Theory of Random Signals and Noise*, McGraw-Hill, New York.

Davenport, W. B., Jr. (1970): *Probability and Random Processes, An Introduction for Applied Scientists and Engineers*, McGraw-Hill, New York.

DiFranco, J. V., and W. L. Rubin (1968): *Radar Detection*, Prentice-Hall, Englewood Cliffs, New Jersey.

Dubes, R. C. (1968): *The Theory of Applied Probability*, Prentice-Hall, Englewood Cliffs, New Jersey.

Dwight, H. B. (1961): *Tables of Integrals and other Mathematical Data*, 4th ed., Macmillan, New York.

Gray, R. M., and L. D. Davisson (1986): *Random Processes: A Mathematical Approach for Engineers*, Prentice-Hall, Englewood Cliffs, New Jersey.

Helstrom, C. W. (1984): *Probability and Stochastic Processes for Engineers*, Macmillan, New York.

James, H. M., N. B. Nichols, and R. S. Phillips (1947): *Theory of Servomechanisms*, M.I.T. Radiation Laboratory Series, vol. 25, McGraw-Hill, New York.

Klapper, J., and J. T. Frankle (1972): *Phase-Locked and Frequency-Feedback Systems*, Academic Press, New York.

Korn, G. A., and T. M. Korn (1961): *Mathematical Handbook for Scientists and Engineers*, McGraw-Hill, New York.

Lathi, B. P. (1968): *An Introduction to Random Signals and Communication Theory*, International Textbook, Scranton, Pennsylvania.

Marcum, J. I., "Studies of Target Detection by Pulsed Radar," Special Monograph Issue *IRE Transactions on Information Theory*, vol. IT-6, no. 2, April, 1960.

Marcum, J. I., "Table of Q Functions," U.S. Air Force Project RAND Research Memorandum RM-339, January 1, 1950 (ASTIA Document AD116551).

McFadden, M. (1963): *Sets, Relations, and Functions*, McGraw-Hill, New York.

Melsa, J. L., and A. P. Sage (1973): *An Introduction to Probability and Stochastic Processes*, Prentice-Hall, Englewood Cliffs, New Jersey.

Middleton, D. (1960): *An Introduction to Statistical Communication Theory*, McGraw-Hill, New York.

Miller, K. S. (1974): *Complex Stochastic Processes, An Introduction to Theory and Application*, Addison-Wesley, Reading, Massachusetts.

Milton, J. S., and C. P. Tsokos (1976): *Probability Theory with the Essential Analysis*, Addison-Wesley, Reading, Massachusetts.

Mumford, W. W., and E. H. Scheibe (1968): *Noise Performance Factors in Communication Systems*, Horizon House-Microwave, Dedham, Massachusetts.

Papoulis, A. (1962): *The Fourier Integral and its Applications*, McGraw-Hill, New York.

Papoulis, A. (1965): *Probability, Random Variables, and Stochastic Processes*, McGraw-Hill, New York. See also the 2d ed., 1984.

Parzen, E. (1962): *Stochastic Processes*, Holden-Day, San Francisco, California.

Pearson, K., editor (1934): *Tables of the Incomplete Gamma Function*, Cambridge University Press, Cambridge, England.

Peebles, P. Z., Jr. (1976): *Communication System Principles*, Addison-Wesley, Reading, Massachusetts.

Prabhu, N. U. (1965): *Stochastic Processes, Basic Theory and its Applications*, Macmillan, New York.

Rosenblatt, M. (1974): *Random Processes*, 2d ed., Springer-Verlag, New York.

Ross, S. M. (1972): *Introduction to Probability Models*, Academic Press, New York.

Spiegel, M. R. (1963): *Theory and Problems of Advanced Calculus*, Schaum, New York.

Thomas, J. B. (1969): *An Introduction to Statistical Communication Theory*, Wiley, New York.

van der Ziel, A. (1970): *Noise: Sources, Characterization, Measurement*, Prentice-Hall, Englewood Cliffs, New Jersey.

Wilks, S. S. (1962): *Mathematical Statistics*, Wiley, New York.

Wozencraft, J. M., and I. M. Jacobs (1965): *Principles of Communication Engineering*, Wiley, New York.

Ziemer, R. E., and W. H. Tranter (1976): *Principles of Communications Systems, Modulation, and Noise*, Houghton Mifflin, Boston, Massachusetts.

A posteriori probability, 18
A priori probability, 18
Algebra of sets, 7
Ambiguity function, 272
Amplitude modulation (AM), 192, 275–279
Antenna temperature, 231
Arcsine density, 84, 333
Associative law of sets, 7
Autocorrelation function:
　bound on, 153
　of complex process, 160–161, 193
　defined, 149, 152–154
　of derivative of a process, 166, 246
　derivatives of, 203, 246
　Fourier transform of, 177, 180, 193
　Fourier transform of time average of, 177, 180
　of linear system response, 212–213
　measurement of, 156–158
　at product device output, 191
　properties of, 153
　relationship to power spectrum, 177, 179–181, 193
　of stationary process, 149
　time, 151
　time average of, 177, 179–180
　of wide-sense stationary process, 150, 153
Autocovariance function:
　of complex process, 160
　defined, 155, 160
　Fourier transform of, 196
　of wide-sense stationary process, 156, 158

Available power gain, 232
Average effective noise temperature, 237
Average effective source temperature, 237
Average operating noise figure, 236
Average standard noise figure, 236
Axioms of probability, 10

Band-limited random process, 219
　properties of, 221–223, 224–226
Band-limited white noise, 188
Bandpass random process, 219
Bandwidth:
　noise, 210, 217–218, 238
　of a power spectrum, 177–179
　rms, 177–179, 198, 201
Bayes, Thomas, 16*n*.
Bayes' theorem, 16
Bernoulli density, 331
Bernoulli, Jacob, 25*n*.
Bernoulli trials, 25
Bessel function, 299, 331
Beta density, 333
Binary communication system, 17
Binary process, semirandom, 162
Binary symmetric channel, 17
Binomial coefficient, 26, 33
Binomial density function, 46, 332
　characteristic function of, 85
　mean of, 81
　variance of, 81
Binomial distribution function, 46

Bivariate gaussian density, 124, 335
Boltzmann, Ludwig, 227n.
Boltzmann's constant, 227
Bound(s):
  on autocorrelation function, 153
  on cross-correlation function, 154, 155
  on linear system response, 210
Bounded variation, 310n., 321n.
Butterworth filter, 243

Cartesian product space, 22n.
Cauchy, Augustin Louis, 61n.
Cauchy density function, 61, 333
Cauchy distribution function, 62
Cauchy random variable, 61
Causal system, 210
Central limit theorem, 105–108
Central moments, 70, 82, 120–122
  related to moments about origin, 82
Certain event, 10
Channel, 17, 276, 281
  binary symmetric, 17
Characteristic function:
  of binomial density, 85
  defined, 72, 122
  of Erlang random variable, 85
  of exponential random variable, 72–73
  of gaussian random variable(s), 83, 138
  joint, 122–124, 138
  marginal, 123
  moment generating property of, 72, 123–124
  of Poisson random variable, 85
  for several random variables, 124
  for two random variables, 122
Chebychev's inequality, 85
Chebychev, Pafnuty L., 85n.
Chi-square density, 85, 331, 334
Class of sets, 3
Classification of process, 144–146
Coefficient of skewness, 71
Colored noise, 190
Combined experiment, 22
Combined sample space, 22
Communication system(s):
  amplitude modulation (AM), 275–279
  binary, 17
  frequency allocations, 275–276, 280
  frequency modulation (FM), 280–284
  three-symbol, 33
Commutative law of sets, 7
Complement of set, 6
Complex random process, 160–161
  (See also Random process)
Complex random variable, 133

Conditional density function, 52, 62
  with interval conditioning, 99
  with point conditioning, 96–99
  properties of, 53
Conditional distribution function, 52, 62, 96
  with interval conditioning, 99
  with point conditioning, 96–99
  properties of, 52
Conditional expected value, 69, 139
Conditional probability, 13
Conditioning event, 55–57
  interval, 99
  point, 96–99
Continuous random process, 145
Continuous random sequence, 146
Continuous random variable, 37
Convolution integral, 103, 207, 223
Convolution of density functions, 103, 105
Correlation:
  of independent random variables, 119, 134
  of orthogonal random variables, 119, 134
  of random variables, 119, 134
  of uncorrelated random variables, 119, 134
Correlation coefficient, 121, 136
Correlation functions:
  autocorrelation, 149, 152–154, 160–161,
    212–213
  autocovariance, 155–156, 160–161
  cross-correlation, 150, 154, 213–214
  cross-covariance, 156
  of derivative of a process, 166, 246
  time autocorrelation, 151
  time cross-correlation, 152
  (See also listings by specific types)
Correlation integral, 268
Correlation receiver, 269
Countable set, 3
Countably infinite set, 3
Covariance:
  of independent random variables, 121
  of orthogonal random variables, 121
  of random variables, 121, 134
  of uncorrelated random variables, 121, 134
Covariance functions (see Correlation functions)
Covariance matrix, 127
  after linear transformation, 132
Crest-factor, 280, 306
Crossover frequency, 308
Cross power, 183
Cross-correlation function(s):
  bounds on, 154, 155
  of complex process, 160–161, 193
  defined, 150, 154
  of derivative of process, 166, 203, 246

Cross-correlation function(s):
  Fourier transform of, 184, 185–186, 193
  of independent processes, 154
  of jointly wide-sense stationary processes, 150, 154
  measurement of, 156–158
  of orthogonal processes, 154
  properties of, 154
  real and imaginary parts of, 199
  relationship to cross-power spectrum, 184, 185–186, 193
  of response of linear system, 213–214
  time, 152
  time average of, 184, 185–186
Cross-covariance function:
  of complex processes, 160–161
  defined, 156
  of independent processes, 156
  of uncorrelated processes, 156
  of wide-sense stationary processes, 156
Cross-power density spectrum:
  of complex processes, 193
  defined, 183
  of linear system, 217
  properties of, 184
  real and imaginary parts of, 199
Cross-power density spectrum:
  relationship to cross-correlation function, 184, 185–186, 193
Cross-power spectrum (*see* Cross-power density spectrum)
Cumulative probability distribution function (*see* Probability distribution function)

Damping factor, 295
Decibels, 240*n.*
Delta function (*see* Impulse function)
De Morgan, Augustus, 7*n.*
De Morgan's laws, 7, 32
Density function (*see* Probability density function)
Derivative of random process, 166, 203
  power spectrum of, 177
Detection probability, 304–305
Deterministic random process, 147
Deterministic signal, 173
Difference of sets, 5
Dirichlet, Peter Gustav Lejeune, 321*n.*
Dirichlet conditions, 321*n.*
Discrete random process, 145
Discrete random sequence, 146
Discrete random variable, 37, 79
Disjoint sets, 3
Distribution function (*see* Probability distribution function)

Distributive law of sets, 7
Domain, 58
Duality principle, 8

Effective noise temperature, 229, 233, 237
Efficiency, 279, 305–306
Elementary event, 12
Elements of a set, 3
Empty set, 3
Ensemble, 143
Ensemble average, 143
Ensemble member, 143
Envelope detector, 278
Equal sets, 5
Ergodic random process, 152, 157*n.*
Ergodic theorem, 152
Erlang, A. K., 85*n.*
Erlang density, 334
Erlang random variable, 85
Event(s):
  certain, 10
  on combined sample space, 23–24
  defined, 9
  elementary, 12
  impossible, 10
  joint, 89
  mutually exclusive, 9
  pairwise independence of, 19
  probability of, 10
  statistically independent, 18–22
Excess available noise power, 233
Expectation (*see* Expected value)
Expected value (expectation):
  conditional, 69, 139
  of a function of a random variable, 68, 133
  of a function of several random variables, 117–119
  of random process, 143, 153, 160
  of a random variable, 67–68, 133
  of a sum of random variables, 118
Experiment(s), 8
  combined, 22
  independent, 25
  mathematical definition of, 8, 11–12
Exponential density function, 49, 68, 70, 71, 72–74, 334
Exponential distribution function, 49, 334

False alarm, 302
Filter (system):
  Butterworth, 243
  gaussian, 241
  idealized, 209–210
  matched, 258–262

Filter (system):
  prediction, 263
  smoothing, 263
  Wiener, 262–266
Finite set, 3
First-order stationary random process, 148
Fourier, Baron Jean Baptiste Joseph, 320n.
Fourier series:
  coefficients, 324, 325
  defined, 324
  Fourier transform of, 324, 325
Fourier transform(s):
  of autocorrelation function, 177, 180, 193
  of cross-correlation function, 184, 185–186,
    193
  defined, 72, 173, 320
  existence, 321
  inverse, 72, 173, 320
  multidimensional, 323
  properties of, 321–323
  table of pairs of, 329
  of time-averaged autocorrelation function, 177
  of time-averaged cross-correlation function,
    184
  uniqueness, 161
Frequency-domain analysis, 172
Frequency modulation (FM), 280–284
Fresnel integrals, 273

Gamma density, 335
Gamma function, 85, 330, 331
Gauss, Johann Friedrich Carl, 43n.
Gaussian density function:
  conditional, 112, 128
  marginal, 125, 128
  mean value of, 81, 124
  N-dimensional (N-variate), 127, 128
  of one random variable, 43, 60, 61, 335
  two-dimensional (bivariate), 112, 124, 335
  variance of, 81, 124, 336
Gaussian distribution function:
  of one random variable, 44, 335
  table of, 315
Gaussian filter, 241
Gaussian random process, 156, 158–160
  variance estimate of, 167
Gaussian random variable(s):
  characteristic function of, 83, 138
  conditional density function of, 128
  defined, 43, 124, 127
  linear transformation of, 77, 130–133
  marginal density function of, 125, 128
  mean value of, 81, 124
  moment generating function of, 83

Gaussian random variable(s):
  properties of, 128
  variance of, 81, 124, 336

Idealized system, 209–210
Impossible event, 10
Impulse function:
  defined, 41, 310–312
  limiting forms of, 176, 195
  N-dimensional, 311
  relationship to unit-step function, 41, 311, 312
  two-dimensional, 311–312
Impulse response of linear system, 206, 207
Incomplete beta function, 331
Incomplete gamma function, 330, 331
Incremental available power, 227
Incremental noise power, 227
Independent events, 18–22
Independent experiments, 25
Independent random processes, 148
Independent random variables, 100–102,
    133–134
  joint characteristic function of, 122–124, 138
Infinite set, 3
Integrals:
  Fresnel, 273
  table of definite, 318–319
  table of indefinite, 316–318
Integrator, 157
Intersection of sets, 6
Interval conditioning, 99
Inverse Fourier transform (see Fourier transform)

j, 60n.
Jacobi, Karl Gustav Jakob, 129n.
Jacobian, 129
Joint central moments, 120–122
  of two discrete random variables, 136
Joint characteristic function, 122–124
  of independent random variables, 138
  of two gaussian random variables, 83, 138
Joint event, 89
Joint moments, 119–120
  of two discrete random variables, 136
Joint probability, 13
Joint probability density function (see Probabil-
    ity density function)
Joint probability distribution function (see Prob-
    ability distribution function)
Joint sample space, 88
Jointly ergodic random processes, 152, 157n.
Jointly gaussian random processes, 156, 160
Jointly wide-sense stationary random processes,
    150, 160–161, 213

Khinchin, A. I., 180

Laplace, Marquis Pierre Simon de, 59*n*.
Laplace density function, 59, 336
Laplace transform, 173
Leibniz, Gottfried Wilhelm von, 76*n*.
Leibniz's rule, 76
Likelihood, 9
Linear system:
  causal, 210
  cross-correlation function, 213–214
  cross-power spectrums, 217
  defined, 206
  idealized, 209–210
  impulse response, 206, 207, 210
  noise temperature, 239
  optimum, 257
  output autocorrelation function, 212–213
  output mean-squared value, 212
  output mean value, 211
  output power, 212, 216
  output power spectrum, 215
  physically realizable, 210
  response to deterministic signal, 206–209
  response to random signal, 211
  stable, 210
  time-invariant, 207
  transfer function, 208
  white noise evaluation of, 214–215
Liquid helium, 200
Log-normal density function, 62, 86, 336
Log-normal distribution function, 62, 336

Mapping, 35
Marcum's Q-function, 305, 331
Marginal characteristic function, 123
Marginal density function (*see* Probability
  density function)
Marginal distribution function (*see* Probability
  distribution function)
Maser, 188*n*., 200
Matched filter, 258
  for colored noise, 259–260
  impulse response, 260, 261, 267
  maximum signal-to-noise ratio of, 260, 268
  output signal from, 268
  for rectangular pulse, 261–262
  transfer function, 260
  for white noise, 260–262
Matrix, covariance, 127
Mean frequency, 178
Mean (expected) value:
  from autocorrelation function, 153
  conditional, 69, 139

Mean (expected) value:
  estimate of, 167
  of function of a random variable, 68, 133
  of function of several random variables,
    117–119
  of process sample function, 151
  of random process(es), 143, 153, 160
  of random process derivative, 166
  of random variable, 67–68, 133
  of sum of random variables, 118
  of system response, 211
Mean-squared error, 263–266
Mean time between failures (MTBF), 64
Measurement of correlation functions, 156–158
Median of random variable, 61
Mixed random variable, 37
Mode of random variable, 61
Modulation index, 306
Moment generating function, 73
  of gaussian random variable, 83
Moments:
  central, 70, 82, 120
  from characteristic function, 72, 123–124
  interrelationships, 82
  joint, 119–122
  from moment generating function, 73
  about origin, 69, 82, 119–120
  of two discrete random variables, 136
Monotonic transformations:
  decreasing, 75, 76
  increasing, 75–76
Mutually exclusive events, 9
Mutually exclusive sets, 3

N-order stationary random process, 151
Narrowband gaussian process, 220
Narrowband random process, 220
Natural frequency, 295
Natural numbers, 80
Noise, 1
  arbitrary source of, 228
  colored, 190
  in control system, 285–288
  in phase-locked loop, 288–295
  resistive source of, 227
  thermal, 187, 227
  types of, 226*n*.
  white, 187
Noise bandwidth, 210, 217–218, 238
Noise figure:
  of attenuator, 238
  average, 236
  average operating, 236
  average standard, 236

Noise figure:
 average system operating, 240
 incremental, 234
 operating spot, 235, 249
  spot, 234–235
 standard spot, 235, 250
Noise performance:
 of amplitude modulation system, 278–279
 of frequency modulation system, 281–284
 of phase-locked loop, 292–295
 of radar, 301–305
Noise temperature:
 of antenna, 231
 of attenuator, 238
 average, 237–238
 average effective input, 237
 average effective source, 237
 effective, 228
 effective input, 233
 standard, 235
 system, 239
Nondeterministic random process, 147
Nonstationary random process, 147
Normal density function (*see* Gaussian density
  function)
Normal distribution function (*see* Gaussian distribution function)
Normalized second-order moment, 121
Null set, 3

Open-loop transfer function, 285
Operating spot noise figure, 235, 249
Operator, 206
Order of moment, 119, 120
Orthogonal random process, 154, 161
Orthogonal random variables, 119, 134
Outcome of experiment, 8
Overmodulation, 278*n.*, 306

Parseval, M. A., 323*n.*
Parseval's theorem, 323
Pascal, Blaise, 332*n.*
Pascal density, 332
Phase detector, 290
Point conditioning, 96–99
Poisson, Siméon Denis, 47*n.*
Poisson density function, 47, 85, 332
 mean of, 81, 332
 variance of, 81, 332
Poisson random variable, 47, 140
Power:
 from autocorrelation function, 153
 incremental available, 227
 from power density spectrum, 175

Power:
 in a random process, 153, 173, 175
 in response of linear system, 212, 216
 from second moment of process, 175
Power density (*see* Power density spectrum)
Power density spectrum, 153, 173
 bandwidth of, 177–179
 of complex process, 193
 defined, 175
 of derivative of process, 177
 inverse Fourier transform of, 177, 180
 of output of linear system, 215
 of product device output, 191
 properties of, 176–177
 relationship to autocorrelation function, 177,
  179–181
 relationship to autocovariance function, 196
Power gain, available, 232
Power spectral density, 173*n.*
Power spectrum (*see* Power density spectrum)
Power transfer function, 215
Prediction filter, 263
Probability, 9
 a posteriori, 18
 a priori, 18
 axioms of, 10
 conditional, 13
 defined, 10
 of detection, 302, 304–305
 of false alarm, 302, 303
 joint, 13
 of miss, 302
 total, 15
 transition, 18
Probability density function:
 arcsine, 84, 333
 Bernoulli, 331
 beta, 333
 binomial, 46, 81, 332
 Cauchy, 61, 333
 chi-square, 85, 331, 334
 conditional, 52, 96–99
 defined, 40, 93, 148
 of discrete random variable(s), 41, 94
 Erlang, 334
 existence, 40
 exponential, 49, 334
 gamma, 335
 gaussian, 43, 60, 61, 112, 124–128, 335
 joint, 93
 Laplace, 59, 81, 336
 log-normal, 62, 86, 336
 marginal, 95, 96
 of *N* random variables, 94

Probability density function:
  normal, 43*n.*
  Pascal, 332
  Poisson, 47, 85, 332
  properties of 42, 53, 94
  of random process(es), 148*n.*
  Rayleigh, 50, 336
  Rice, 299, 337
  of sum of random variables, 102–105
  transformation of, 74–80, 128–133
  triangular, 43
  of two random variables, 93
  uniform, 48, 337
  Weibull, 85, 338
Probability distribution function:
  arcsine, 333
  Bernoulli, 331
  beta, 333
  binomial, 46, 332
  Cauchy, 62, 333
  chi-square, 334
  conditional, 52, 62, 96–99
  defined, 37, 89, 147–148
  of discrete random variable(s), 38, 90
  Erlang, 334
  exponential, 49, 334
  gamma, 335
  gaussian, 43–45, 335
  joint, 89, 90, 148
  Laplace, 336
  log-normal, 62, 336
  marginal, 92, 93
  of one random variable, 37
  Pascal, 332
  Poisson, 47, 332
  properties of, 37–38, 91
  of random process, 147*n.*, 147–148
  Rayleigh, 50, 337
  Rice, 337
  of several random variables, 90
  of sum of random variables, 102–105
  of two random variables, 89
  uniform, 48, 337
  Weibull, 338
Process (*see* Random process)
Product device, 191–193
Product of sets, 6
Product space, two-dimensional, 88
Pulse repetition frequency (PRF), 301

Quantizer, 58

Random point, 88

Random process(es):
  autocorrelation function, 149, 152–154, 160
  autocovariance function, 155, 160
  band-limited, 219
  bandpass, 219
  classified, 144–146
  complex, 160–161
  continuous, 145
  cross-correlation function, 150, 154, 160–161
  cross-covariance function, 156, 160–161
  cross-power spectrum, 182–185, 193
  defined, 142, 160
  density function of, 148
  derivative of, 166, 246
  deterministic, 147
  discrete, 145
  distribution function of, 147–148
  ergodic, 152
  estimate of, 168
  first-order stationary, 148–149
  gaussian, 156, 158–160
  jointly ergodic, 152, 157*n.*
  jointly wide-sense stationary, 150, 160–161, 213
  mean of, 143, 153, 160
  mean value estimate of, 167
  N-order stationary, 151
  narrowband, 220
  nondeterministic, 147
  nonstationary, 147
  orthogonal, 154, 161
  periodic component in, 153
  power of, 153, 173, 175
  random telegraph, 169
  sample function of, 143
  second-order stationary, 149
  semirandom binary, 162, 296
  stationary, 147
  statistically independent, 148
  strict-sense stationary, 151
  time correlation functions of, 151–152
  uncorrelated, 156, 161
  variance of, 156
  wide-sense stationary, 150, 160–161
Random sequence:
  continuous, 146
  discrete, 146
Random signal, 1
Random telegraph process, 169
Random variable(s):
  auxiliary, 140
  Cauchy, 61
  central moments of, 70, 82, 120–122
  characteristic function of, 72, 122

Random variable(s):
  coefficient of skewness, 71
  complex, 133
  conditions for, 36
  continuous, 37
  correlation of, 119
  correlation coefficient, 121, 136
  covariance of, 121, 134
  defined, 34, 88, 89
  discrete, 37
  Erlang, 85
  gaussian, 43, 81, 119n., 124–128, 130–133
  independent, 100–102, 133–134
  joint central moments of, 120
  joint characteristic function of, 122–124, 138
  joint moments of, 119, 123, 124
  lattice-type, 106n.
  marginal characteristic functions of, 123
  mean of, 67–68, 133
  median of, 61
  mixed, 37
  mode of, 61
  moments of, 69–70
  N-dimensional, 89
  orthogonal, 119, 134
  Poisson, 47, 140
  Rayleigh, 50, 81
  skew, 71
  standard deviation, 70
  transformation of, 74–80, 128–130, 130–133
  uncorrelated, 119, 134
  uniform, 48, 59
  variance of, 70, 133
  vector, 88
  Weibull, 85
Random vector, 88, 89
Range, 58
Range sample space, 88
Rational power spectrums, 267
Rayleigh, Lord (John William Strutt), 50n.
Rayleigh density function, 50, 336
  maximum value of, 61
  mean of, 81
  median of, 61
  mode of, 61
  variance of, 81
Rayleigh distribution function, 50, 337
Rayleigh random variable, 61
Realizable linear system, 210
Realization of a process, 143
Rice density, 337
rms bandwidth, 177–179, 198, 201

Sample function, 143

Sample space, 9
  combined, 22
  continuous, 9
  discrete, 9
  joint, 88
  range, 88
Sampling function, 326
Schwarz, Hermann Amandus, 259n.
Schwarz's inequality, 259
Second-order stationary random process, 149
Semirandom binary process, 162, 296
Series, table of, 319
Set(s):
  algebra of, 7
  associative law of, 7
  class of, 3
  commutative law of, 7
  complement of, 6
  countable, 3
  countably infinite, 3
  defined, 3
  difference of, 5
  disjoint, 3
  distributive law of, 7
  elements of, 3
  empty, 3
  equality of, 5
  finite, 3
  infinite, 3
  intersection of, 6
  mutually exclusive, 3
  null, 3
  product of, 6
  sum of, 6
  uncountable, 3
  union of, 6
  universal, 4
  Venn diagram of, 5
Set theory, 2
Skew, 70
Skewness, 70
Smoothing filter, 263
Spectrum (Fourier transform), 72, 173
Spot noise figure, 234–235, 249, 250
Stable linear system, 210
Standard deviation, 70
Standard noise temperature, 235
Standard source, 235
Stationarity (see Random process)
Statistical average (see Expected value)
Statistical independence:
  of events, 18–22
  of experiments, 25
  by pairs, 19

Statistical independence:
  of random processes, 148
  of random variables, 100–102, 134
Subexperiments, 22
Subset, 3
Sums of random variables:
  density function of, 102–105
  distribution function of, 102–105
  mean of, 118
  variance of, 122
System noise temperature, 239

Tables:
  of definite integrals, 318
  of finite series, 319
  of Fourier transform pairs, 329
  of gaussian distribution function, 314
  of indefinite integrals, 316–318
  of trigonometric identities, 315–316
Telegraph process, 169
Thermal noise, 187
Thevenin, Léon, 227n.
Thevenin voltage source, 227
Threshold, 284, 302, 303
Time autocorrelation function, 151
Time average, 151
Time cross-correlation function, 152
Time-domain analysis, 172
Time-invariant system, 207
Total probability, 15
Transfer function:
  open-loop, 285
  of phase-locked loop, 290–292
  of system, 208, 215
Transformation of random variable(s), 74–80,
    128–130
  linear, 77, 130–133
  monotonic, 75–77
  nonmonotonic, 77–80
  square-law, 78–79
Transition probability, 18
Trial, 8

Trigonometric identities, table of, 315–316
Two-dimensional product space, 88

Uncorrelated random processes, 156, 161
Uncorrelated random variables, 119, 134
Uncountable set, 3
Uniform density function, 48, 139, 337
  mean of, 80, 337
  variance of, 80, 337
Union of sets, 6
Unit-impulse function (see Impulse function)
Unit-step function:
  defined, 38, 311, 312
  related to impulse function, 311, 312
  two-dimensional, 311–312
Universal set, 4

Variance:
  estimate of, 167
  of random process, 156
  of random variable, 70, 133
  of sum of random variables, 122
Vector random variable, 88
Venn, John, 5n.
Venn diagram, 5
Voltage density spectrum, 173

Weibull density, 85, 338
Weibull, Ernst H. W., 85n.
White noise:
  autocorrelation function of, 187
  band-limited, 188
  defined, 187
  in system evaluation, 214–215
  power spectrum of, 187
Wide-sense stationary random process, 150,
    160–161
Wiener, Norbert, 180, 262n.
Wiener filter, 262–266, 288, 307
  minimum mean-squared error of, 266
  transfer function of, 265
Wiener-Khinchin relations, 180